# Capturing the Universe

## A photographer's guide to deep-sky imaging

Chris Woodhouse

Routledge
Taylor & Francis Group
NEW YORK AND LONDON

First published 2021
by Routledge
52 Vanderbilt Avenue, New York, NY 10017

and by Routledge
2 Park Square, Milton Park, Abingdon, Oxon, OX14 4RN

*Routledge is an imprint of the Taylor & Francis Group, an informa business*



*Library of Congress Cataloging-in-Publication Data*
A catalog record for this title has been requested

ISBN: 978-0-367-36654-4 (hbk)
ISBN: 978-0-367-36653-7 (pbk)
ISBN: 978-0-429-34744-3 (ebk)

Typeset in Adobe Minion Pro and Myriad Pro
by Chris Woodhouse

*Publisher's Note*
This book has been prepared from camera-ready copy provided by the author.
Visit: *https://www.digitalastrophotography.co.uk* for additional book resources

# Contents

*A book for those aiming their camera upwards.*

night hides the World but
reveals the Universe

*ancient Persian proverb*

# Preface

*A book designed for those entering the wonderful world of astrophotography*

I catapulted myself into astrophotography in 2011 after buying the entire kit from another, who was selling up. Over the next few years I experienced both success and failure first hand, as I fathomed out things the hard way. This period was both challenging and huge fun. A large part was spent researching and building up an expertise that resulted in the first and second edition of my book, *The Astrophotography Manual*. This book addresses a gap in the market and is deliberately aimed at those who are already more than casual astrophotographers, drawing directly from my first-hand experiences while they were still fresh. The complimentary reviews confirmed the style and the pitch was just right for the aspiring and intermediate astrophotographer and in particular, the cogent write-ups on important topics, commonly skipped over by other publications. The fact it was up to date with the latest trends made it even more useful. *The Astrophotography Manual* is not for everyone though and the content is a little daunting for a complete novice. I did promise my better half that it would be the last book for a while. The vacuum, however, was not filled by postponed domestic projects and after only a few months I started in a new direction.

The result is a book that fills another void. Most astrophotography books pitch at the beginner and too many fill their pages with images from equipment manufacturers' websites and on the surface, cover all the bases. When you read them in more detail, however, seeking real practical information to solve a particular problem, say on how to set up an autoguider, it only then becomes apparent that the content is superficial.

This book is different, its approach is to keep things simple, practical and within a modest budget. It is deliberately aimed at keen amateur photographers who are familiar with DSLRs or mirrorless cameras and it purposefully kicks off using their existing photographic equipment and skills. After these early examples, it progressively introduces new processes and equipment for more challenging deep-sky objects. (Deep-sky objects, by the way, are defined as those beyond our solar system.) This approach introduces and explains the more complex topics when needed. In the same manner, I bare my soul, sharing some of my first imaging attempts, as well as more recent ones, to show what is likely in the beginning and possible, after a few years' practice.

I avoid using expensive or complicated software and make the most of free utilities and budget applications. Many readers of *The Astrophotography Manual* particularly liked the practical chapters and found the processing flow diagrams very useful. This book continues with this hands-on approach. There are the usual traditional chapters on setting up equipment, taking and processing images. After some general introductions to what is coming, these topics are explained in more detail, as required, to support progressively more demanding case studies. This approach is rather like peeling away the layers of an onion and some topics are revisited a few times, each time at a more advanced level. The end game aspires to the more automated system of an advanced amateur and blends with the content in *The Astrophotography Manual*.

I only write about things with which I have direct experience. Regurgitating data sheets is not my style. Ironically, it is much harder to write a beginner's guide than a more advanced manual since it requires constant vigilance and restraint to avoid it becoming too technical, too quickly. This is, however, an unavoidably complicated hobby. Writing this book also required me to turn the clock back and abandon my automated observatory, the luxury of a permanent installation and high-end software. It presented new problems to solve, new software to evaluate and gizmos to design. It was an interesting experience, that highlighted just how many acquisitions and upgrades had occurred over the last few years; it is easy to lose sight of what it is like to start out (or the sum total of one's outlay, which can be scary).

In this book I have kept extravagancies to a minimum and sought out value-for-money propositions. In some cases this avoids "obvious" choices due to their expense or steep learning curve. This includes buying used equipment where practical and using low-cost upgrades to improve performance and reliability. In addition, there are a few construction projects too, supported by my website. 'Clear Skies'.

*chris@digitalastrophotography.co.uk*
*www.digitalastrophotography.co.uk*

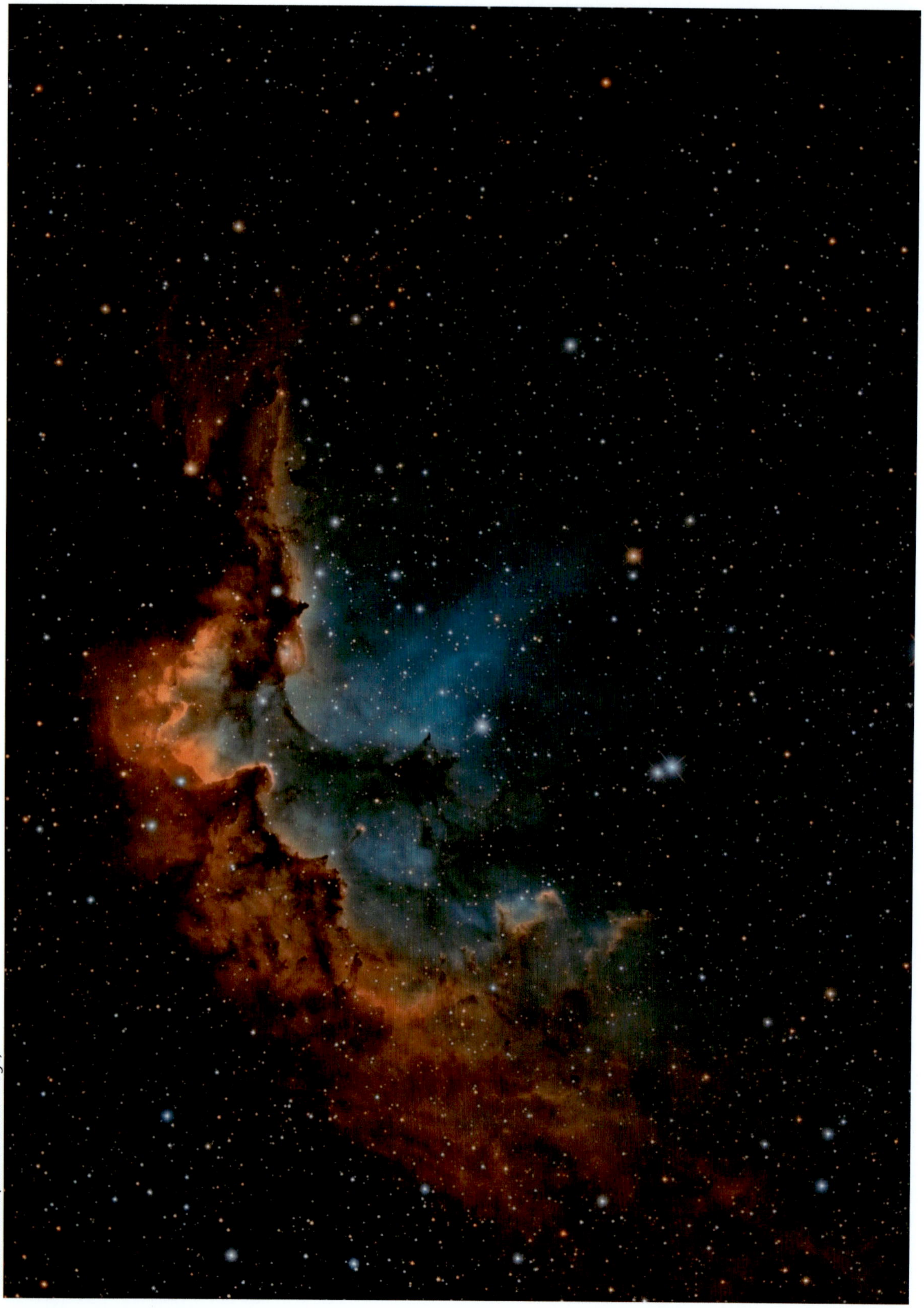

*Wizard Nebula (narrowband image)*

# About The Author

*I promised my wife I would not write another book, but then I also said I did not need another telescope.*

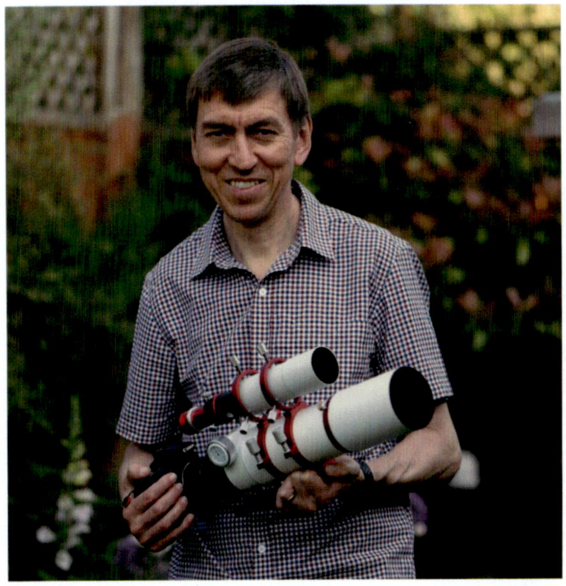

Chris was born in England and from his teenage years was fascinated by the natural sciences, engineering and photography, all of which he found more interesting than football. At the weekend he could be found building or designing some gadget or other. At school he used a slide-rule and log books for his exams at 16. Two years later, scientific calculators had completely displaced them. He studied Electronics at Bath University and by the time he had completed his masters degree, the computer age was well under way and 8-bit home computers were common. After a period designing military communication and optical gauging equipment, as well as writing software in Forth, Occam, C++ and Assembler, he joined an automotive engineering company.

As a member of the Royal Photographic Society, he gained LRPS and ARPS distinctions and pursued a passion for all forms of photography, mostly using traditional monochrome techniques. Not surprisingly, this hobby, coupled with his professional experience led him to invent and patent several highly regarded f/stop darkroom meters and timers, still sold throughout the world. During that time digital cameras evolved rapidly and photo ink-jet printers slowly overcame their annoying limitations. Resisting the temptation of the early optimistic digital promises, he authored a book on traditional monochrome photography, *Way Beyond Monochrome*, to critical acclaim and followed with a second edition to satisfy the ongoing demand. Digital monochrome appeared to be the likely next avenue for his energy, until an eye-opening presentation on astrophotography. This renewed a dormant interest in astronomy now made possible by affordable digital camera technology. Astrophotography is the perfect fusion of science, electronics and photography.

Like many before, his first attempts ended in frustration and disappointment, but he quickly realized the technical challenges of astrophotography responded well to a methodical and scientific approach. He found this, together with his photographic eye and decades of printing experience, were an excellent foundation to produce beautiful and fascinating images from a seemingly featureless sky.

The outcome was *The Astrophotography Manual*, acclaimed by many readers as the best book on the subject in the last 15 years, followed by a ground-breaking second edition in 2017. Chris was accepted as a Fellow of the Royal Astronomical Society, founded in 1820, for his work with outreach events and for sharing his passion with the younger generation.

**Acknowledgements**

This book and the intensive research that it requires would not have been possible without the ongoing support of his wife Carol (who even dug the footings for his observatory) and the wider on-line community. Special thanks also goes to Andrew Duffield for loaning equipment and for his contributions to some chapters.

It is one of the pleasures of this hobby to share problems and solutions with other hobbyists and this edition builds upon the knowledge and wisdom of many astrophotographers. This hobby is a never-ending journey of refinement, knowledge and development. It is a collaborative pursuit and I welcome any feedback or suggestions for this book or the next adventure.

Chris Woodhouse ARPS, FRAS

# Introduction

*Digital sensors, more than any other technology, have made high-quality amateur astrophotography what it is today.*

If you have ever stared up at the night sky from a dark site, it is easy to imagine how the stars and planets have been a source of wonder and speculation for millennia. In modern times and with our familiarity with satellite-based telescopes, mobile phone applications and computerized mounts, it is hard to comprehend how so much was fathomed by the ancients with no optical aids, computers or reliable timepieces. The greats, including Ptolemy, Hipparchus, Copernicus, Kepler, Hooke and Newton worked it all out by patient observation and mathematics. In this age of instant gratification, I'm considered patient for recording a dim deep-sky image over 40 hours, yet these men observed for a lifetime to shape their conclusions. Ironically, astrology was the driving force behind astronomy for centuries, followed by navigation and then more scientific study. Up until the 1920s, we only believed there was one galaxy; our Milky Way. Today, we routinely image thousands of galaxies from land and space-based telescopes.

Astrophotography started with a Daguerreotype of the moon in 1840 and slowly progressed along with the improvement of light-sensitive emulsions and optics. The first astrophotograph taken with a digital sensor was in 1974, though it would not win any prizes. By 2000, digital sensors had largely displaced film and astrophotography, as we know it, was born.

The pace of innovation since then now allows amateurs to take images from their backyard that rival those of professional observatories from the previous decade. Amateurs can even now detect unseen planetary objects (exoplanets) around distant suns with surprisingly modest means. The history of astronomy is fascinating. The concept that exoplanets may have the potential for extraterrestrial life is a long way from the controversial observations and theorems of Galileo, who defied the established religious doctrine of 17th century Italy and dared to support Kepler's sun-centered solar system. (The Vatican publicly admitted that Galileo's views of the solar system were correct in 1992!)

Astronomy and astrophotography are unavoidably technical subjects, which for many is a large part of their allure and off-putting for others. It is all too easy to lose oneself in techno-babble and in this book, I try to include both basic principles as well as more challenging concepts for the curious. This is aimed to give context to case studies or to back up practical recommendations. In addition, although there are many friendly forums that can offer help and advice, it helps to know the common terminology to ask the right questions, as well as to understand the answers!

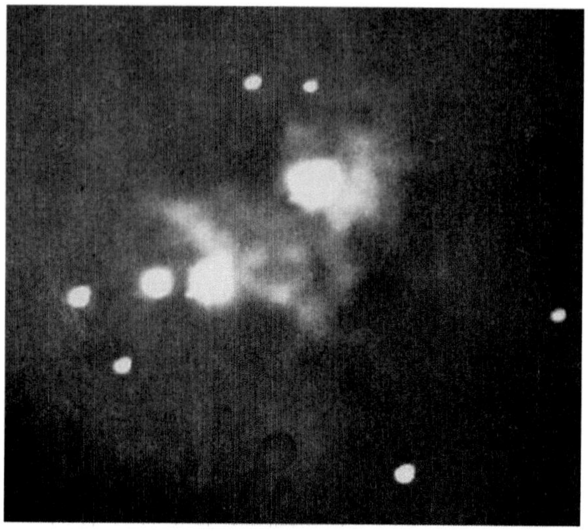

*fig.1 The first photograph of the Orion Nebula, taken by Professor Henry Draper, in 1880.*

### New Horizons

The pace of development in both hardware and software continues unabated. The "bang for the buck" continues to increase with keen competition and improved manufacturing techniques. Consumer digital cameras have reached a saturation point, where their performance improvements are arguably irrelevant to the amateur photographer. Dedicated cameras for astrophotography are a niche product, however, with a longer life cycle and the popular cameras based on mature CCD sensors from the previous decade are being progressively replaced by less expensive CMOS equivalents. In this book, we see just how close the technologies have become with improvements in optical efficiency and low-noise electronics.

Computing prices continue to fall and the use of smartphones coupled, by WiFi or Bluetooth, to enabled accessories provide a tantalizing glimpse into the future, which may make the traditional PC or laptop redundant. For those pioneering users who like a challenge, Raspberry Pi offers an affordable interesting alternative. By the time of publishing, solutions using these may be more user-friendly and reliable enough to make mainstream applications. For now though, although it is interesting, it is beyond this entry-level guide.

### Scope of Choice

Consumer choice is overwhelming, making it hard to keep things simple and easy to misplace spending priorities within a strict budget. My own systems have moved on in the last few years and as a result, writing this book was a salutary lesson in humbleness. It required me to return to my roots and use my experience to make the best out of least (and once again, find the mosquito repellent). It is not always the case, but less expensive equipment often requires additional care and nurturing to produce good results. In this book, my aim is to encourage the reader with early successes and minimum outlay. One can always spend a shed-load of money on equipment but without understanding the fundamentals, one soon becomes all the more frustrated. I witness this on the forums from time to time. A new member, often newly retired, thinking that they can simply buy a high-end system and it will simply turn on and work. Turn-key systems are still a dream, and, if we are being honest, dull. Although astrophotography is not easy to do well, it is not out of reach either.

### About This Book

While *The Astrophotography Manual* is a fast track to intermediate and advanced astrophotography, this book reins that in and is designed for an amateur photographer who wants to do astrophotography. This is more challenging to write than an advanced guide as it is all too easy to assume the user has a scientific or engineering background and lose them in the first chapter. The concept is to bring the subject down to earth, with realistic goals and assumptions. A title on my shelf advertises itself as making images with affordable equipment and software. The author then goes on to describe a fully-fledged system in a permanent, rotating observatory costing thousands!

This book makes the most of free or value-for-money software and using modest equipment. It assumes the reader has a digital camera, tripod and some basic image editing skills. For that reason, it avoids the heavyweight applications and uses less intimidating software titles. If we are being true to our frugal goal it should exclude Adobe Photoshop too. That may be a step too far for some but I will showcase some worthy alternatives that offer better value.

As mentioned before, this book takes a unique and progressive approach through the first steps in astrophotography. It does not follow the traditional approach; with dry theoretical chapters that review sensors, telescopes, mounts and so on in an almost abstract manner and with perhaps a brief practical example or two at the end. This book, after the briefest of introductions to the world of astrophotography, takes the reader on a journey of increasing refinement. This is punctuated with increasingly demanding practical assignments. At the beginning, these require no more than an existing digital camera and tripod and slowly build up to a modest telescope, mount and a dedicated astrophotography camera. Specialist techniques are introduced along the way, just when they are needed so that by the end, the reader has a small gallery of images to amaze their friends, a functional system and a hunger to continue.

The case studies introduce unique challenges and methods by which they are acquired and processed. A worked example is often a wonderful way to explain things and these case studies deliberately use a variety of equipment, techniques and software. There are too many to try them all and I have chosen the most popular ones that, at the same time, are reliable and good value for money. New software titles are being released each year. They can seem attractive at first but, in reality, it usually takes a few years for them to settle down and become a practical proposition.

Practical examples are more valuable if they are imperfect and we learn from them. Similarly, the full-page images are a selection of my first-time images, attainable images and later on, more aspirational ones. Some case studies are deliberately experimental and present an opportunity to discuss alternative promising techniques. Fixing problems can be fun, until they waste too many clear nights. This is when a helping hand is welcome and I share my top "things gone wrong" and simple ideas to track down gremlins in increasingly sophisticated imaging systems. Astrophotography and astronomy, in general, lend themselves to practical invention. To that end, simple practical tips to make life easier are sprinkled through the book, along with recommendations for further reading (a rare thing these days) and useful resources, many of which are on the Internet.

# New Horizons

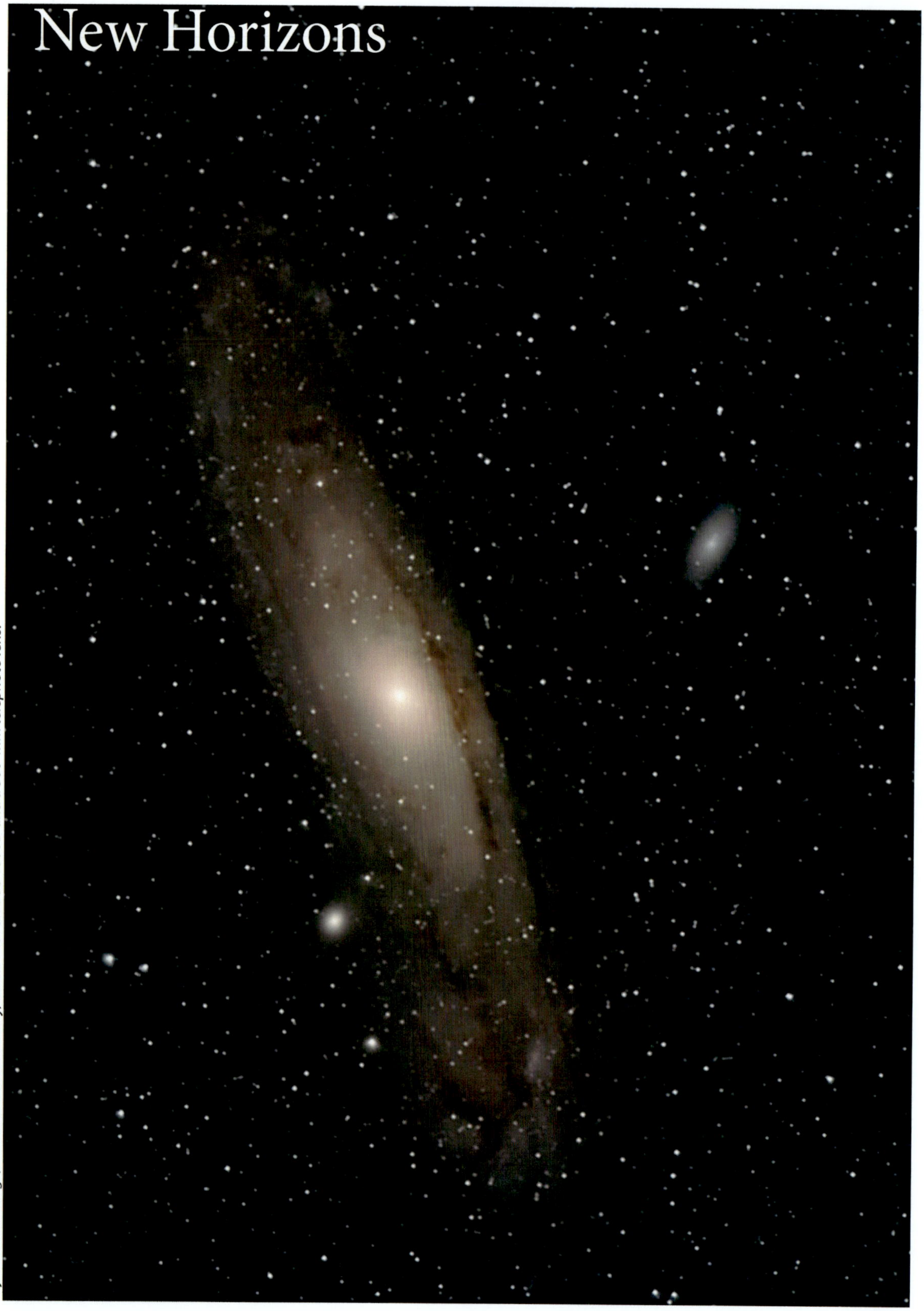

*My first ever image; the Andromeda Galaxy, taken with an EOS 1000D and a 300-mm telephoto lens.*

# The Diverse Universe

*A hobby that will fill you and others with wonder, satisfaction and frustration in equal measure.*

Astrophotographers have the luxury of many specialties to pursue but in the main, the images that adorn the multitudinous websites consist of stars, special events, planets and deep-sky objects. This chapter and the following give an astronomical grounding in the various objects in space and the systems we use to characterize, locate and measure them. It is not essential to understand astronomy to obtain good pictures, but I think it helps to decipher the lingo as well as add to the enjoyment and appreciation of our own and others' efforts.

At the same time, it is important to realize that over the last decade, amateur astrophotography has elevated itself beyond making pretty pictures, and makes significant contributions to scientific research. Those contributions include discovering supernovae, potential exoplanet candidates, discovering new nebulae, directly observing an impact on Jupiter and detecting new comets. One reason for this is that the handful of professional large observatories cannot cover the entire sky at any one time and so the contribution from thousands of amateurs is invaluable, especially when it comes to identifying transient events. Although I might happen upon something in my lifetime, I am content with making high-quality images and inspiring others to do so too.

Astrophotography is not one hobby, but many; not only are there some specialty topics, individual circumstances strongly influence what one image. Depending on viewing conditions, equipment, budget and available time, amateur astronomers can vary from occasional imagers using a portable setup, to those with a permanent installation capable of remote control and operational at a moment's notice. The subjects are just as numerous too; from high magnification planetary, and deep-sky imaging, through medium and wide-field imaging in broad or selective wavelengths. Then there is lunar and solar photography as well as landscape astrophotography, which creates wonderful starry vistas. As with any hobby, there is a law of diminishing returns and once the fundamentals are in place, further enhancements often have more to do with convenience and reliability than raw performance. I have a permanent setup and a few portable configurations, all of which are limited by my location. Any further purchase would do little to increase my enjoyment. (Well, that is what I said to my better half!)

### A Public Health Warning

This chapter touches on some of the more common forms of astrophotography and the likely setups. Unlike digital photography, one-upmanship between astrophotographers is rare but even so, once you are hooked, it is tempting to pursue an obsessive frenzy of upgrades and continual tuning. It is important to realize that there is a weak link in the imaging chain. Mundanely, it is more likely to be your location, prevailing light pollution, weather, atmospheric conditions, obscuration and family commitments than the latest camera model. Suffice to say, I did warn you!

### Size is Important

Staring up at the night sky one simply sees a generally monochrome scene, with the brightest stars and the moon. Taking a long exposure of the night sky with a digital camera reveals faint details that simply do not register with our eye and brain, even when viewed through a modest telescope. I have a dozen prints of deep-sky objects arranged around my desk at work. (When we refer to deep-sky objects, we are talking about those objects beyond our solar system and comprise of stars, nebula and galaxies.) Colleagues commenting on my prints are amazed by the vivid color and weird shapes and almost always think they are imaged through a colossal telescope. It is a surprise to them when they realize that they are not. Many of these objects dwarf the moon in apparent size and if they were considerably brighter, would likely have been worshipped by ancient civilizations. Of course, there are many very tiny objects too that require long focal length lenses to resolve them at a sensible scale. This enormous range of scale equally requires a large range of equipment. Fig.1 shows a selection of common astrophotography targets, printed at a size according to their relative scale. It is easy to see that no one telescope can deliver all these objects. Of note, is the large difference between the largest planet and even the smaller nebulae. Less obvious in these

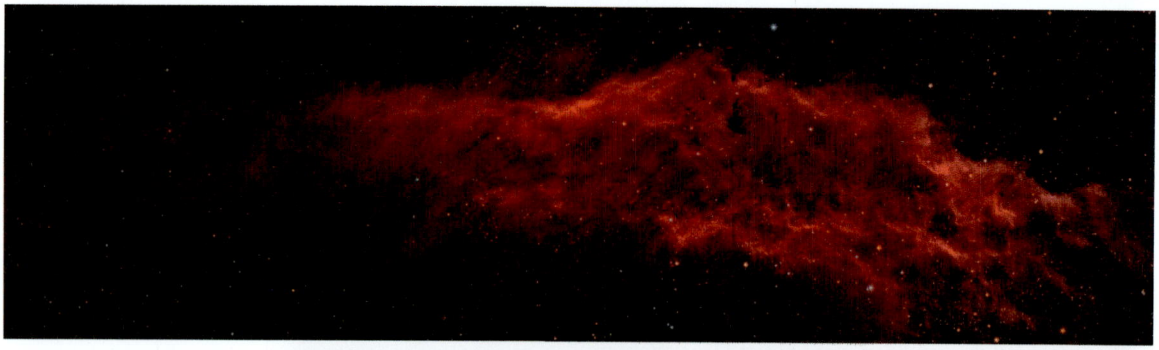

*fig.1* *A selection of common astrophotography objects, reproduced at approximately the same scale, with the Moon as a useful reference. Note the size of Jupiter (below)!*

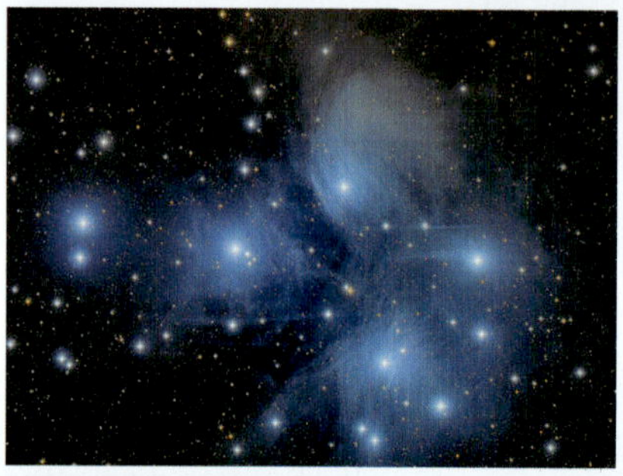

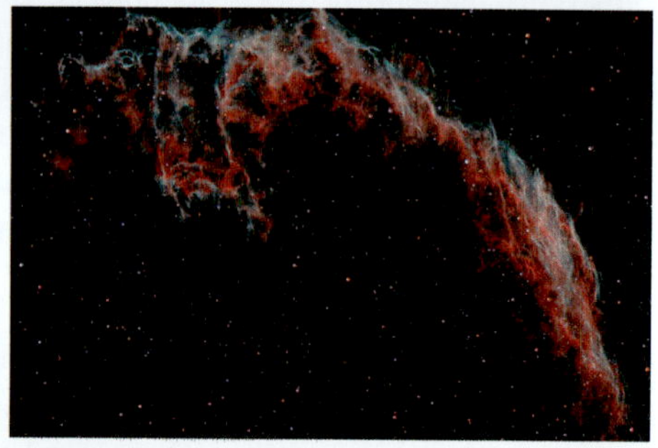

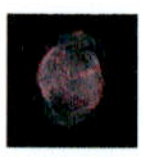

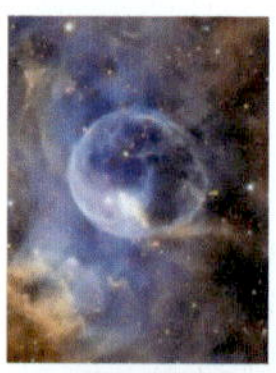

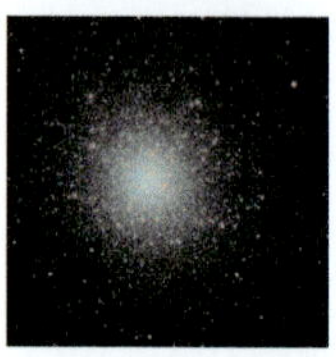

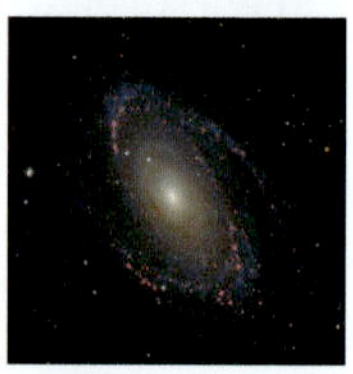

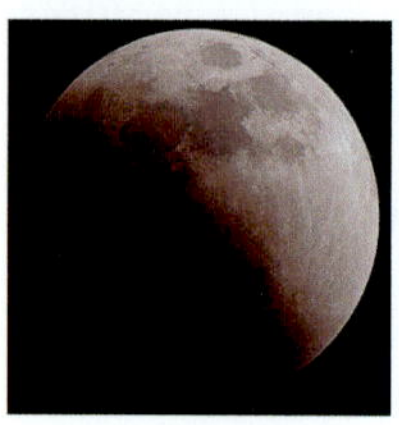

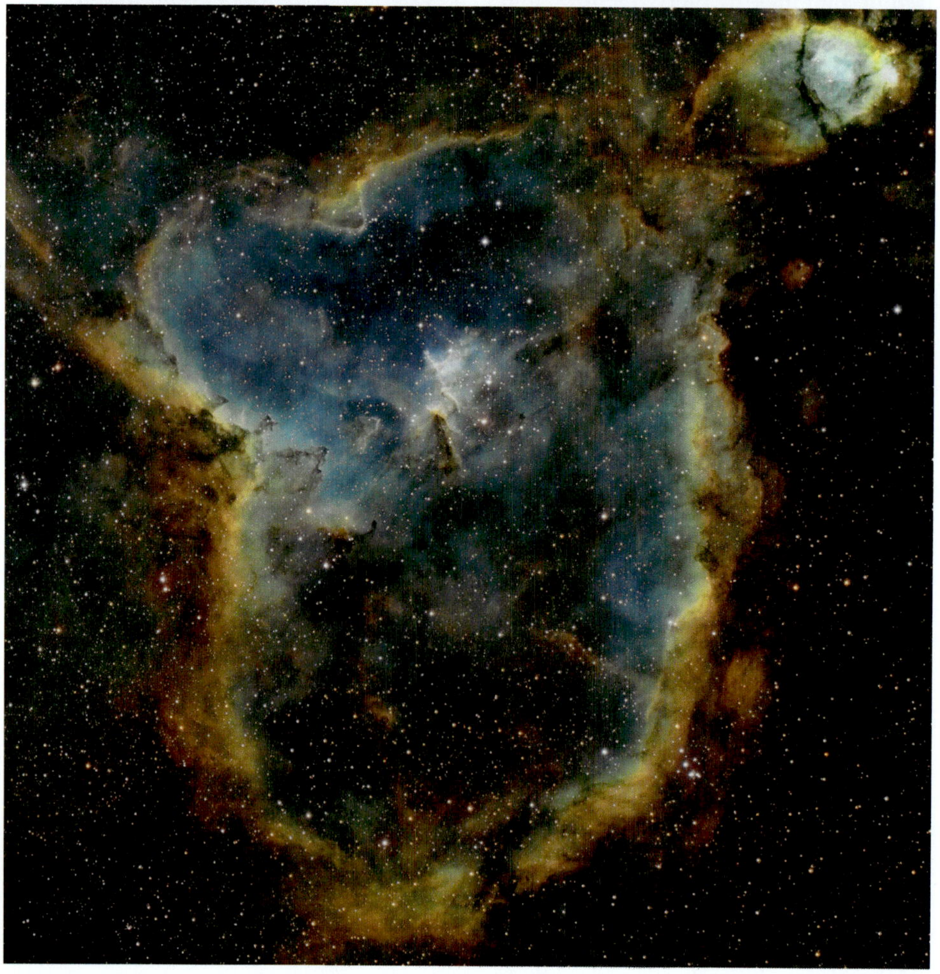

images is the equally wide subject brightness range. This is accommodated by different exposure lengths and special image processing but when objects of very different brightness appear in the same image, this becomes quite a challenge.

In fig.1, Jupiter is considerably smaller than the deep-sky objects. While many deep-sky objects can be imaged with the same general purpose short and medium focal length instrument, the planets require much longer focal lengths. Solar and lunar imaging also use unique methods and to do them justice, beyond the scope of this book. There are plenty of other avenues to explore; however, the following list covers most, requiring increasing magnification. We will go on to explain the technical terms a little later on.

### Landscape Astrophotography

These images are astronomy-related and typically involve the surrounding landscape. Examples include images of aurora, meteorite showers, eclipses or a wide-angle shot of the Milky Way overhead. Digital SLRs and those cameras with larger sensors make ideal cameras for these images and a great place to start astrophotography with little additional cost. The low image magnification makes it possible to take images with an inexpensive telescope mount, or with care, just a tripod. Overall image sharpness is challenged by a stationary foreground and stars that circle overhead by 1 degree every four minutes. The most successful images require careful planning to provide an interesting foreground and sky; a combination of being in the right place at the right time. At a dark site, a panorama of the Milky Way makes a fantastic image.

### Wide-Field Deep-Sky Astrophotography

In common with many visitors to my gallery, I once thought our moon was the biggest object in the night sky. One of the most impressive is the Andromeda Galaxy, thought to be a nebula up until the 1920s. Under a dark sky, one may just discern its center with the naked eye but the entire object's span is six times the width of the Moon. It is interesting to ponder what ancient civilizations would have made of it had they perceived its full extent.

Such objects are within the grasp of an affordable short focal-length lens in the range 350-500 mm. Imaging at lower image magnifications is less demanding of accurate tracking and even in light polluted areas, it is possible to use special filters to reduce the effect of the ever-present sodium street light. These deep-sky images use dedicated CCD/CMOS cameras

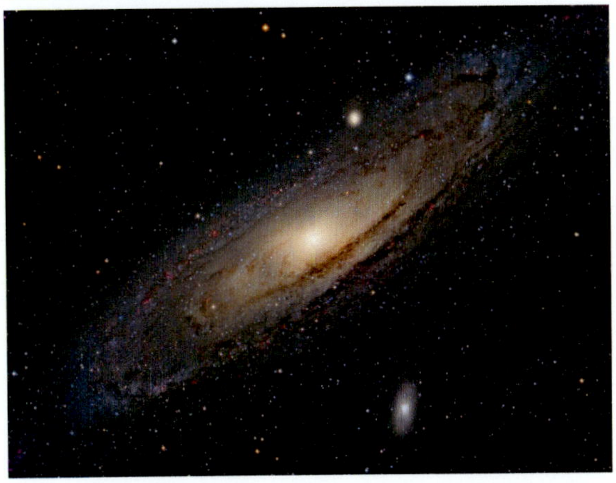

*fig.2* *The Andromeda Galaxy is a magnet for beginners. My first ever image is shown on page 10. The image above shows the improvement after a few years imaging. It was acquired over several weeks, using a short telescope, a monochrome CCD sensor and a filter wheel with red, green and blue filters.*

or consumer digital cameras with interchangeable lenses, either coupled to the back of a short telescope or with a camera telephoto lens. Typically, the camera system fits to a motorized telescope mount and multiple exposures of a few seconds to 20 minutes each. Short focal length telescopes by their nature have short lengths and smaller diameters with correspondingly lightweight focus tubes. There are some practical challenges associated with this type of photography, these include achieving fore-aft balancing and the rigidity of the focus mechanism and tube as a result of a heavy camera hanging off its end. If you live under a regular flight path, a wide field brings with it the increased chance of aircraft trails across your images.

Deep-sky images, as they come out of the camera, unlike those for lunar, solar or planetary images, mostly appear as featureless black, with a few pinpricks of light. It takes special processing to transform the faint shadow details into billowing clouds of glowing gas, dust and thousands of faint stars.

### Narrow-Field Deep-Sky Astrophotography

The smaller objects, typically galaxies and small nebulae require a longer focal length to make meaningful images, starting at around 800 mm. As the magnification increases, the image brightness reduces, unless the aperture increases at the same rate. More likely, it requires longer imaging times and the need for more specialized cameras, with high sensitivity and low noise. This quickly becomes a lesson in practicality and economics. Affordable refractor telescopes at the time of writing have typically a 5-inch or smaller aperture and at the same time, reflector telescopes have between 6- and 10-inch apertures. Larger models do exist, to 16 inches and beyond, but come with the inherent risk of an overdraft and a hernia. The longer exposures required for these highly magnified objects benefit from patience, good tracking and a cooled CCD camera. At higher magnifications, the effect of atmospheric turbulence is noticeable and it is usually the weakest link in the imaging chain. These objects are more demanding on equipment and technique; their long exposures require accurate tracking and image focus, which in turn typically require autoguiding and autofocus systems.

### Lunar Imaging

The Moon is the most obvious feature of the night sky and easily passed over for more unusual objects. Several astronomers, including the late Sir Patrick Moore, specialized in lunar observation and photography. Being a large and bright object, it does not mandate extreme magnifications or an expensive cooled CCD camera. (The Moon is smaller than you may imagine and, for instance, a 200-mm lens fitted to an APS-C camera renders a small image.) Many successful lunar photographs use a modest refractor telescope with a small video or digital camera adapted to fit into the eyepiece holder. It requires high magnification and the short resulting video image wobbles and shimmers. This video is only a starting point. Many frames are not sharp; subsequent processing discards these and aligns and combines the remainder to make a detailed image. Increasingly, digital SLRs are used for lunar photography, either in the increasingly popular video modes or take individual stills at high shutter speeds. The unique aspect of the Moon, and to some extent some planets too, is that their appearance changes from night to night. As the Moon waxes and wanes, the interesting boundary between light and shade, the terminator, moves and reveals the details of a different strip of the lunar surface. No two nights are precisely the same and many are drawn to its enigmatic lure, though the lack of color is made up for by special events such as eclipses.

### Solar Imaging

Solar imaging is another specialist activity, especially during the summer months, and provided it is practiced with extreme care, conventional telescopes can be employed using a purpose-designed solar filter fitted to the telescope. Specialist solar scopes which feature fine-tuned filters, maximize the contrast of the Sun's features. These are expensive devices though and cannot be used for any other form of imaging. The resulting bright image is usually photographed with a high-speed video camera or a still camera.

### Planetary Imaging

The larger and brighter planets, Jupiter, Saturn, Venus and to a lesser extent Mars, share some of the challenges of lunar imaging. These bright objects require short exposures but with considerably higher magnification, often achieved with a long focal length telescope with the equivalent of a teleconverter lens. The camera of choice usually has a small sensor size, a high frame-rate (or video mode) and a fast interface. In these situations the small chip size is ideally matched to the image size, although there is nothing stopping one using any digital camera with video capability. I have made pleasing images of Jupiter and Mars using just a refractor with a focal length of just over 900 mm combined with a high-quality 5x teleconverter and an adapted webcam. Planetary imaging has its own practical challenges; it requires extreme magnification to show sufficient

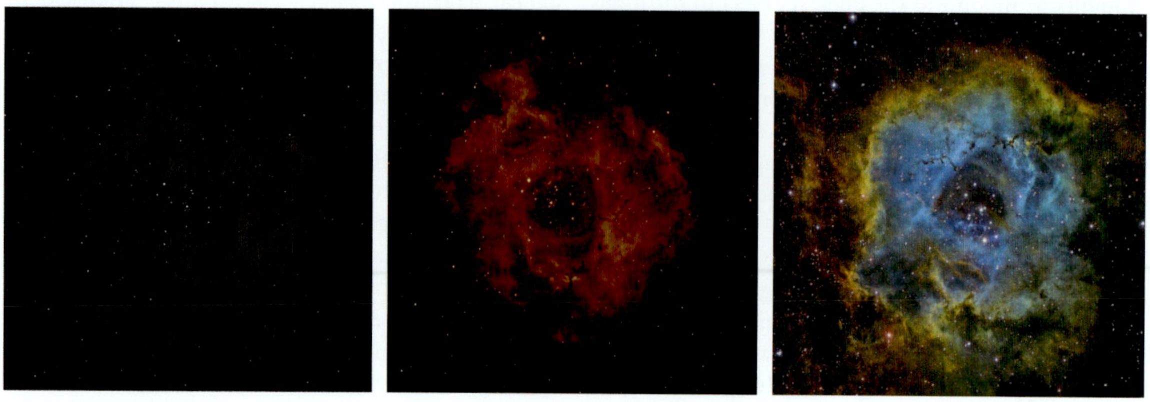

*fig.3 The Rosetta Nebula, as it might appear through an eyepiece, in more realistic colors using a conventional camera and finally, imaged with a cooled CCD camera, over many hours, through narrowband filters and processed in "false-color" for pictorial effect.*

surface detail and as a consequence is also more difficult to locate onto the small sensor using a portable setup.

At high magnification, all focus and atmospheric problems are more obvious. The unique exposure and processing techniques are quite different from those used in deep-sky astrophotography and lunar, solar and planetary imaging are considered specialty pursuits for an established astrophotographer.

#### Other Imaging Activities

Spectroscopic analysis, supernova hunting, asteroid, minor planet, exoplanet, comet and satellite tracking are further specializations for some. Supernova hunting requires a computer-controlled mount directing a telescope to briefly image multiple galaxies each night, usually following a programmed sequence. Each image, in turn, is compared with prior images of the same object. The prize is not a pretty image but the identification of an exploding star that suddenly outshines its host galaxy. Each of these specialties has interesting technical challenges associated with object location, tracking and imaging.

### Astrophotography Subjects

As you can see, astrophotography is not one pursuit, but many. We have already introduced a number of different terms that may be new to you, so, by means of a quick introduction, what follows is a brief description of the different objects and subjects that make up the world of astrophotography.

#### Stars

The points of light that we see in the night sky are stars, well, almost. Our own Sun is a star, but the planets of our solar system are not, they merely reflect our own Sun's light. Every star is a gravitationally bound luminous sphere of plasma; a thermonuclear light bulb. Our eyes slowly adapt to dark conditions and after about 30 minutes in dim light, we can see up to 3,000 on a dark night with the naked eye. That number decreases rapidly with light pollution. Most stars have their own solar system, but their distance and brightness is such that we cannot directly observe any orbiting planets, even with the help of space-borne telescopes. Not all stars are equal; they can be a range of masses, temperatures and brightnesses. Stars have a sequence of formation, life and decay spanning billions of years. Most stars, including our own sun, follow a predictable life cycle. There are exceptions; intensely dense white dwarfs and the huge red giants, some of which are so large, we could fit our entire solar system within their boundary. There are countless stars in a galaxy but at the end of a star's life, if it explodes and briefly becomes a supernova, it can outshine its entire parent galaxy. In our own Milky Way galaxy, documentary evidence suggests on average, there are about three supernova events per century.

From a visual standpoint, although stars may be different physical sizes, they are so distant from Earth, they become singular points of light. The only star to be resolved as something other than a point of light, and only by the largest telescopes, is the red giant Betelgeuse in the constellation Orion. It is puzzling then that stars appear in photographs and through the eyepiece in varying sizes, in relation to their visual intensity. This is an optical effect which arises from light scatter and diffraction along the optical path through our atmosphere, telescope optics and the sensitivity cut-off of our eyes or imaging sensor. Stars as single objects are not the most interesting objects to photograph, although there is satisfaction from resolving double stars and specific colored stars, such as the beautiful Albireo double.

When imaging stars, the main consideration is to ensure that they all are circular points of light, all the way into the corners of the image, sharply focused and with good color. This is quite a challenge since the brightness range between the brightest and dimmest stars in the field of view may be several orders of magnitude. In these cases, the astrophotographer has to make a conscious decision on which stars will oversaturate the sensor and appear as pure white blobs and whether to make a second or even third image set, with shorter exposure times, for selective combination later on. Very few images are "straight".

#### Constellations

Since ancient times, astronomers have grouped the brighter stars as a means of identification and order. In nontechnical terms, we refer to them as constellations but strictly speaking, these star patterns are asterisms and the term constellation defines the bounded area around the asterism (fig.4). These are irregular in shape and size and together they form a US state-like jigsaw of the entire celestial sphere. This provides a convenient way of dividing the sky and referring to the general position of an object. The 12 constellations that lie on the path of our companion planets' orbits (the ecliptic) have astrological significance and we know them as the constellations of the Zodiac.

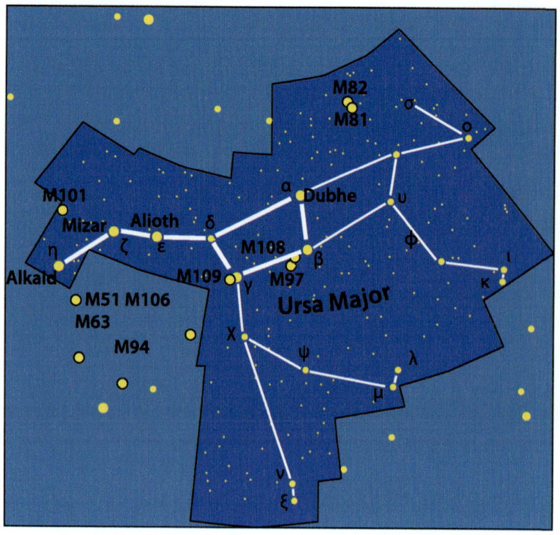


*fig.4 The "Plough" is one of the most identifiable asterisms, also commonly known as the Big Dipper in the US, being part of the constellation Ursa Major, the extent of which is shown in dark blue. Many of the stars are named and additionally use the Greek alphabet to designate their order of magnitude. This constellation is also home or close to several well-known galaxies, here indicated by the M-designation from the Messier catalog.*

### Star Names

Over thousands of years, each culture has created its own version of the constellations and formed convenient join-the-dot depictions of animals, gods and sacred objects. It has to be said that some stretch the imagination more than others. Through international collaboration, there are now 88 official constellations. The brightest stars have been named for nearly as long. Many, for instance, "Arcturus" and "Algol", are ancient Arabic in origin.

For some time a simple naming system has been used to label the bright stars in a constellation: this comprises two elements, a consecutive letter of the Greek alphabet and the possessive name of the constellation or its abbreviation. Each star, in order of brightness, takes the next letter of the alphabet: for instance, in the constellation Centaurus, the brightest star is Alpha Centauri or αCen, the next is Beta Centauri or βCen and so on.

Beyond the limits of the Greek alphabet, the most reliable way to define a star is to use its coordinates. As the number of identifiable stars increased, incremental catalog systems were introduced to identify over 1 billion objects in the night sky. The various catalogs overlap considerably and occupy gigabytes on my PC.

## Deep-Sky Objects

A deep-sky object is a broad term referring to anything in the sky apart from singular stars and objects in our solar system. They form the basis of most astrophotography subjects and include clusters, nebulae and supernova remnants.

### Star Clusters

As stars appear to be randomly scattered over the night sky, one would expect there to be several groups of apparently closely packed stars. Clusters are strictly groups of stars in close proximity in three dimensions. They are characterized into two groups: those with a loose sprinkling of approximately 100 to 1,000 younger stars, such as the Pleiades, are called open clusters. These may also have glowing gas and dust associated with them. Those with 10,000 or more densely packed stars are older and are referred to as globular clusters of which M13 and Omega Centauri in the Northern and Southern Hemispheres are wonderful examples.

In modern times we have detected clusters in neighboring galaxies but they are too distant to resolve as individual stars. The clusters we image are located in our own Milky Way galaxy. As well as being beautiful objects, clusters contain some of the oldest stars in the galaxy and are subject to intense scientific study too. An image of a star cluster is a showcase for good technique as it requires a combination of good tracking, focus, exposure and resolution. They are a challenging subject for the beginner.

### Double and Binary Stars

A double star describes a distinguishable pair of stars that appear visually close to one another. In some cases, they really are, with gravitational attraction, and these are termed visual binaries. Binary stars are one stage on, a pair of stars revolving around a common center of gravity but appearing as one star. Amazingly, scientists believe that over 50% of Sun-like stars have orbiting companions. Most binary stars are indistinguishable but sometimes with eclipsing binaries, the light output is variable, with a defined periodicity. These have more scientific interest than imaging purposes. Resolving well-known double stars provides a practical assessment of optical resolution.

### Variable Stars

Variable stars also have more scientific significance than pictorial. A class of variable star, the Cepheid Variables, unlocked a method to gauge distances by a chance discovery: in the early 20th century, scientists realized that the period of the pulsating light from many Cepheid Variables in our neighboring galaxy, the Small Magellanic Cloud, showed a strong correlation to their individual average brightness. By measuring other variable stars' period and intensity in another galaxy, scientists were able to ascertain their relative distance.

### Nebulae

A nebula is an interstellar (between stars) cloud of dust, hydrogen, helium, oxygen, sulfur, nitrogen or other ionized gas. In the beginning, before Edwin Hubble's discovery, galaxies beyond the Milky Way were called nebulae. Nebulae form the backbone of many unusual and beautiful astrophotographs. Nebulae are classified into several types; diffuse nebulae and planetary nebulae.

### Diffuse Nebulae

Diffuse nebulae are the most common and as the name suggests, have no distinct boundaries. They can emit, reflect or absorb light. Those that emit light are formed from ionized gas, which as we know from sodium, neon and xenon lamps, radiates distinct colors. This is particularly significant for astrophotographers since the common hydrogen, oxygen, sulfur and nitrogen emissions do not overlap with the common sodium and mercury vapor lamps used in city lighting. As a result, even in heavily light-polluted areas, it is possible to image a faint nebula through specific "narrowband" filters with little interference. Diffuse nebulae can also be very large and many fantastic images are possible with short focal-length optics. The Hubble Space Telescope has made many iconic false-color images using "The Hubble Palette," comprising narrowband filters tuned to ionized hydrogen, oxygen and sulfur emissions which are assigned to green, blue and red image channels in an RGB image.

### Planetary Nebulae

These amazing objects are expanding glowing shells of ionized gas emitted from a dying star. They are often faint and tiny in comparison to diffuse nebulae and require higher magnifications for satisfactory images. They are not visible to the naked eye and the most intricate details require space-telescopes operating in visible and non-visible electromagnetic spectrums. The first planetary nebula to be discovered was the Dumbbell Nebula in 1764 and its comparative brightness and large 1/8th-degree diameter render it visible through binoculars (included in fig.1 and fig.5). Only the Helix Nebula is bigger or brighter.

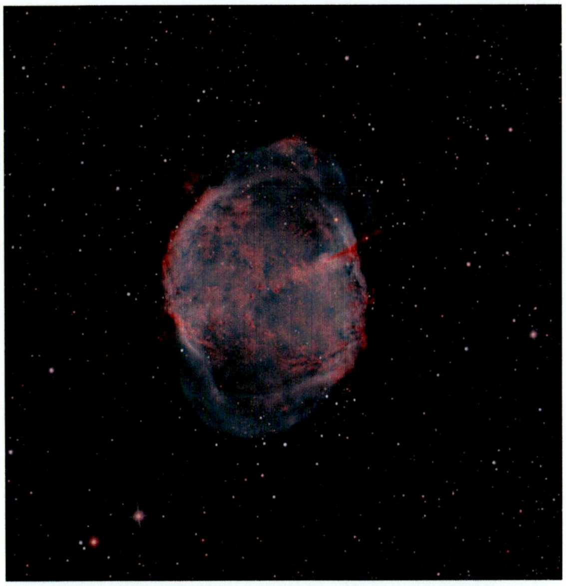

*fig.5 This image of the Dumbbell Nebula is an example of a planetary nebula, the first to be discovered. What we see here is an expanding, glowing shell of ionized gases, ejected from a red giant star during its later life.*

### Supernovae Remnants

One other fascinating nebula type forms when a star collapses and explodes at the end of its life. The subsequent outburst of ionized gas into the surrounding vacuum emits highly energetic radiation in visible and invisible wavelengths. The Crab Nebula is a notable example, originating from a stellar explosion (supernova) recorded by astronomers around the world in 1054. Amazingly, by comparing recent images with photographic evidence from the last century, astronomers have shown the nebula is expanding at the rate of about 1,500 kilometers per second. After certain classes of supernova events, there is a gravitational collapse into an extremely dense, hot neutron star. Astronomers have detected a neutron star at the heart of the Crab Nebula. They often give off gamma and radio waves but have also been detected visibly.

### Galaxies

As mentioned earlier, the existence of other galaxies outside our own was a late realization in 1925 that fundamentally changed our view of the universe. Galaxies are gravitationally bound collections of millions or trillions of stars, planets, dust and gas and other particles. At the center of most galaxies, scientists believe there is a super-massive black hole. There are billions of galaxies in the observable universe but terrestrial astrophotography concerns itself with the brighter ones. There are approximately 200 brighter than magnitude 12, but at magnitude 14 the number rises to over 10,000. The brightest is the Large Magellanic Cloud, a neighbor to our Milky Way and easily visible to the naked eye by observers in the Southern Hemisphere. Charles Messier in the 18th century cataloged many other notable examples and this catalog is a ready-made who's who.

Galaxies come in all shapes and sizes, making them beautiful and fascinating. Most of the imaging light from galaxies comes from their stars, though there is some contribution from ionized gases too, as in nebulae. Imaging galaxies require good seeing conditions and low light pollution since they are in general, less luminous than stars or clusters, have less distinct boundaries and emit a wider range of colors.

Quasars may appear as stars, but in fact they are the bright cores of very distant galaxies and are the most luminous things in the universe. They were first identified through their radio wave emissions and only later linked to a faint, visible, heavily red-shifted dot. The extreme energies involved with their emissions are linked to the interaction of gas and dust spiraling into a black hole. A few quasars are visible from Earth and within the reach of the amateur astrophotographer's equipment. These are some of the most distant objects we can see from Earth and an image of one is more of academic interest, or to claim the T-shirt.

### Solar System Objects

The prominent planets in our solar system were identified thousands of years ago. The clue to how is in the name. Derived from the ancient Greek, "planet" means wanderer, and in relation to the background of wheeling stars, Mercury, Venus, Mars, Jupiter, and Saturn appeared in different positions each night. Unlike the continual annual stately cycle of star movement, some planets perform U-turns at certain times in the calendar. Those planets closer to the Sun than the Earth are called inferior planets (Venus and Mercury) and correspondingly Mars, Jupiter, Saturn, Uranus and Neptune are called superior planets. By definition, planets orbit a sun and need to be a significant distinct ball-like shape of rock, ice and gas. The definition is a bit hazy and as such Pluto was demoted in the 20th century, after much debate, to a minor planet (of which there are many).

The Keplerian and Newtonian laws of motion amazingly predict the precise position of our planets in the night sky. Within planetarium programs, their position has to be individually calculated but from an imaging standpoint, for the short duration of an exposure, their overriding apparent motion is from the Earth's rotation, which is adjusted for by the standard (sidereal) tracking rate of a telescope.

From Earth, some planets change appearance: planets appear larger when they are close to "opposition" and closest to Earth. Mercury and Venus, being closer to the Sun than the Earth, show phases just as our Moon does, and Jupiter, Saturn and Mars change their appearance with planet rotation and tilt.

The massive Jupiter spins very quickly and completes a revolution in about 10 hours. This motion sets a limit on the exposure time of a photograph to about 90 seconds at medium magnifications and less with more. Above this time, its moons and the surface features, including the famous bands and red spot become increasingly blurred. Saturn, whose iconic ring structure has inspired astronomers since the first telescopic examination, has an interesting cycle of activity. These rings, which in cosmic terms are unbelievably thin at less than 1 kilometer, have an inclination that changes over a 30-year cycle. In 2009, the rings were edge-on and were almost invisible to Earth and reached a maximum of 30° inclination during 2016–17.

Mars rotates at a similar rate to Earth. Terrestrial photographs of Mars show some surface details as it rotates. In addition, there are seasonal changes caused by the axial tilt and its highly eccentric orbit. From an imaging standpoint, this affects the size of its white polar ice cap of frozen carbon dioxide during the Martian year (lasting about two Earth-years). Its size is under $1/120^{th}$ degree and requires high magnification and stable atmospherics for good results. It is a challenging object to image well.

### Asteroids, Satellites and Meteorites

At various times, these too become subject to photographic record (more commonly by accident). Of these, asteroids are perhaps the least interesting to the pictorialist until they fall to Earth. These lumps of rock or ice are normally confined to one of our solar system's asteroid belts, but in our prehistory, may have been knocked out of orbit by collisions or gravitational interactions of the planets. On rare occasions, asteroids pass closer to Earth than the Moon.

Satellites, especially when they pass in front of the Moon or Sun in silhouette, are visually interesting and require forward planning. More commonly, satellite images are indistinct reflections of sunlight against a dark sky. There are thousands of man-made satellites circling the Earth. The most well known have published orbital data which can be used within planetarium programs to indicate their position or line up a computer-controlled telescope. They orbit from 180 km or more away at a variety of speeds, depending on their altitude and purpose.

Meteorites are not in themselves special objects, merely the name we give natural objects when they make it to the Earth's crust. During their entry into our atmosphere, their extreme speed and the friction with the air heats them to extreme temperatures, leading to their characteristic blazing light trail and occasional mid-air explosions. The larger ones are random events, but there are regular meteor showers as the Earth passes through a stream of debris from a comet. This debris is usually made up of fine particles, smaller than a grain of sand and they burn up in our atmosphere. These events are named after the constellation from which the streaks appear to emanate. Famous meteor showers are the Perseids in August and the Leonids in November, which produce many streaks per hour. Often the most spectacular photographs make use of a wide-angle lens on a static camera and repeated exposures on a time-lapse for later combination of the best ones. Apple iOS devices and others have astronomy applications that provide a timely reminder of the regular meteor showers, location and activity. It is estimated that 60 tonnes of cosmic dust make it to Earth each day.

## Special Events

Over thousands of years, astrologers have attached significance to special astronomical events. The most well known, yet strangely unproven, is the "Star of Bethlehem" announcing Jesus' birth, which may have been a supernova explosion. These events include special causes, like a supernova, where an individual star can achieve sufficient short-lived intensity to be visible during the day, or the sudden appearance of a bright comet. Many other events consider the relative positions of a planet and the Sun, the Moon and the Sun, the phases of the Moon or the longest day or night. Modern society has disassociated itself from Astrology, but the rarity of some events encourages astronomers and physicists to travel the world to study eclipses, transits or another one-off event. The good news for astronomers is that, apart from a supernova, everything else is predictable. (Edmond Halley realized that most comets too have a predictable orbit and appearance.) For an imaging standpoint, the luck and skill of capturing a rare event add to the satisfaction of the image. As they say, "chance favors the prepared mind" and astrophotography is no different.

### Exoplanets

In recent years, some amateurs have joined in the search for exoplanets, made feasible by low-noise CCD cameras and high-quality imaging equipment. This is a highly specialized area and particularly difficult to generate reliable results. Although imaging is involved, the exoplanets themselves are not revealed by any visual means but their likely presence is detected by minute changes in their host star's light intensity as the exoplanet crosses (transits) its surface.

### Comets

Comets become interesting when they pass close to the Sun. In space, they are lumps of ice and rock circling in enormous orbits. As their orbit passes close to the Sun, the characteristic tail and tiny atmosphere (coma) develop. The tail points away from the Sun and arises from the effect of solar radiation and wind on the comet's volatile contents. Short-period comets, with orbits of less than 200 years, are widely predicted, and with a little luck, can be photographed in good conditions. More occasional visitors are often detected by the various near-Earth object telescopes long before they become more readily visible. A bright comet discovered in September 2012, name ISON, passed too close to the Sun in January 2014 and melted, robbing us of a spectacular imaging opportunity. A photograph

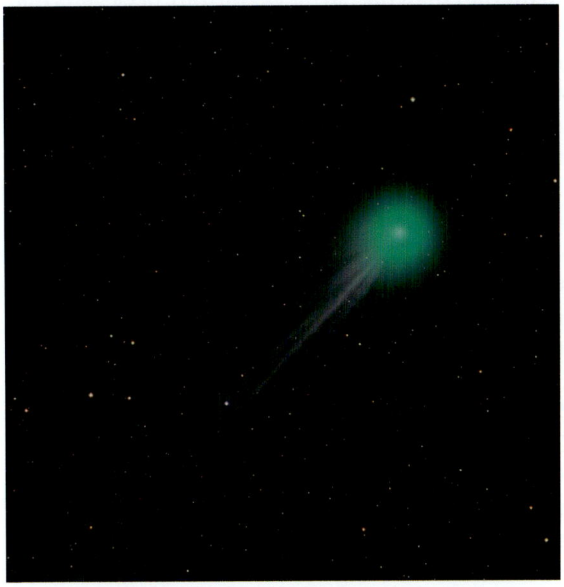

*fig.6 Some comets are regular visitors and some, like this one, take us by surprise. It was first detected in August 2014 but I was too engrossed in book writing to notice the January headlines. Thankfully, a friend pointed it out to me and I was able to grab this image before the clouds rolled over.*

of a comet is a wonderful thing, but to image it as it passes by another landmark site, such as a star cluster, makes it memorable.

### Lunar Eclipses

A lunar eclipse occurs when the Moon, Earth and Sun are in a direct line and the Moon is in Earth's shadow. We can still see the Moon, which is illuminated from scattered sunlight that passes through our atmosphere and normally appears red.

### Solar Eclipses

A solar eclipse occurs when the Earth, Moon and Sun are in a direct line and the Moon blocks our view of the Sun. By an amazing coincidence, the Moon and Sun have the same apparent size, and eclipses may be partial, where the Moon clips the Sun, or total, which provides a unique opportunity to image the solar corona safely. A total solar eclipse will only be visible from a select 100-kilometer wide tract of the Earth's surface, and avid observers will travel to far-flung corners of the world to get the best view of a "totality". My friend Andrew Duffield traveled to the US to photograph a total eclipse with a small telescope (fig.7).

### Planetary Transits

Mercury and Venus, the "inferior" planets, lie closer to the Sun than the Earth. On the rare occasions that they pass in front of the Sun, they are in transit. Man-made satellites also transit the Moon and Sun for a few seconds. Photographing the Sun during a transit requires the same mandatory precautions as any other form of solar photography. Transits occur when the nearer object is smaller than the more distant object. (Occultations occur when it is the other way around and it is possible to get transits and occultations between planets too.) In 2065, Venus transits Jupiter and in 2067, Mercury occults Neptune. Maybe something for the younger reader.

### Superior and Inferior Conjunctions

These are general terms for line-ups of astronomical bodies from an observer's standpoint. These may be between planets, a planet and the Moon or Sun or other combinations. From an imaging standpoint, it is interesting when one can make an image of two close important bodies, though the brightness difference makes it challenging. Planetarium programs are very adept at predicting these events and can produce timetables for their occurrence. Certain combinations add an interesting and unique dimension to say a landscape astrophotograph.

### Opposition

Another particular event, opposition, occurs when two bodies are on opposite sides of the sky from an observed position. This is of most significance to astrophotographers since when a superior planet is in opposition, it generally is at its closest point to Earth and hence its apparent size will be a maximum. Jupiter increases its apparent size by 66%. Mars' change is more extreme and its apparent diameter increases by more than 600%. It makes sense to image planets when they are at their closest to Earth.

### Equinoxes and Solstices

These regular events occur when the Earth is at a specific point in its orbit around the Sun. In the case of the equinox, the tilt of the earth's axis is tangential to the Sun and it has the unique characteristic that night and day are of equal length. It does not have any major imaging significance, but it does for our celestial coordinate system. There are two equinoxes per year (spring and autumn) and the celestial coordinate system uses the Sun's position at the spring equinox to define an absolute reference point for measuring right ascension. There are also two solstices each year, in winter and summer. These mark the shortest and longest day and occur when the tilt of the Earth's axis is in line with the Sun. Their significance for astrophotography mostly relates to the number of available hours for imaging.

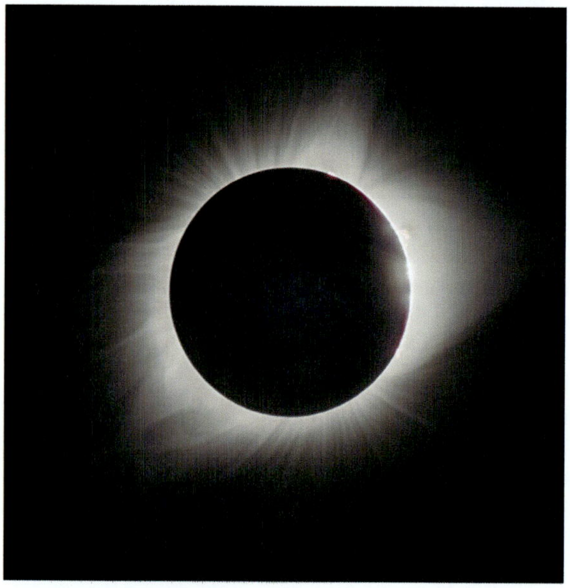



*fig.7 A brilliant image, taken in the US, of the total solar eclipse in 2017. The obscuration reveals the extended corona and the earthshine, which illuminates the lunar surface.*

# Terms of Reference

*Knowing your way around the various reference systems help identify and locate interesting objects in the sky.*

When you first start astronomy and astrophotography, it is not uncommon for the terminology to be impenetrable for a while. A broad understanding of the popular terms is useful before we move on too far, as well as understanding the conversations on the online forums and manufacturer's websites. In particular, I found the multiple references to objects, time and coordinate systems extremely confusing as well as the general proliferation of strange acronyms. It took a little while to appreciate these terms. Of course, there are many other technical terms too, explained in later chapters, as needed, as well as some others in the glossary.

### What is in a Name?

A part of the language of astronomy are the names we use to refer to various objects of interest. The most popular may have common and even fanciful names. All are documented in a more scientific manner in various astronomical catalogs. These catalogs are the à la carte menu of the cosmos and a valuable resource to the astrophotographer. It is not uncommon to have multiple names and reference numbers for the same object.

The first astronomers recorded the brightest visible stars onto ornate charts, giving stars individual names. These names appear foreign to Westerner's because they are; as many have Arabic origins. One of the earliest catalogs dates from the first millennium and lists more than 1,000 stars in detail, and interestingly included the fuzzy outlines of the Andromeda Galaxy and the Large Magellanic Cloud. As soon as telescopes were used to methodically survey the heavens, the number of objects increased exponentially. This created the need for systematic catalogs by type, position and brightness.

### Classification

As observations became more sophisticated, it was necessary to find ways of classifying stars and organizing them in logical ways. Johann Bayer started the convention of prefixing the constellation name with a letter from the Greek alphabet, in the order of their brightness, a system that is still in use today. John Flamsteed, in his star atlas of 1725, listed stars using numbers combined with the constellation in the order of their right ascension. (John Flamsteed was the first Astronomer Royal at the Greenwich Observatory, London. The observatory was built on the meridian and his telescopes pivoted in altitude only and so it was convenient for him to label stars in the order they crossed the line of sight.) Singular stars are not particularly engaging for imaging purposes but star catalogs do have an important use for accurate telescope pointing.

In 1781 the French astronomer Charles Messier published "Nebulae and Star Clusters." Crucially, this was not a star catalog but one of the deep-sky objects. He used a simple index, each prefixed with "M" for Messier to identify these objects; for example, M31 is the Andromeda Galaxy. Since observations with a telescope at that time only showed the most obvious deep-sky objects, it follows that these objects, in turn, are prime subjects for amateur astrophotography. The Messier catalog is very convenient and arguably the backbone of amateur astrophotography in the Northern Hemisphere. Indeed, at star parties "The Messier Marathon" is a challenge to see how many of his catalog items (there are 110) one can view in a single night.

One hundred years on, another significant catalog, the New General Catalog (prefixed NGC and compiled by J. Dreyer) listed about 8,000 objects, stars and deep-sky objects and remains a useful comprehensive catalog. It is astonishing to realize that these early catalogs were compiled by hand, without the help of computers or photographic records, but by patient observation and often in poor conditions.

The "Guide Star Catalog" (prefix GSC) is another important catalog, initially compiled to support the Hubble Space Telescope and is also used by amateurs with plate-solving software. (Plate-solving is a technique that recognizes the relative positions and brightness of stars in an image and matches it to a catalog database to calculate the actual image position, as well as its scale and rotation.)

In the following century, as telescopes continued to improve and crucially, photography allowed astronomers to see fainter objects, the catalogs continued their exponential expansion. In the early 20th century the Henry Draper Catalog listed more than a quarter of

| Catalog | Date | Objects | Notes |
|---|---|---|---|
| Messier "M" | 1771 | 110 | Deep space objects, including galaxies, nebulae and clusters, visible from Northern Hemisphere |
| Hickson "HCG" | 1982 | 100 | Compact groups of galaxies |
| New General Catalog "NGC/IC" | 1888<br>1908 | 7,840<br>5,386 | Nebula and star clusters |
| Abell "Abell" | 1958-89 | 4,073 | Galaxy clusters and planetary nebula |
| Rodgers, Campbell & Whiteoak "RCW" | 1960 | 182 | Hα regions, principally in Southern Hemisphere |
| Barnard "B" | ~1923 | 349 | Dark nebulae |
| van de Bergh "vdB" | 1966 | 158 | Reflection nebulae |
| Sharpless "SH2-" | 1953-59 | 313 | Hα and planetary nebula and supernova remnants |
| Caldwell "C" | 1995 | 109 | 109 deep space bright objects missed by Messier or in Southern Hemisphere, by Sir Patrick Caldwell Moore |
| Lynds' Bright Nebula "LBN" | 1965 | 1,117 | Bright nebulae |
| Gum "Gum" | 1955 | 85 | Bright Hα regions |
| Cederblad "Ced" | 1946 | 330 | Diffuse nebulae |

*fig.1 A selection of common catalogs from which one will find the majority of imaging targets. (These are not star catalogs.) They were compiled with other things in mind and one has to be selective, choosing those objects of sufficient size for the imaging field of view. There is also considerable overlap between these catalogs; with many objects appearing in more than one.*

million stars, and later still, using satellite imagery, the Tycho-2 catalog identifies positions and color information of 2.5 million stars in the Milky Way.

In practice, many common objects have several names, corresponding to their listing in each of the popular catalogs, and in addition, descriptive names based on their appearance. Thankfully, we do not need to pore over large books of numbers but can use planetarium programs on computers, smartphones or tablets to select objects for viewing or imaging, display its image and display its relative size and brightness. Many planetarium programs can also command a telescope to point to an object via a number of connections; from external RS232 serial, Bluetooth, WiFi, wired Ethernet or remotely over the Internet.

Today the main catalogs are available in digital formats and are freely available; for example from US and European Space Agency websites, though many are automatically included with planetarium programs. Clearly, in the early days, as new objects were identified the catalogs expanded and overlapped previous editions. Subsequently, as measurement techniques improved, those with more accurate measurements of position, brightness and color replaced earlier surveys. Even so, stars and galaxies are on the move, relative to Earth and to each other and so any catalog's accuracy will change in time. This perhaps has less significance for the amateur but for scientific use, renewed surveys are required to update their databases.

## Too Much Data?

To avoid the proverbial needle in a haystack issue, we often use simplified compiled catalogs which select objects for specific purposes, for instance, all stars or galaxies brighter than a certain magnitude. Even with digital computers, too much data can obscure or slow down the search and display. The various catalogs on my PC occupy over 20 GB.

## Catalogs for Astrophotographers

The catalogs of interesting objects are the most useful for astrophotographers (fig.1). For deep sky objects, subsequent to the ubiquitous Messier catalog, Sir Patrick Moore generated a supplementary hit list of 109 objects in his Caldwell Catalog. He noticed that Messier had excluded objects that were only visible in

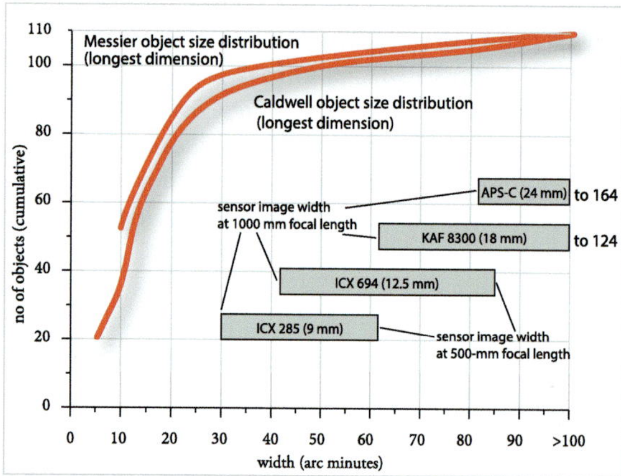


*fig.2 This graph shows how many objects are smaller than a certain field of view for several sensor sizes and focal lengths. Only 2 Messier objects are wider than 100 arc minutes.*

the southern hemisphere and had missed quite a few interesting bright deep-sky objects too. Since Messier had already taken the "M" prefix, Moore used his middle name Caldwell and used "C" instead. His catalog is listed in numerical order of degrees away from Polaris (declination). Most of the Messier objects are less than 2 degrees wide. The graph in fig.2 compares the field of view of common sensor/telescope combinations with the object size.

In addition to Caldwell and Messier, a group of astronomers selected 400 deep-sky objects from the 5,000 listed in John Herschel's Catalog of 1864, all of which are observable from mid-northern latitudes and with a modest telescope. It is called the Herschel 400. About 60 objects in the Herschel 400 also repeat in the Messier or Caldwell catalogs. Another set, by Stewart Sharpless, lists 313 areas of bright nebulosity. These objects are prefixed with Sh2-. These objects are less commonly imaged and include some overlooked treasures. Others select galaxy groups, double stars and so on.

The astrophotographer has more objects to photograph than a lifetime of clear nights. The choice is bewildering and thankfully many planetarium programs offer recommendations for a given night. The huge astrometric databases are of more importance to the scientific community but can be used for improving pointing accuracy and supernova detection in amateur systems. Most are available as free downloads from the Internet and most planetarium programs are able to load and access them selectively. If too many catalogs are enabled at the same time, the star map is cluttered with multiple names for each object. To add to the fun, several popular objects have multiple common names and their catalog number is used to remove ambiguity. Fig.1 shows the principle catalogs used for imaging. The star catalogs shown in fig.3 are principally used for location; including plate solving and modeling purposes.

### What to Image?

At first, the instinct will be to make "me-too" images. The most popular targets are often the brightest and largest, though, as it turns out, not necessarily the easiest to do well. There is nothing wrong with this at all and the many images of these familiar objects on the Internet provide a good yardstick to measure yourself against. The early astronomers could only detect the brightest objects and is one reason why the Messier catalog is the staple for many beginners. There are plenty of things outside this modest list and the challenge is to locate them and work out when one can image them. Many can be simply identified by using a planetarium program (with the most common catalogs loaded). Here, one peruses the evening sky, looking East, for likely candidates. This approach is instructive but I find that only the more popular diffuse nebulae are shown.

| Catalog | Date | Objects | Notes |
|---|---|---|---|
| General Star Catalog "GSC1" "GSC2" | 1989 | 20M 1B | Star catalog to magnitude 15 and 21 for space telescope navigation and plate solving. |
| Tycho 2 "TYC" | 1997 | 2.5M | Star catalog with revised proper motion, brightness and color data. |
| Hipparcos "HIP" | 1993 | 120,000 | Star catalog with extremely accurate position and motion data. |
| USNO A2 | 1997 | 526M | Extensive astrometric catalogs, down to magnitude 18, useful for plate solving in narrow fields. Recommended over UCAC4 for plate solving. |
| UCAC4 | 2012 | 113M | Extensive astrometric catalog down to magnitude 16, superseding prior UCAC catalogs. |
| NOMAD | 1997 | 526M | Merged compiled database from common sources above. Not used for critical scientific applications. |

*fig.3 A selection of common star catalogs, mostly used by astrophotographers for location purposes. Several of these are used by plate-solving applications, that correlate an image to one of these star catalogs, to determine the precise center coordinates of this image. Knowing the target and the image coordinates allows an application to issue a pointing correction.*

To find more unusual objects, other resources, including specific software packages, list catalog entries that comply with certain criteria (azimuth, altitude, size, brightness, type etc.). A paper atlas is useful too; some are dedicated to astrophotography and concentrate on imaging targets rather than stellar ones. Charles Bracken's *The Astrophotography Sky Atlas* is very useful for this purpose. Another good book, with excellent suggestions for each month, is the popular *The 100 Best Astrophotography Targets* by Kier Ruben. These and other books are listed in the bibliography.

## Object Location

Object location in a moving system requires a sense of time and space. As the famous quote goes, "time is an illusion" and as it happens, so too are coordinate systems. Consider the lone astronomer, sitting on a planet observing billions of stars and galaxies floating around in space, all in constant motion with respect to each other and their planet, which is spinning and rotating around its solar system that is, in turn, rotating around its host galaxy. With such a dynamic universe, one can start to appreciate the dilemma that faces anyone who wants to make a definitive time and coordinate-based system to locate objects in the universe. It is quite a challenge and requires some pragmatic assumptions.

The solution is to agree on a suitable reference for space and time. Even something as simple as the length of an Earth day is complicated by the fact that although our Earth spins on its axis at a particular rate, since we are simultaneously moving around the Sun, the length of a day, as measured by the Sun's position, is different by about 4 minutes. An Earth-based coordinate system for measuring a star's position is flawed since the Earth is spinning, oscillating and orbiting its solar system, galaxy and so on. In practice, it requires some practical assumptions to form a basis to which corrections are made for lesser effects. Our Earth's daily rotation is almost constant and the tilt of the axis about which it rotates varies very slowly over 26,000 years (over an angular radius of 23°). (I find it incredible that this slow shift was detected and measured by Hipparchus in 125 BC , without the aid of a telescope or accurate timepiece.) The name given to the change in the orientation of the Earth's axis is "precession" and the position of the North Celestial Pole (NCP) moves against the background of stars. Currently, Polaris is a good convenient approximation (about 0.75° away) but in 3,200 years, the star Gamma Cephei, currently 12.5° away, will be closer.

The practical upshot of all this is that there are several coordinate and time systems, each optimized for a specific purpose. The accuracy requirements will be different for science-based study, versus more humble, down-to-earth systems employed by amateur astronomers. Having said that, the amateur is impacted by the gradual change in precession and a good polar scope, designed to align a telescope to the NCP, will have several concentric rings to show the changing position of Polaris over a decade or so.

## Time Systems

### Local Time (LT)

This is the time on our watch, designed for convenience. Most countries make an hour correction twice a year (daylight saving) to make the daylight hours fit in with sunrise and sunset. As one travels around the Earth, the local time in each country is designed to ensure that the daylight hours and the Sun's position are aligned.

### Local Sidereal Time

Local sidereal time is a system designed for use by astronomers. It is based on the Earth's rotation and does not account for its orbit around the Sun. Its "day" is 23 hours, 56 minutes and 4.1 seconds and allows one to form an accurate star clock. If you look at the night sky at a given LST each night, the stars appear in the same position. It is the basis of the Equatorial Coordinate system described later on.

### Universal Time (UT)

Perhaps the most common time system used by amateur astronomers is Universal Time. This is another clock time, based on the local time on the north-south meridian, which passes through Greenwich, London. It has a number of different names, including Greenwich Mean Time (GMT), Zulu Time and Coordinated Universal Time (UTC). It is synchronized with the Earth's rotation and orbit and is accurate enough for practical purposes. Each night at a given time, however, a star's position will change. This is attributable to the 4-minute time difference between a 24-hour day and a sidereal day.

### Atomic Time

Time systems based on astronomical events are ultimately flawed. The most stable time systems are those based on atomic clocks; over the course of a decade, small changes in the Earth's rotational speed add up. Atomic clocks use the ultra stable property of Cesium or Rubidium electronic transitions. If you use Global Positioning Satellite (GPS) signals to locate and set your time, you are also benefitting from the stability of atomic clocks.

### Barycentric or Heliocentric systems

For academic interest rather than practical use for the amateur, this system uses the Sun as the reference point for observation, rather than the Earth. This removes the sub-second errors incurred by the change in Earth's orbit between measurements. One use of this system is the timing of eclipsing binary stars.

## Other Time References

There are some other time references that you may meet along the way. Of these, the most common are:

### Julian Dates (JD)

Julian dates are a day-number system that allows users to calculate the elapsed time between two dates. The formula converts dates into an integer that allows one to quickly work out the interval. For example, the 22nd

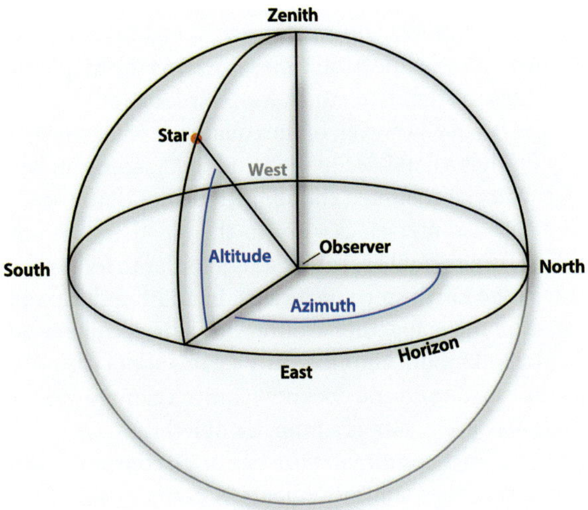


*fig.4 Horizontal coordinates are convenient for locating an object in the sky, from its altitude and a simple compass bearing (azimuth). Unfortunately for a single object, its Alt/Az coordinates constantly change with time and with the observer's position on the Earth.*

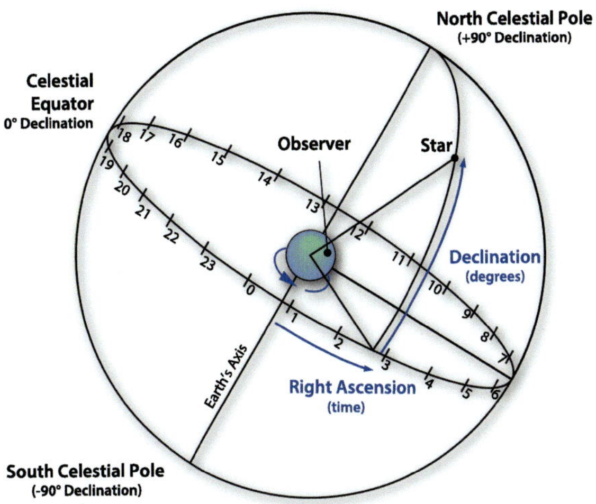


*fig.5 Equatorial coordinates are at first sight more confusing but each object has a unique coordinate that is unchanging with season, time or the observer's position. These coordinates are used by catalogs and telescope mount systems.*

January 2013 is JD 2456315. (A similar idea is used by spreadsheet programs to encode dates.) An example of an online calculator can be found at:

*http://aa.usno.navy.mil/faq/index.php*

## Epoch

An epoch is a moment in time used as a reference point for a time-changing attribute, for instance, the coordinate of a star. Star position data (called astrometric data) often references the epoch of the measurement or coordinate system. One common instance, often seen as a check-box in planetarium and telescope control software, is the choice between J2000 and JNow. These are the coordinate system as defined in 2,000 AD and today. As the years progress, the difference and selection will become more significant. In most cases, the underlying software in the applications or device drivers translate coordinates between epochs and is transparent to the practical user. I say in most cases but even the most experienced users sometimes are caught out by a new application which mysteriously consistently points to the wrong region of the sky or fails to center on a star after repeated corrections.

## Coordinate Systems

There are several fundamental coordinate systems, each with a unique frame of reference and designed for very different purposes.

## Horizontal Coordinates

Perhaps the most well known and easily understood system uses the astronomer's time and position on Earth, with a localized horizon and the zenith directly above. The position of an object is measured with a bearing from North (azimuth) and its elevation (altitude) from the horizon, as shown in fig.4. These terms are often abbreviated to AZ and ALT. This system works like a battleship gun and is embodied in altazimuth telescope mounts, which are the astronomy equivalent of a pan and tilt tripod head and also abbreviated to "Alt-Az mounts".

While it is easy to judge the position of an object in the night sky with this information, an object's ALT/AZ coordinates change with time and location. In the image-planning stage, horizontal coordinates, say from a time-synchronized planetarium program, are a useful reference for determining the rough position of the subject, but the inconvenient truth is that horizontal coordinates are only temporarily useful.

## Equatorial Coordinates

Unlike horizontal coordinates, a star's position, as defined by equatorial coordinates, is a constant for any place and time on the Earth's surface. (Well, as constant as it can be in the context of star's relative motion and Earth's motion within its galaxy.) For a given epoch, planetarium programs or the handset on a programmable telescope mount will store the equatorial coordinates of planets, galaxies and many stars. It requires a little maths, using the additional information of your local time and location on the Earth for a computer to convert any star's position into horizontal coordinates or display on a computer screen.

Equatorial coordinates are tricky to explain; as with horizontal coordinates, they have two reference points for two angular values. The first reference is the Celestial Equator, a sort of tilted horizon, as shown in fig.5, according to the tilt of the Earth's axis of rotation. If you can imagine the stars as points on a glass globe around the spinning Earth, a star's declination or DEC is the angular measure from the equator. For instance, the pole star (Polaris) is very close to the North Celestial Pole and has a declination of +89.5°. If you observe the sky from the North Pole, you would see a fixed set of stars endlessly going around in a perfect circle with the center directly above you. In this special case, a star's declination is equal to its altitude. If the observer is on the equator, the stars rise in the East, pass straight over and set in the West, like a universal comb-over.

The second reference point lies on the Celestial Equator, from which the star's bearing is measured in hours, minutes and seconds (for historical reasons) rather than degrees. Unlike the azimuth value in horizontal coordinates, which is measured clockwise from true north, the star's bearing (right ascension or RA) is measured counter-clockwise from the zero-hour reference point. (The reference point is arbitrary and corresponds to a special event, on the occasion of the Spring Equinox, where the Sun, moving along the ecliptic, crosses the Celestial Equator.)

From an observer's standpoint, say at the latitude of the UK or North America, the North Celestial Pole is not at the zenith but some 30–40° away, and the stars wheel around, with many appearing and disappearing across the observer's horizon. (From these latitudes, the North Celestial Pole is directly above the North Pole and hence Polaris has been used as a night-time compass for thousands of years, which is just as well since magnetic North is inaccurate, shifts with time and changes with one's position on the Earth.)

The equatorial coordinate system is quite confusing for an observer unless they are equipped with an aligned telescope to either celestial pole; unlike horizontal coordinates, the right ascension for any given direction is continually changing. Even at the same time each night, the right ascension changes by 4 minutes; the difference between a day measured in universal and sidereal time. Unlike the horizontal coordinate system, an astronomer armed with just a compass and equatorial coordinates would be unable to locate an object without the additional information of their latitude and local time (as well as the magnetic declination from true north for their location).

The beauty, however, of the equatorial system is that any star has a fixed declination and right ascension and an equatorial-mounted and aligned telescope only needs to rotate counter-clockwise (in the Northern Hemisphere) on its right ascension axis in order to follow the star as the Earth spins on its axis. In addition, since all the stars move together around this axis, a camera on an aligned mount automatically rotates about the pole, just as the stars do and does not require a camera rotator to resolve every star as a pinprick of light.

Equatorial coordinates are not quite constant, even if one discounts star movements. A comparison of the readouts of a star's position in successive years shows a small change, due to the Earth's precession mentioned earlier, and serves as a reminder that the absolute position of a star requires its coordinates and epoch. In practice, the alignment routine of a computerized telescope mount or as part of the imaging software soon identify the initial pointing error and make adjustments to their reference model. Linked planetarium programs accomplish the same correction through a "sync" command that compares the theoretical and actual target and compensates with a calculated adjustment to the telescope position and a small nudge.

## Other Reference Terms

### Galactic Coordinates

Galactic coordinates are used for scientific purposes and remove the effect of the Earth's orbit by using a Sun-centered system, with a reference line pointing towards the center of the Milky Way. By removing the effect of Earth's orbit, this system improves the accuracy of measurements within our galaxy.

### Ecliptic, Meridian and Celestial Equator

There are a couple of other terms that are worth explaining since they come up regularly in astronomy and astrophotography. The ecliptic is the apparent path of the Sun across the sky, essentially the plane of our solar system. The planets follow this path closely too and planetarium programs have a view option to display the ecliptic as an arc across the sky chart. It is a useful aid to locate planets and plan the best time to image them.

The meridian is an imaginary north-south divide that passes through both celestial poles, the zenith and the north and south points on the observer's horizon. This has a special significance for astrophotographers since with many telescope mounts, as a star passes across the meridian, the telescope mount has to stop tracking and perform a "meridian flip." This flips the telescope end-to-end and side-to-side on the mount so that it can continue to track the star without the telescope colliding with the mount's support. At the same time, the image turns upside down and any guiding software has to change its polarity too. During the planning stage, it is useful to display the meridian on the planetarium chart and check to see if your object is going to cross the meridian during your imaging session so that you can intervene at the right time, perform a meridian flip and reset the exposures and guiding to continue with the exposure sequence. The more advanced systems can accomplish this without manual intervention.

The Celestial Equator has been mentioned briefly before in the discussion on equatorial coordinates. The plane of the celestial equator and our Earth's equator are the same, just as the North Celestial Pole is directly above the North Pole. The effect of precession, however, means that as the tilt of the Earth's axis changes, so does the projection of the Celestial Equator and the stars will appear to shift in relation to this reference plane.

### Degrees, Minutes and Seconds

Most software accepts and outputs angular measures for longitude and latitude, arc measurements and declination. This may be in decimal degrees (DDD.DDD) or in degrees, minutes and seconds. I have encountered several formats for entering data and it is worthwhile to check the format being assumed. Common formats might be DDDMMSS, DDD° MM' SS" or DDD:MM:SS.

In each case, a minute is 1/60th degree and a second is 1/60th of a minute. In astrophotography, the resolution of an image or sensor (the arc subtended by one pixel) is measured in arc seconds per pixel and the tracking error of a telescope may be similarly measured in arc seconds. For instance, a typical tracking error over 10 minutes, without guiding, may be ± 15 arc seconds but a sensor will have a much finer resolution of 0.5–2 arc seconds per pixel.

| distance | km | AU | ly | pc |
|---|---|---|---|---|
| Earth to Moon | $3.8 \times 10^{5}$ | $2.5 \times 10^{-3}$ | 1.2 lsec | $1.2 \times 10^{-8}$ |
| Earth to Sun | $1.5 \times 10^{8}$ | 1 | 8.3 lmin | $4.8 \times 10^{-6}$ |
| Sun to nearest star | $4.0 \times 10^{13}$ | $2.7 \times 10^{5}$ | 4.2 ly | 1.3 |
| Sun to center of Milky Way | $2.6 \times 10^{17}$ | $1.7 \times 10^{9}$ | $2.8 \times 10^{4}$ ly | $8.2 \times 10^{3}$ |
| nearest galaxy | $2.1 \times 10^{19}$ | $1.4 \times 10^{11}$ | $2.2 \times 10^{6}$ ly | $6.8 \times 10^{5}$ |
| furthest we can see | $1.2 \times 10^{23}$ | $8.0 \times 10^{14}$ | $1.3 \times 10^{10}$ ly | $3.8 \times 10^{9}$ |

*fig.6 A selection of objects and their distances from Earth in a variety of units. It helps to use different units to express distances more usefully as they become mind-numbingly big.*

## Distance

The fourth dimension is distance. Knowing the distance of an object is not required for imaging purposes but for completeness, it is interesting to see how scientists use different systems to measure the mind-boggling vastness of space. Again, several units of measure are commonly in use, with scientific and historical origins, which work in increasingly large scales. The vastness of space is such that it is cumbersome to work with normal measures in meters or miles. Larger units are required, of which there are several.

### Light-Years

Light-years are a common measure of stellar distances and as the name suggests, is the distance traveled by light in one year, approximately $9 \times 10^{15}$ meters. When we know the distance of some cosmic event, such as a supernova explosion, we also know how long ago it occurred. Distances in light-years use the symbol "ly".

### Astronomical Unit

The astronomical unit or AU for short is also used. An AU is the mean Earth-Sun distance of about $150 \times 10^{9}$ meters. It is most useful when used in the context of the measurement of stellar distances in parsecs.

### Parsecs

A distance in parsecs is determined by the change in a star's angular position from two positions 1 AU apart. It is a convenient practical measure used by astronomers. In practice, a star's position is measured twice, 6 months apart. A star 1 parsec away would appear to shift by 1 arc second. It has a value of approximately 3.3 light-years. The parsec symbol is "pc". The further the star's distance, the smaller the shift in position. The Hipparcos satellite has sufficient resolution to determine stars up to 1,000 pc away.

All these measures of large distances require magnitude uplifts; hence kiloparsec, megaparsec, gigaparsec and the same for light-years. It is easy to lose a sense of scale and to feel totally insignificant as these distances become ever larger. This sense of awe only increases as you take up this amazing hobby and realize the vastness and complexity of the cosmos.

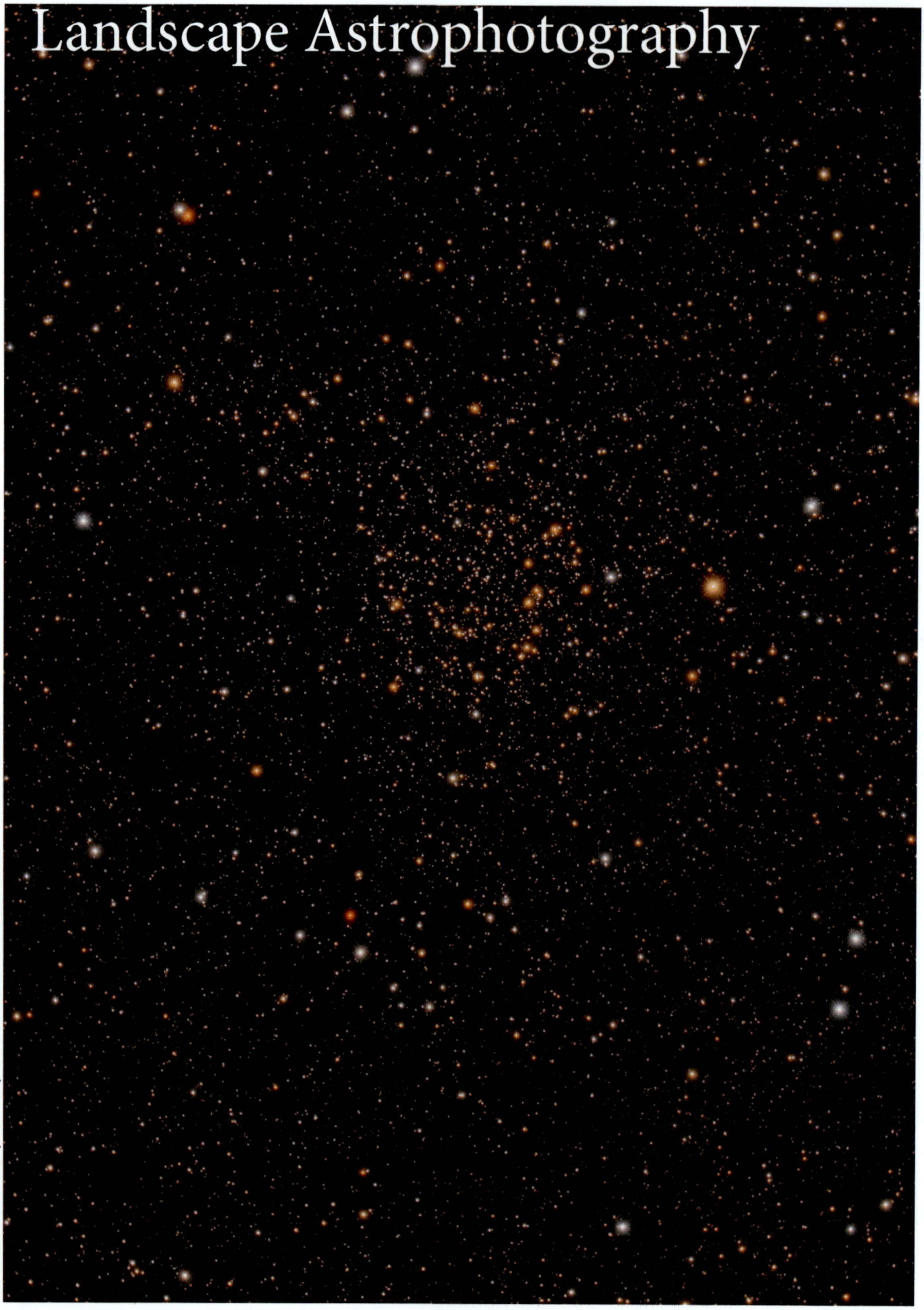


*Caroline's Rose cluster (NGC7789) taken with a color CMOS camera.*

# Getting Started

*A gentle introduction, without using specialized equipment.*

One of the best ways of learning is to just give it a go. The first two practical assignments consider landscape astrophotography. This is a logical starting point, in so much that this genre does not require specialized equipment but, at the same time, does introduce some new challenges. Most likely, your existing camera and tripod system is all that is needed, with perhaps a little insider knowledge and some patience. Some photographers specialize in this field and go to great lengths (and distances) to create spectacular images. For what initially seems to be a straightforward assignment can quickly turn into a complex technical process. We will keep it simple and then add in some complexities later on.

The subject is a diverse one; the astrophotography element covers anything from wide-field views of the Milky Way to comets, aurora, star trails or special solar-system events. The landscape part is equally important and some photographers travel to distant continents to find interesting foregrounds and context for their image. As such, the best images are carefully planned to select the most photogenic location and season, as well as the precise timing and image-taking position, leaving nothing to chance. For example, framing the Milky Way between silhouetted trees or the rising Moon close to a church spire. For the rest of us, the location may be local or a lucky chance during a vacation. Either way, there are several things we can do to make the most of each opportunity.

Our practical assignments take on two common subjects: star trails and wide field vistas. At first glance, landscape astrophotography appears straightforward; one just attaches a camera to a tripod, leaves it on automatic, takes a picture of the land and sky and there you have it. Taken from a suitably dark location you may have a record of the scene but it will unlikely have much "wow" value. To create something more compelling often requires a more thought and effort. There are several approaches; some follow the path of "take the best possible single-image exposure," usually using an expensive, fast-aperture lens and a full-frame DSLR. While it is a valid approach, it is not in keeping with the concept of this book and each assignment will introduce new techniques that are commonplace in digital astrophotography to substantially improve image quality and using lesser means.

## Challenges and Considerations

Before we start, it is worth considering the overriding practical challenge; the inescapable problem of operating with very little light and light pollution. At low light levels, the imperfections of sensors and the nature of light itself conspire to make images with a proportionally high level of image noise. To make matters worse, we often require a boost to the shadow detail to make the interesting details more obvious. At the same time, the shadows are where the apparent noise level is at its highest. Astrophotographers are perpetually finding clever ways of reducing the visible noise levels in their images by improving the signal and lowering the noise levels. Their ratio, commonly referred to as the Signal to Noise Ratio, is abbreviated to SNR. The higher the SNR number, the cleaner the image and the more it can withstand image manipulation before becoming ugly. As you will soon come to appreciate, a large part of astrophotography is the perpetual struggle to conquer noise and imperfections, at the performance limit of electronic sensors and in the presence of light pollution.

There are also some basic considerations that apply to all kinds of astrophotography and to start off on the right foot, it requires a brief acquaintance with each before we try out our first assignment.

- camera stability
- noise reduction options
- file formats
- sensor pixel size
- image resolution
- image exposure
- image calibration
- light pollution (lunar and man-made)

### Camera Stability

The need for a stable camera support increases with focal length and exposure time. A good tripod is an essential investment for any photographer and I use a selection of carbon fiber models, as they offer a good performance-to-weight ratio and the legs soak up small vibrations. The tripod head is equally important and especially the coupling to the camera. Those couplings with soft rubber or cork are best avoided since, although their purpose is to grip and not scratch a baseplate, the compliant material acts as a springboard. I prefer smooth flat plates, with an anti-rotation ledge or ones with very shallow hard pads, all of which have a dovetail fitting compatible with Arca Swiss. I am not fan of those camera bodies with a rounded polycarbonate base. These rock on a compliant coupling, no matter how tight the tripod screw is secured. Arca Swiss plates are compatible with a number of high-quality heads and I have several substantial ball heads, each with a panoramic base plate (fig.1). If I am doing vertical shots, I use a right-angle plate which allows the camera mass to be held centrally over the tripod head rather than leaning over to one side (fig.2).

If the camera is secured to a stable tripod, camera shake from an SLR mirror and shutter should not be an issue during a long exposure, as any vibration is short-lived. Human-induced vibration is more of an issue and some form of remote release, intervalometer or delayed start is useful to avoid blurred results. An intervalometer facilitates a series of fixed-length exposures at regular intervals. Some cameras have an internal mode, or one can use a handheld unit, as shown in fig.3. This model uses a 2.5-mm jack plug system that works with my Fuji and Canon systems. This model conveniently uses inexpensive AAA cells. On the camera itself, remember to disable any image stabilization modes on the body or lens. They are more likely to cause image movement.

### Noise Reduction Options

Noise is the enemy. By taking photographs in the dark, we create a situation where the generated noise in the sensor and from the light itself are greater than the faint details of the object. There are a few settings on a camera, however, that allegedly improve image noise: most digital cameras have a long-exposure noise-reduction option. This automatically takes a single dark frame exposure, immediately after the image exposure and for the same exposure time, before subtracting from the image. This removes the hot pixels but doubles the overall exposure time and interrupts a multiple image sequence. There may also be high-ISO noise-reduction mode too. Often, this is at the expense of image resolution and is not sufficient for high-quality astrophotography. In both cases, it is better to *disable* these options, as there are more effective methods for reducing image noise during image processing, with less image degradation.

Astrophotographers use multiple exposures combined with statistical techniques to reduce image noise in dim images. This important concept is fundamental to how astrophotographers achieve such amazing results from apparently an empty sky and is given a full explanation in the two *Improving Quality* chapters. Lastly, to reduce thermal noise, we keep the camera as cool as possible (for instance, by turning off any "Live View" or LCD display mode whenever possible).

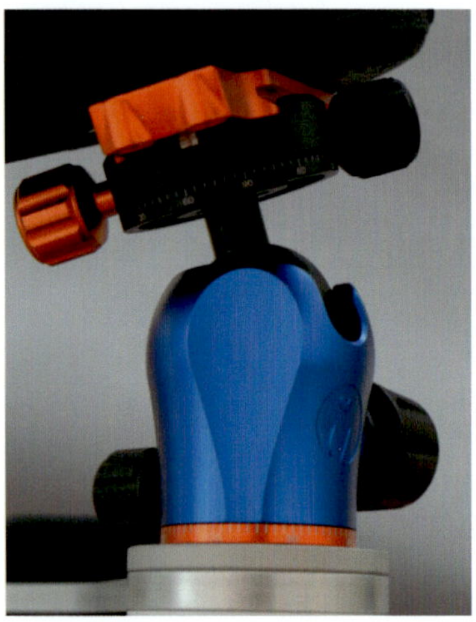

*fig.1 A good solid tripod head is essential. This model has a panoramic base and top plate, which help with checking balance and making panoramas.*

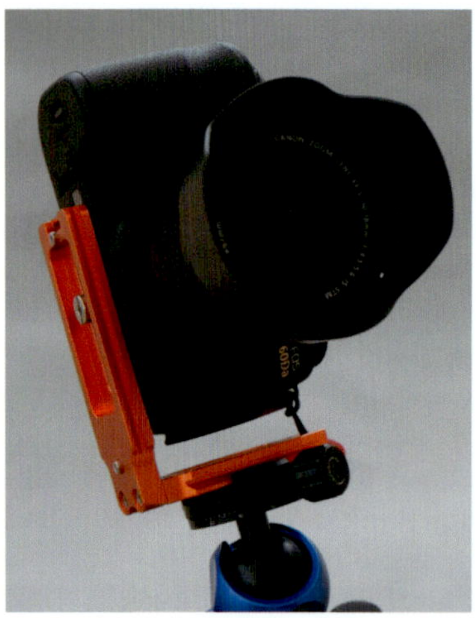

*fig.2 To improve balance, it is better to use a right-angle plate to support the camera in portrait mode. The long plate will also come in useful later on, for camera balancing.*

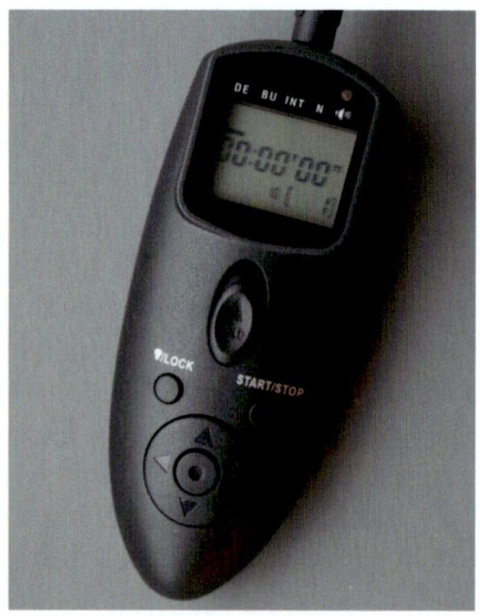


*fig.3 An intervalometer makes a simple sequence easy and vibration free. It also facilitates long exposure durations with the camera shutter set to "B.".*

### File Formats

The best file format for astrophotography is the camera's RAW format. If the camera has a compressed RAW format, it is best to avoid it, to maximize compatibility with image processing applications. A JPEG file may be sufficient for a casual image but it is less tolerant to typical image processing methods. Although RAW files are not affected by various camera modes (such as noise reduction, sharpening, film effects and tone curves) they are not quite as unadulterated as we are led to believe. All the cameras I have tried have to some extent altered the RAW data from the sensor. This undisclosed manipulation is often detected when one tries image calibration. (In addition, user reports on the Internet allege early Nikon DSLR and some Sony models consume faint stars, believing them to be image noise. With so many models in circulation, it is a good idea to search the astro forums for issues, especially before making a purchase.)

The other reason to use a high bit-depth is that image processing, especially with deep-sky images, severely distorts the tonal range to reveal faint detail and requires the highest tonal resolution it can muster. Most digital cameras have 12- or 14-bit resolution in their sensor electronics. These produce $2^{12}$ (4,096) to $2^{14}$ (16,384) light levels each for red, green and blue pixels, stored in a 16-bit file format. JPEG files, on the other hand, store 8-bits (256) per color and more readily show artifacts after image processing (such as banding or posterization). Although more universally convenient, they unnecessarily compromise image quality.

Landscape astrophotography is a hybrid activity with little specialist equipment and the images are typically stored on the memory card. In the longer term, and with full-on astrophotography it is more likely you will use an acquisition program for remote control and capture, which provides the opportunity to store the files on a computer in RAW or FITS formats. The FITS format is a scientific file format used by astronomy programs and which also permits even higher-bit depths than TIFF and RAW. Astrophotography has very singular needs and the specialist capture programs are optimized for this purpose, preserving the original data and storing it in 16- or 32-bit file formats.

While on the subject of file formats, there is a lesser-known benefit of specialist image manipulation applications that occurs during the conversion from the separate color sensor photosite values to an RGB color image. Digital camera RAW (and FITS) files require processing (de-Bayering) to create a color image from the individually filtered photosite values. The standard RAW converters used by photographers combine several (typically 4) adjacent sensor photosite values, softening the result. Some cameras additionally have an anti-alias filter, a sort of mild diffuser, in front of the sensor. In general photography, an anti-alias filter trades resolution for smoothness and is most needed to reproduce straight lines without jagged edges or strange colored banding. While this is important for normal pictures it is less so for astrophotography, as there are no straight lines. Several astrophotography programs use their own de-Bayer algorithms (especially optimized for star-fields) and convert the individual RGB photosite data into a 16- or 32-bit composite file.

### Sensor Pixel Size

Astrophotography turns tables on the never-ending pursuit of more sensor pixels. Less, in this case, is often more. In landscape astrophotography, in particular, one reads claims that full-frame DSLRs have less noise than APS-C format ones. The quality gain comes from the larger physical aperture of an equivalent lens. At the sensor level, at the same pixel size, there is little appreciable difference. My $3,000 cooled CCD astrocamera uses a 15 year-old four-thirds sensor of just 8 megapixels. Its pixels are 5.4 µm apart and in practice, most of my images do not benefit from any finer pitch. As the pixel size decreases, sensor efficiency and noise performance generally worsen (though micro-lenses and back-illuminated sensor technology may partially compensate for the losses). When you double the pixel count in any give area, the pixel dimensions only decrease by 40%. By way of comparison, the APS-C sensors in my 12-megapixel Canon EOS 1100D, 18-megapixel EOS 60Da and 24-megapixel Fuji X-T20 have 5.2, 4.3 and 3.9 µm pixel pitch respectively.

### Image Resolution

When it comes to lenses, we soon learn how image resolution is limited by the optic's physical aperture. This is not to be confused with the common f/ratio (the ratio of the focal length divided by the aperture diameter). Diffraction from a circular aperture creates a circular "haze" around a star. Likewise, an irregular shaped aperture causes an irregular shaped haze. Camera lenses typically have between 5 and 9 aperture blades that approximate a circle at any setting. At anything other than full aperture, when the blades are not in the light path, the diffraction from each of the segments causes a radial blur. When the object is a bright light source (for example, a star), the outcome is a faint star burst shape of the fairy-tale variety. This is emphasized by over-exposure or any stretching operation during image processing and may quickly become objectionable (fig.4).

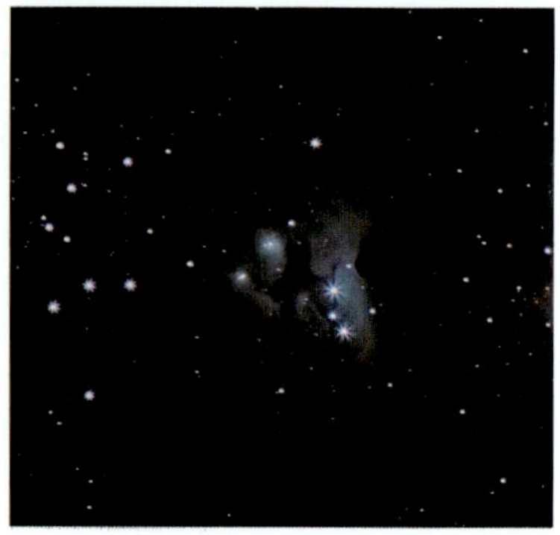

*fig.4 Irregular diffraction occurs around any discontinuity in the aperture. Here, starbursts from the 8-bladed aperture of a Canon EOS 300 mm f/4 L lens.*

### Image Exposure

No matter what camera you use, image quality improves with total exposure as it, in turn, improves the signal to noise ratio. Unfortunately, highlight clipping sets the upper limit to any single exposure. To improve image quality further the principle method used by astrophotographers is to take multiple images and combine them during image processing. The exposures are combined by averaging, to reduce the random noise level in the image. This still leaves behind consistent issues, like hot pixels, sensor row or column irregularities, dust shadows and vignetting. These issues can be fixed too, before averaging, using a process called image calibration.

### Image Calibration

Image calibration is commonplace in full-blown astrophotography but it is seldom mentioned by conventional photographers who stray into landscape astrophotography. Image calibration is an advanced topic and one that will dominate later chapters, but suffice to say that a full set of calibration images include short-exposure images with the lens cap on, images of the same duration as the image exposure (again, with the lens cap on) and images taken through a diffuser. In astrophotography parlance, these are called bias, dark and flat frames. These are used to correct the consistent errors, present in each image exposure, before averaging.

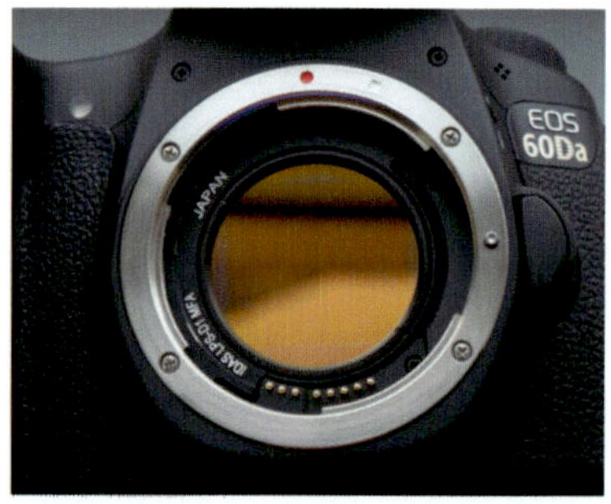


*fig.5 The popular IDAS light-pollution filters are now available in a lens throat mount for EOS cameras. This model, however, cannot be used with EF-S lenses but works fine with EF lenses and my Contax adaptor.*

### Light Pollution (lunar and man-made)

Last but by no means least, urban light pollution increasingly ruins this hobby. Even if it is not immediately apparent, a 30-second exposure of the sky reveals a murky orange mush when low-pressure sodium lights are prevalent. Light Pollution (LP) filters can help, in so much that they not only reduce the orange glow but, at the same time, removing this background light also removes the random noise that comes with it. There are several versions, all of which have complex coatings that block select wavelengths corresponding to principle street light lamps (including the latest LED street lighting, which has a broader spectrum). LP filters are only a partial solution and are increasingly expensive in larger sizes. They are becoming available as clip-in versions, that fit in front of the mirror (fig.5) or sensor. Canon DSLRs are well catered for and lately, other makes too. LP filters do not work as well with wide-angle lenses or wide aperture lenses. This is because dichroic filters rely on precisely controlled coatings, the optical path length of which determines the filter response. When light passes through at extreme angles, the effective optical path length increases and the filter response shifts, producing an uneven effect over the image.

After the Sun has set, the ambient light level continues to fall and true astronomical darkness starts when it is typically about 18° below the horizon (typically 1 hour after sunset or before sunrise). It is easy to be caught out by the Moon too. A full Moon illuminates the upper atmosphere and the scattered light from the airborne dust and aerosols washes out faint stars and hence, much of the Milky Way. Even if the Moon is not in the picture, its effect will be and my advice is to avoid periods when the Moon is more than one quarter illuminated and high in the sky. It does

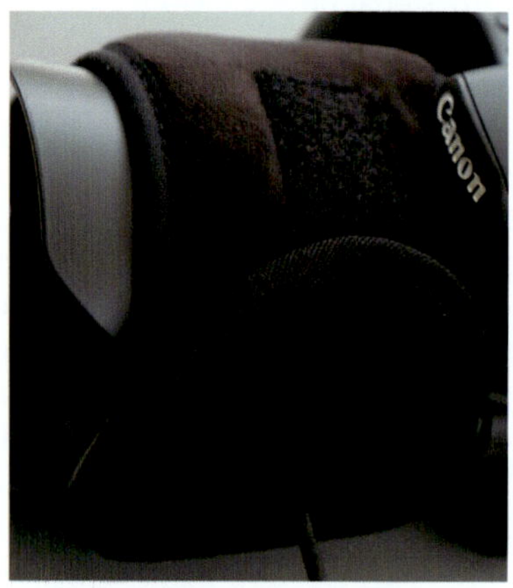


*fig.6 This innovative dew heater tape is allegedly derived from a soup warmer! It is powered by a 5-volt USB connection.*

not have to be left to chance; there are numerous free smartphone applications and online resources to take out the guesswork. To anticipate other issues, it is best to avoid areas under busy flight-paths or close to busy roads, both of which may add a few unwanted light trails. For a landscape photo, this just requires us to be vigilant during the exposures. In the assignment on star trails, which require extended exposure, it may be easier said than done as the overall image-taking period can be several hours.

## Equipment for Landscape Astrophotography

### Camera Considerations

Within reason, one can use any camera for landscape astrophotography but for best results, an APS-C or larger format camera, with interchangeable lenses is preferred. These models have better sensors, a RAW file option, better manual controls and permit a wider choice of focal lengths and larger apertures. Astrophotography is not kind to equipment; dew and low temperatures are a hostile operating environment to many and some choose an old model for the purpose. This is OK up to a point but it is worth pointing out that some older cameras have an intrusive issue called "amp glow" that occurs during prolonged exposures. The effect is reminiscent of light flare along the edge of the frame in a film camera and it is not trivial to remove. It is also likely that extended operation will exceed the capacity of one battery (especially in the cold) and spare batteries are a must, or preferably a battery eliminator; a dummy battery that fits into the camera permitting an external DC power source and removes the need to touch the camera during the exposure sequence.

### Focus Considerations

Framing and focusing a dark sky is a frustrating business, especially with a wide-angle lens. Autofocus will fail and it is virtually impossible to manually focus through a viewfinder. I found an articulated screen, or an HDMI port that can be connected to a portable video monitor are invaluable, especially when they are used with the camera's manual-focus screen magnification and preview features. The lens infinity marking and camera readout are seldom accurate and I use various alternatives with camera lenses: focusing on a planet, the Moon, a distant street lamp or simply taking a series of images at different focus positions and looking at the result up close on the screen with the preview magnifier. In later chapters we explore alternative focusing strategies that work with higher magnification systems.

### Lens Considerations

The chances are you will be reaching for a wide-angle lens (rather than a fish-eye lens) for landscape purposes. These focal lengths are less likely to show unintentional star-trailing and can show a larger sweep of the Milky Way. The lens choice itself does not alter perspective (only your feet do that) but wide angles do enable some dynamic views, encompassing dramatic sweeping vistas. Wide-angle zooms are particularly complex affairs, with many optical elements. The better models give excellent results over most of the image but still struggle at the outer edges and increasingly so at the longer end of the zoom range. This is not an issue when there is little image detail (say a featureless sky) but one quickly realizes that stars are particularly unforgiving and show up every optical issue.

It used to be the case that prime lenses were superior to any zoom lens. Zoom lenses have come a long way and the performance differences are less obvious between modern designs. There have been some outstanding optics over the years and the taking lens does not necessarily have to come from the camera manufacturer. It is not uncommon to see all manner of DSLR and film SLR lenses fitted to DSLR bodies via an adaptor and even more on mirrorless cameras on account of the shorter distance between their lens mount and the sensor. I use old Contax Carl Zeiss lenses on my Canon and Fuji. These lenses are rugged, inexpensive and focus manually. Optically, the longer focal lengths fare better than the wide-angles, even on the smaller APS-C sensor but are outperformed by modern, premium zoom and prime designs. For super-wide focal lengths suitable for vistas, the independent brands offer some interesting models. An example is Samyang, who sell 10-, 12-, 14- and 16-mm focal lengths for multiple mounts. Whichever lens you choose, its maximum aperture is an important consideration, the wider apertures give you more exposure latitude; f/4 is good, f/2.8 or wider is better, but at a price and they may also cause light-pollution filters to be less effective.

Depending on the ambient conditions, some form of dew prevention may be necessary. Condensation on the front optical surface will ruin any image. Long lens hoods are an effective passive means but are not applicable for a wide-angle lens. In the end, some form of gentle heating is required (to prevent condensation rather than evaporate it.) For occasional use, a pocket hand-warmer, strapped around the lens with an elastic band may suffice, or better still, a wrap-around heated band, called a dew heater tape, that is plugged into a 5- or 12-volt power supply (fig.6). This warms the lens elements by a few degrees, sufficient to prevent condensation. Since they usually wrap around the entire lens barrel, one thing to consider when using these devices is how do you reach the focusing ring for fine tuning.

## Exposure (Land and Sky)

Returning to the immediate challenges of landscape astrophotography, it is quickly apparent that the foreground and sky exposures may have different exposure requirements. The automatic features of digital cameras do not work as well in dim light and it is best to set exposure, focus and white balance to manual. In the case of a single exposure, it is usually optimized for the sky and short enough to keep star elongation to a minimum, leaving the landscape in silhouette.

If you are up for more of a challenge, you might additionally consider the problem in two halves, with separate exposures for land and sky and then combined later with a mask, during editing. Even star trails images, where the camera is static for the exposure, may need some nimble manipulation to visually balance the two elements of the image.

### Foreground Matters

Often taken at night, or at least in very dim lighting, the purpose of the foreground is to give some interesting context to the sky image and provide an image base that gives a notion of scale to the sky. Typically, its role is to support the image of the sky, rather than detract from it. This image may be a simple black silhouette or a dimly-lit alluring landscape. The most interesting ones have a range of object distances, giving a sense of depth and drawing the eye into the frame. That requires a careful choice of focus and aperture to ensure sufficient depth of field. For APS-C cameras, lenses are typically at their best in the aperture range f/5.6 to f/8 and 1 f/stop smaller for full-frame cameras. At true dark sites, without any light pollution, an interesting option is to experiment with light-painting during the exposure, using a hand-held torch or flash to emphasize some interesting nearby feature. The foreground exposure therefore typically has the following requirements:

- sufficient exposure to render a dimly-lit or darker foreground and potentially employ multiple exposures to minimize image noise with stacking
- stopped-down and focused to provide sufficient depth of field (or use focus bracketing)
- shutter speed chosen to either freeze or blur motion (e.g. rendering moving water or branches)
- ISO set to be compatible with chosen shutter speed and aperture (for single exposure)
- focal length chosen for sufficient field of view (or one can combine several exposures taken in a panoramic sweep)
- avoids man-made light sources in the foreground image as the deep exposure emphasizes the dimmest light into a distracting highlight
- manual white balance (daylight, cloudy or tungsten settings are common choices)

The general aim is to render the landscape in muted tones. If it is metered directly, this requires deliberate under-exposure of about 1–2 stops, or alternatively darkened during image processing. (One alternative to consider is exposing the foreground image at twilight and darken these tones during image processing using a combination of editing adjustments. It is quite easy to do and reduces image noise at the same time.) Noise is still a consideration; at the optimum aperture and focus point, which traditionally is the hyperfocal distance, the sensor sees less light intensity than at full aperture.

The hyperfocal distance $H$ is the focusing distance which provides the best overall focus coverage and at which infinity is still acceptably in focus. It varies with focal length, aperture and format. Traditional lenses have a depth of focus scale that indicates the hyperfocal distance. Many modern lenses have no scale and we have to use an aide memoir. An example of hyperfocal distances for an APS-C camera with different lenses set to f/5.6 is shown in fig.7. It is derived from the ancient equation below, where $f$ is the aperture number, $f_L$ is the focal length and c is the Circle of Confusion (CoC).

$$H = \frac{f_L^2}{c.f}$$

The CoC is the diameter of blur that is considered acceptable. It is highly dependent upon the resolution of the eye and camera and lens manufacturers, who use a range of values in their assumptions. As the sensor (or film) format increases, the degree of image enlargement for the same print size reduces and the size of the acceptable blurriness can increase. Based on standard vision, which assumes 1 arc minute resolution, the CoC is 0.022 mm for 35-mm film. The depth of field scale on my Carl Zeiss Contax lenses assumes a generous 0.04 mm. These figures are optimistic but similar to those adopted by other OEMs. On older lenses, a common rule of thumb is to use the depth of field markings on the lens for one f/stop larger than the taking aperture. For instance, if the aperture is set to f/8, use the f/5.6 markings. The smaller and more common APS-C format is about 30% smaller and the common CoC is equally smaller; in the range of 0.025–0.015 mm. With somber landscape tones, our eyes are less able to discern fine details and the larger CoC values may suffice.

On APS-C lenses, the optimum aperture is around f/5.6. Smaller apertures give more depth of field but slowly lose resolution due to diffraction. The example in fig.7 assumes a CoC of 0.025 mm for common APS-C lenses. These values are best used for the landscape part of the image. The sky exposure requires perfect infinity focusing, for pin-prick stars and to minimize optical aberrations in the image margins. In very dark surroundings, the foreground exposure can be up to 20x more than the sky exposure and stopping down will reduce light intensity and make it noisier. If the aim is to take a single exposure, the ISO setting, shutter speed and aperture need to create a dim image. If you choose to average multiple exposures, you can be more creative.

It is important to realize that a high ISO setting on a camera does not make the sensor any better; it largely boosts the signal (and noise) by about the same amount and may make it worse. There are a few variables in play here; if a 2-second image taken with a low ISO is stretched to resemble that of a 2-second exposure at a higher one, the image will look a little noisier but if the overall exposure value is kept the same, the low ISO image will be better, since the sensor has received more light, from a longer exposure time and with the same aperture setting (figs.8, 9).

White balance in astrophotography is a controversial subject; although there is a technically correct light balance, using stars of known color, it is tricky to accomplish and our aim is to make an aesthetically pleasing image and a "true" setting may not be agreeable for the landscape. For digital camera users, white balance is fairly arbitrary, but a manual white balance is a sensible precaution for consistency. Surely, I hear you say, if you are using RAW files, the white balance setting is irrelevant? The answer should be yes but depending on the image processing software you are using, it may assume the white balance setting that is stored in the RAW file's metadata. A simple precaution is to set a consistent white balance, especially if you combine images in processing. A daylight setting or a custom white point, metered off a neutral white card in daylight (including any light-pollution filter) is a good starting point in areas of low light pollution. For those less fortunate locations, a tungsten lamp setting offsets the reddish glow to some extent.

### The Sky is the Limit

The sky exposure plan depends on the purpose; a panorama of the Milky Way may have a different treatment to a narrower view of an eclipsed moon or conjunction and different again to star trails or aurora. The sky exposure is a compromise between competing considerations:

- to retain star color, each individual exposure should not cause the star's exposure to reach the maximum pixel value
- since star intensity varies widely, the above condition is a compromise between the number of clipped bright stars and revealing faint ones

| focal length | H (m) | focal length | H (m) |
|---|---|---|---|
| 8 | 0.5 | 23 | 3.8 |
| 10 | 0.7 | 27 | 5.2 |
| 12 | 1.0 | 35 | 8.8 |
| 14 | 1.4 | 50 | 18 |
| 16 | 1.8 | 90 | 58 |
| 18 | 2.3 | 135 | 130 |

*fig.7 Many lenses omit depth of field scales and this example table lists hyperfocal distances for APS-C lenses of different focal lengths (in mm). To use, focus on an object at a distance H and set the aperture to f/5.6.*

*fig.8 This frame is an enlargement of a 2-second exposure at ISO 1600.*

*fig.9 This equivalent exposure is a 16-second exposure at ISO 200. The noise level is considerably lower than the one in fig.8.*

- to give the impression of sharp stars, each exposure should be short, so that the stars are not elongated
- the sky appearance itself can be anything from very deep blue/black to lighter tones and very dependent upon light pollution
- a lens aperture of f/4 or wider is preferred, to maximize light capture and for sharp stars
- images taken at full aperture typically have the poorest corner resolution; stopping down slightly improves things but increases image noise
- images taken at full aperture have more obvious aberrations (though image processing can improve things)
- consider using filters to reduce the effect of light pollution and where to place them
- set a manual white balance for image-to-image consistency
- in the case of star trails, where deliberate elongation requires a longer exposure, it is usual to reduce the ISO setting (and aperture) to retain star color
- take multiple exposures to minimize image noise as well as calibration images (advanced)

It is interesting to contrast the land and sky exposure needs and it is easy to see why some realize an image with two very different sets of exposures and montage them together, since the demands on aperture, shutter speed and the ISO setting conflict (and the focus position too). Fortunately, modern image processing programs have very effective tools that allow you to select the land or sky, feather the edge and blend layers together into a seamless image.

Since stars are in constant motion, to give the impression of sharpness, their elongation in the image needs to be small enough to fool the eye that they are stationary. At a small scale, as on a computer monitor, one can disguise small streaks better than say on a 16x20-inch print. Likewise, it is much harder to disguise star elongation for images taken with a long focal length than a short one.

A general guideline, for a full-frame camera, is to choose a shutter speed up to the value *t*, using the following equation, where $f_L$ is the focal length (mm):

$$t\ (seconds) = 500/f_L$$

and for APS-C cameras, whose sensor requires a higher magnification for the same size output, requires a slightly shorter time:

$$t\ (seconds) = 330/f_L$$

These figures are at the very top end, ideal for a small screen image or print, using a medium quality lens. Ideally, the exposure should be much shorter, as a better lens will have more resolution and fewer optical issues and the more finely rendered stars will show elongation more readily, especially in a full-screen image or a larger print, for display purposes.

## Trailing Stars

In the case of star trails, the compromise is very different, since we want the star to show trailing in each exposure, without necessarily clipping. For this, the exposure is commonly in the range of 30–120 seconds, with an ISO and aperture setting that ensures star color is retained in all but the very brightest stars. On most cameras, an exposure over 30 seconds requires the camera to be set to "B" and activated by an external release or intervalometer, like the one in fig.3. If the exposure is under 30 seconds, things are much simpler. Here, set the camera to "M," and set the exposure time with the drive mode to continuous and use an electronic or mechanical cable release to perform a continuous stream of exposures without pausing. (Again, remember to disable all camera noise-reduction modes.)

The easiest way to settle on an exposure is to experiment; I normally stop down 1 f/stop from full aperture, and try a few test exposures at different ISO settings with a 30-second to 2-minute exposure. To create the impression of a long streak requires multiple exposures, with less than a few seconds' gap between them and processing software that combines them selectively. There are several ways of aligning and combining images to generate a star trail. These include manual methods, in a conventional image processing application, and dedicated software that does the hard work for you.

If you choose the manual route, using Photoshop or Affinity, each image is loaded into an individual layer. Assuming the camera was on a tripod, the images should be aligned (one can easily check by temporarily using a difference blend mode to highlight the image differences for a static part of the scene). To combine, change the blending mode of all the layers (apart from the background layer) to "lighten" or "screen," to combine the moving stars into one image. The blending modes are subtly different and a little experimentation is in order. Unfortunately, this method will also pick out noise, including hot pixels, aircraft, satellite and meteor streaks. The worse frames are ideally edited out and, once complete, one can flatten the individual layers.

The resulting star trail will likely have small gaps in it and as the stars are rotating in the frame, and if the lens has low distortion, it is possible to use the radial blur tool over a few pixels to hide any irregularities. The radial blur tool initially assumes a rotation axis in the middle of the image. Affinity Photo allows you to move the axis directly on the image, making it easy to select the celestial pole. Photoshop is less convenient, and although its tool allows you to move the center, it will require several iterations to find the right point. In both cases, if the image does not contain the celestial pole, extend the image canvas until it does and then crop back after applying the tool. Of course, the radial blur will also attack the landscape image, so an image mask is required to exclude it from the filter selection. At the same time, for some images, a little radial blur will improve the noise appearance in the sky background. As the exposure count increases, however, the manual method soon becomes a tedious task and many resort to a specialist utility.

### Star Trail Software

Apart from the tedium of managing multiple files in Adobe Photoshop or Affinity Photo, at the same time and especially with large pixel counts and 16-bit files, the processing consumes computer memory. While it is feasible to progressively process a few layers, flatten and then add more layers, as the frame count increases, this soon becomes irksome. One alternative is to do an Internet search for a Photoshop of Affinity batch action to automate the same process. Another alternative is to use a dedicated utility application. One favorite is StarStaX, a freeware application for Linux, Windows and OSX, which usefully works on 16-bit TIFF images. After loading the images, its various blending modes can alternatively create a star trail, remove stars and additionally perform dark frame subtraction, to remove warm and hot pixels. It has a convenient feature to fill in small gaps in the star trail too. With these various modes, it is possible to process a set of images twice, once for the sky and a second time for the land, before combining them in Affinity Photo or Photoshop with the aid of a mask. Masking images in either application is a good skill to acquire and there are plenty of on-line resources that explain the various methods, allowing this book to concentrate on the astro-specific processing elements.

## Choosing and Locating a Target

Many astrophotographs concentrate on a singular object or local area. To do so requires a knowledge of their season and location on a particular night. Landscape photography is less selective and the "feature" if any, is often the arc of the Milky Way. The most convenient way to find your bearings and for planning purposes is to use a mobile planetarium application in "compass mode." Here the smartphone or tablet uses GPS to set the location and time and its built in compass and sensors to orientate the screen dynamically with your outstretched arm. The one I typically use is SkySafari, though there are many others, like Luminos (fig.10 ) and TheSky HD. All these applications usefully show the deep-sky objects as well as the Milky Way. As we become more sophisticated, they will connect to a telescope mount and slew it, to point to the object. These applications often have a "tonight's best" selection, which covers interesting events, such as comets, meteor showers, galaxies and common catalog objects. Later on, we consider full-scale planetarium programs for telescope mount control and additionally, for planning purposes.

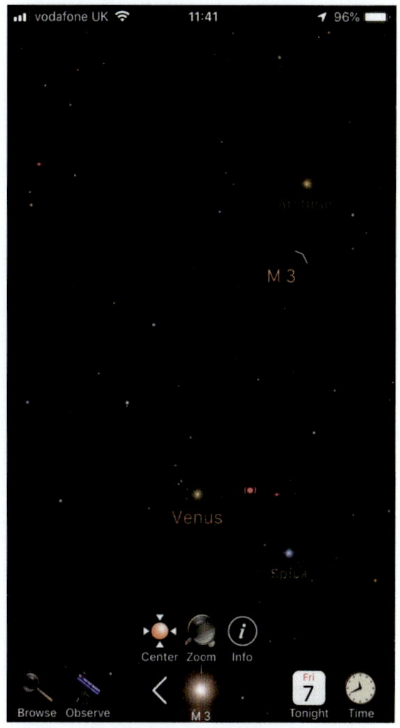


*fig.10 Mobile planetarium apps are an ideal way to find your bearings, choose an interesting part of the sky and plan a night's imaging. This one is Luminos, on an Apple iPhone. As I hold the phone up to the sky and with it angled, to reflect my face, the image moves to align itself with the view. Similar applications are TheSky mobile/HD and SkySafari. With the amount of detail on the screen, a larger screen size, or iPad is more convenient.*

# Star Trails

*Combining long exposures to create a dramatic dynamic.*

***Equipment:***
*Fuji X-T20*
*23 mm f/2 XF*
*tripod with ball and socket head*
*battery adaptor*
*intervalometer*
*mosquito repellent, small torch and head torch*

***Software:***
*RAW converter software*
*Affinity Photo*
*StarStax (or SiriL/ASTAP)*

***Exposure:***
*60 x 120 seconds at ISO 200*

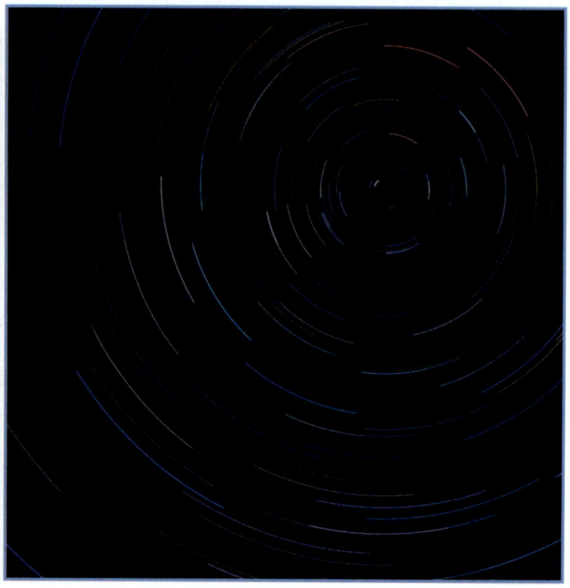

This assignment jostled with *Vista* to be the first up. In the end, it proved simpler to take and process. If you are able to take a successful landscape photograph you can take a star trail.

### Environment

Compared to an image of the Milky Way, a star trail is less sensitive to the environment since the required exposure to show star trails is not as deep (or stretched) as that for showing the subtlety of the Milky Way in all its glory. Having said that, the best results are from sites with dark skies. Not all things go to plan, however, and a winter break in Yorkshire did not produce a single clear night, forcing a more resourceful approach for this assignment.

### Setting Up

Photographing star trails requires the camera to be continuously imaging; a gap of a few seconds may well show up, especially with longer focal lengths. In common with most astrophotography sessions, the likely exposure plan will likely exceed the capacity of a single fully-charged battery, especially in cool conditions. The Fuji just manages 2 hours in freezing conditions but to be on the safe side, I used a battery adaptor. On consumer cameras, these take the form of a dummy cell that fit into the battery compartment and its DC lead passes through a little flap to the side of the battery door (fig.6). These adaptors are powered externally with a DC power supply, usually that plugs into the mains. Welcome to our first problem; you are in the middle of nowhere and even if it was safe to have a mains power supply out in the dew (which it is not) there is nowhere to plug it in. The solution requires a little practical soldering.

### DC Power Supply

For safety reasons, I use low-voltage DC power outdoors and for my EOS and Fuji cameras, I bought a small DC-DC step-down converter for a few dollars and a tiny plastic enclosure to fit it into (described in *Planning for Success*). I wired this as a small in-line voltage regulator, such that I could use a small 12-volt lead-acid battery as the power source (the ones used in burglar alarms are ideal). The required voltage for my adaptors is not stated and although the Fuji and EOS batteries have very similar indicated voltages, around 7–8 volts, these are only a guideline. To find a safe level, I first turned the adjustment screw on the converter so that the output was a few volts less than the camera's lithium battery figure before connecting the adaptor to the camera and turning it on. I slowly increased the voltage until the camera came to life and the battery charge indicator was at 50% or 2 bars. (It does not need to be higher, as the voltage will never change.) The standard power source for astronomical equipment (apart from a few telescope mounts) is 12-volts DC. While carrying a heavy battery in the field is an encumbrance, in this particular assignment, it may also be used for powering a lens warming band, or dew-tape, to prevent condensation forming on the lens surface during the session. Easy to say but more difficult to do in practice, when imaging with a stubby lens, as it gets in the way of the focus and aperture rings.

Star trails themselves take on different pictorial qualities, depending on the camera direction. In the Northern Hemisphere, pointing North will provide circular traces around Polaris. If the camera is pointed East or West, the lines will appear straighter and at an angle (dependent upon the observer's latitude). If the camera is pointed South, the trails will be gentle horizontal arcs. For fun, some photographers use photo editing software to distort the trails for pictorial effect.

### Exposure

A little experimentation is required to settle on the right exposure. If we first consider the shutter speed, it is clear we need a long exposure to register a star trail. Since any star's position moves by 1 degree every 4 minutes, a 30-degree arc requires a 2-hour exposure. A single exposure, however, does not work on several levels: the first being that the sky and land would be significantly overexposed and secondly, if one tried to compensate by using a low ISO setting and stopping down, the moving stars would only leave a faint trail and the long exposure would have significant thermal noise. (With film, one can use a single long exposure and although there is no such thing as sensor noise, it is mandatory to find a site with a really dark sky to avoid the sky and land becoming washed out.)

The solution is to take many shorter exposures, each contributing a small segment of the overall star trail and then selectively combine them. Photo editing software, including Photoshop and Affinity Photo makes it easy to combine several images (using the lighten blend mode) to elongate the star trails. This becomes a chore when the exposure count becomes excessive. For instance, without an intervalometer function, most digital cameras, slowest shutter speed is 30 seconds, representing 0.125 degrees of movement. This generates a lot of files and for a 24 MP camera, 2-hours' worth occupy over 5 GB of space. I prefer to take slightly longer exposures, around 2 minutes and set the aperture and ISO settings to ensure the overall exposure retains star color and renders the night sky and landscape in dark tones. Some sources prefer using shorter exposures, down to 30 seconds, so that a rogue one can be omitted without too much impact.

With a wide-angle lens, however, an aircraft takes longer than 20 seconds to cross the frame and, as my location is 20 miles from two airports, I expected several exposures to have aircraft trails. This is a reality for many of us and part of this assignment was to find and share the best way of removing them during image processing.

As the long exposure is set to create a trail, the aperture selection is a balance between exposure (the

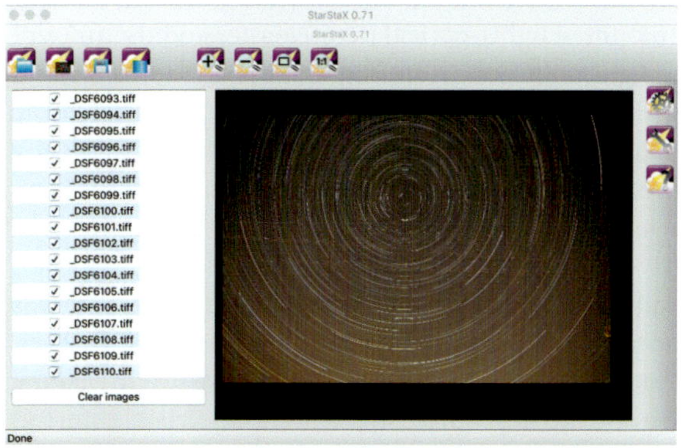


*fig.1 StarStax makes it very easy to combine images to form a star trail. It also has added features for dark frame subtraction and simple stacking too. Here, 70 images have been combined to form an image, to edit in Affinity Photo.*

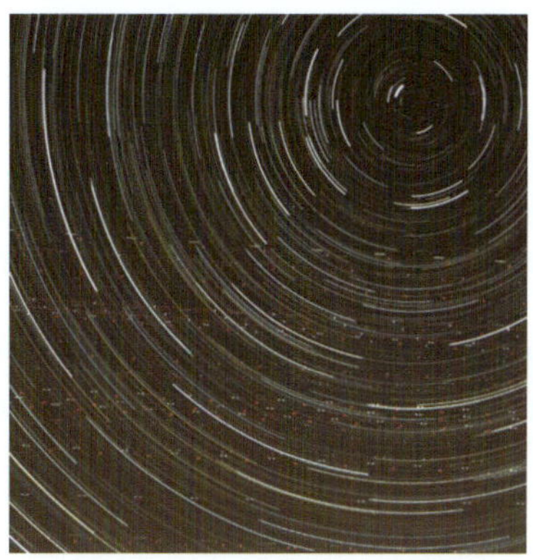

*fig.2 The result of the stack, however, picks up all the hot pixels, including meteor and aircraft trails. These distract from the image and ideally need eliminating.*

amount of light), optical performance and the amount of light pollution. Stopping down a few stops is not a sin in this case since we do not expect to stretch the image much. In this case, the Fuji glass is generally excellent and I decided to set 2-minute exposures on a handheld intervalometer with the lens at f/2.8. A few trial exposures at ISO 200 showed good star color and obvious background light pollution and there was no requirement to increase the ISO setting further. Pictorially, we only need to capture the brighter stars. Ultra-deep exposures not only clip the visible stars to white but at the same time create too many trails from dimmer stars, that coalesce into a white wheel.

### Focusing

We have not discussed focusing up to this point. Accurate focus is essential and it is easy to be caught out by the common wisdom of wide-angle depth of field and the effect of stopping down. Astrophotographers take considerable care in the initial and ongoing focus during any imaging session through a telescope. Stars are probably one of the trickiest things to image and are particularly unforgiving of any focus error. A slight error causes several issues; not only does it bloat star sizes but any lens aberrations are more likely to cause colored fringing near the image borders, or odd-shaped stars from coma or astigmatism. Unfortunately, the classic focus tools that we use for traditional astrophotography are not effective for images taken through wide-angle lenses. That applies both to various focus mask tools and to programs that measure star sizes. The practical upshot is that we have to trust our eyes and the focus magnifier utility on the camera using an electronic viewfinder. In this case, in addition to the 10x focus magnifier option on the LCD screen, I used a portable 7-inch monitor plugged into the camera's HDMI socket.

Before starting, I double-checked the camera's settings and disabled the long exposure noise-reduction mode, set the white balance to daylight and ensured the file format included a RAW format. (Note: in the case of those DSLRs with a mirror lock-up feature, if it requires two remote activations to take an image, then disable this feature when using an intervalometer.)

### Image Processing

Image processing used two OSX applications, StarStax and Affinity Photo, though I could have equally used their PC versions. The flowchart in fig.3 shows the workflow. In the first place, the Fuji's RAW files were placed in a folder and then converted into TIFF files.

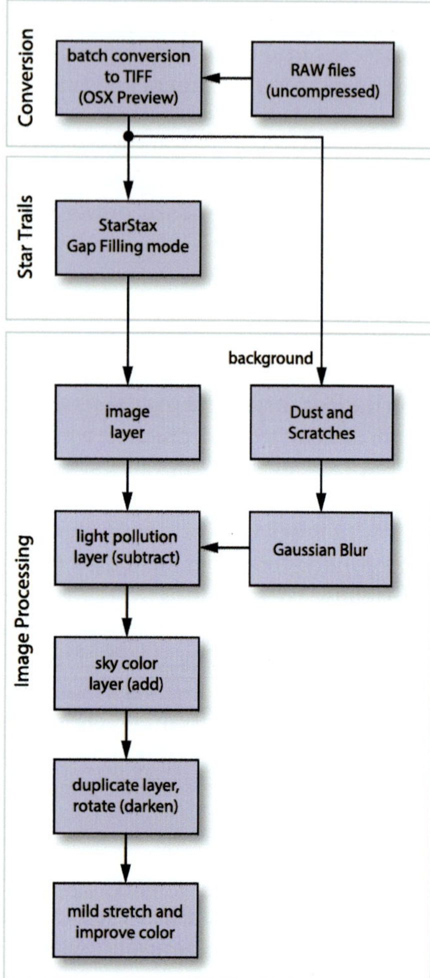


*fig.3 The outline processing workflow, in Affinity Photo and StarStax, both available for Windows and Apple OSX.*

Unfortunately, StarStax cannot read RAW files and the RAW files first required conversion. RAW conversion is a difficult thing for amateur programmers; they are always bringing out new formats and there are just too many to accommodate. Fortunately, in OSX, I found that I could load all the RAW files into the standard Preview application, select all and then export them all as either JPEGs or TIFFs. I loaded these into StarStax. This free application combines the images to form a star trail, very similar to loading all 70 images into layers and using the lighten blending mode, but more conveniently. It has a number of blending modes, including one that fills in gaps between exposures. A few minutes later, it had combined the images to form an image with many colorful 30-degree arcs (fig.1). On closer inspection, apart from the obvious brown light pollution, it was obvious there were a number of

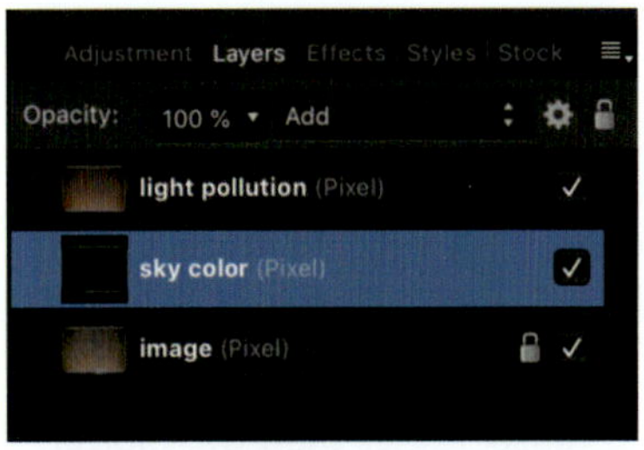


*fig.4 Subtracting a starless image layer and adding a subtle dark gradient, replaces the light pollution with something more agreeable.*

aircraft flights across the field of view and their dotted paths detracted from the arcs of the star trails (fig.2).

To remove the light pollution I used Affinity Photo and its layer blending modes. After loading the star trail image, I loaded one of the original TIFF files into a new layer and used the Dust and Scratches filter to remove all the stars and additionally applied a mild Gaussian blur to form a background image. Changing this layer's blending mode to Subtract immediately removes the light pollution from the image but makes the background too dark. Using the same trick, I added another layer in-between and filled it with a subtle gradient from deep, dark blue at the top to a slightly warmer blue at the horizon. Changing its blending mode to Add, put in a subtle gradient across the image (fig.4).

This gives a more natural look but the pesky aircraft trails remained. Since I could not face the prospect of editing a dozen files, I used a property of the image itself to remove them. The difference between the star and aircraft trails is that one is an arc and the other is a line. It occurred to me to blend duplicated, rotated layers using the Darken blending mode, to eradicate them (fig.5). In practice, I duplicated the star trail layer, selected it and used the transform tool to rotate it by a few degrees. At first, the image was awful, as the two layers were not pivoting about the pole. The trick was to accurately locate the center of rotation. I found the best way of doing this was to temporarily change the blending mode to Difference and move the layer with the mouse and then the cursor keys, pixel by pixel until the star trails all but disappeared. Switching the blending mode back to Darken, chooses the darker pixel, slightly shortening the star trails but removing almost all traces of the aircraft.

My backyard does not have the most interesting of skylines and to give an indication of what you can do, fig.7 shows a composite image of Fountains Abbey with the processed star trails. It is not going to win any awards but as a concept, interesting to consider.

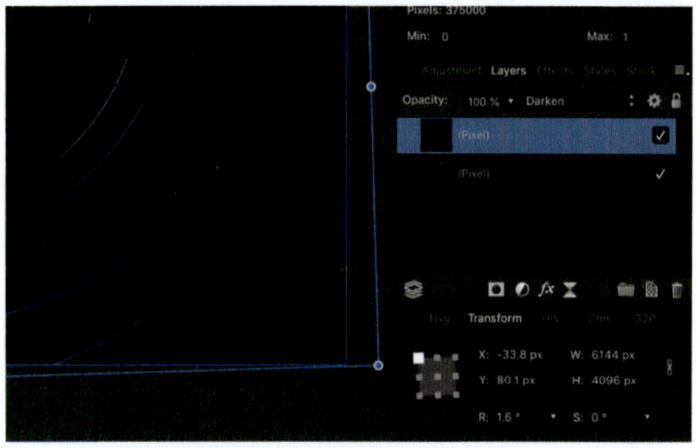


*fig.5 By carefully registering duplicate images and setting one to the Darken blend mode, a small rotation of one about the pole, removes hot pixels and aircraft trails at the same time.*

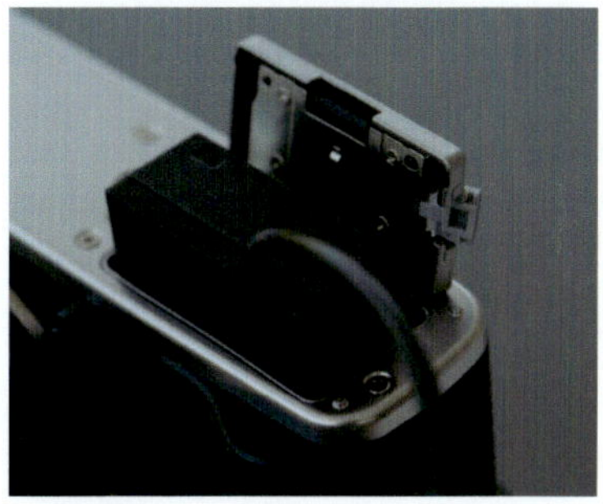

*fig.6 A Fuji W126 battery adaptor. There is either a little flap on the battery lid, or a rubber grommet on the body to allow the cable to poke through. Note: designs move on and this early battery adaptor is not compatible with later bodies, due to the lead's exit position.*

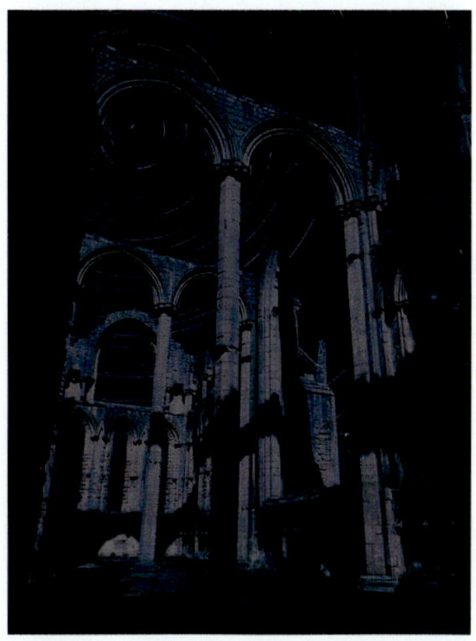

*fig.7 Since an image of a star trail around the pole is similar for alternative locations and seasons, it is plausible to use it as a generic sky image and add a dynamism to an existing twilight image. Here, I took an image of a ruined abbey, masked out the sky and superimposed the star trails.*

# Improving Quality (part 1)

*When it comes to exposure, more is... more.*

With an image under our belt, it is time to introduce some new concepts to improve their quality. Single-exposure images are convenient but, as we discovered, push the boundaries of what sensors can accomplish and favor more expensive DSLR systems. Even with a premium system it is still possible to improve image quality, principally through lower noise levels, using established astrophotography techniques. In this chapter, we consider the first of several of these.

## Astrophotographers Do It Over and Over

Although many online landscape images are carefully crafted from a single exposure, there is no reason why the tricks of the trade, usually associated with long-exposure astrophotography, cannot be applied to shorter exposures too. These improve the image quality, in so much that they reduce the noise level.

What is noise? In a sense, it is the difference between a perfect image and an actual image. Noise is the enemy and, until you try astrophotography for yourself, it may seem this concern is over-stated. It really is not. The thing is, in conventional photography, the likely image manipulations, in terms of boosting local contrast, are modest and in many cases, you are working with a lot of light. When you are working with very few photons, it requires extreme image stretching to reveal faint details, which greatly exaggerate the smallest differences between pixels. When most of the difference comes from noise, rather than the subject, the result is particularly unpleasant. If you can reduce noise, it allows image processing to improve sharpness and local contrast without causing unsightly side-effects.

There is a lot of science behind noise reduction methods but suffice to say that noise can be pigeon-holed into two categories: random noise and pattern noise. These, in turn, are tamed by two principal methods: increasing light exposure and image calibration respectively. Here, we will consider the first, which is in essence, a process of averaging multiple images of the same subject. Calibration is handled in a later chapter.

The averaging process reduces the randomness between frames (i.e. random noise) and at the same time, the subject values stay the same. Later on, we will evaluate other mathematical methods of combining images so that we can eliminate unwanted special events, like meteors, cosmic rays and the 10:40 Boeing 747 departing from New York. To add to the confusion, this combining process is also known by several alternative names, including stacking, integration and averaging. For now, we will stick with simple averaging but there is one obvious problem, particular to landscape astrophotography, that makes imaging a little more challenging, which of course is that the stars are moving and are in a different position in each exposure.

Putting that aside for a moment, a simple test demonstrates the improvement of more exposure: in fig.1,

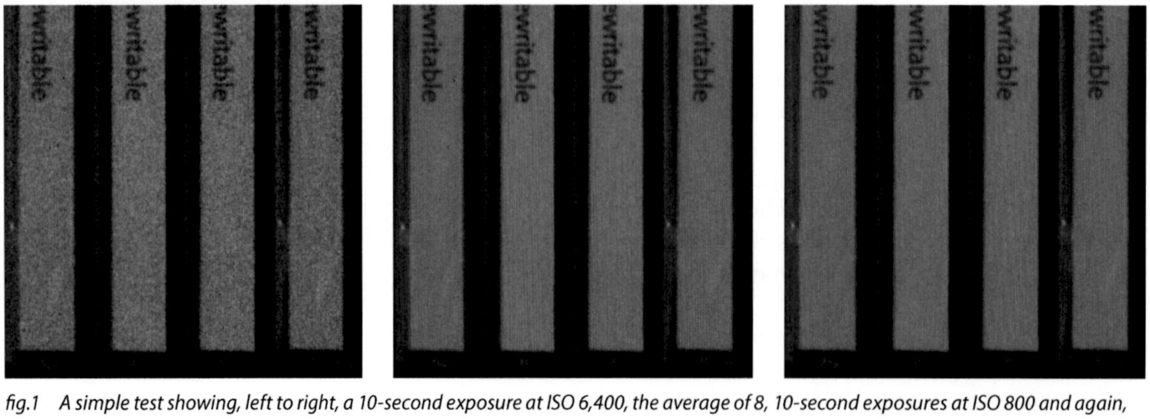

*fig.1 A simple test showing, left to right, a 10-second exposure at ISO 6,400, the average of 8, 10-second exposures at ISO 800 and again, after each image has been calibrated with dark-frame subtraction. The initial improvement is impressive due to the reduction in sensor read noise. With longer exposures, the benefit of calibration increases, as it overcomes the increased levels of thermal noise.*

there is a comparison of a single 10-second exposure at ISO 6,400, with the average of 8, 10-second exposures at ISO 800 and then again, after each of the 8 images has been calibrated. (The averaged images have been stretched in Photoshop to the same intensity as a single shot, taken at ISO 6,400, for a fair comparison.) Even in reproduction, these enlarged sections show a significant improvement with averaging. When it comes to deep-sky imaging, the objects are considerably dimmer and it is common to take many more than 8 exposures.

Another interesting comparison is shown in fig.2. This sequence of shots was taken with the lens cap on, using a variety of exposure times but at the same camera ISO setting. The noise level is shown on each. The last image is the combination of 8, 2-second exposures, which compares favorably with the 16-second exposure. It is hard to over-estimate the significance of this finding. In this case, decreasing the exposure time from 32 seconds to 1/2,000$^{th}$ second only reduces the noise level by 7% but averaging 8, 2-second exposures reduces the noise level by ~50%. As the random noise level reduces, these dark frames show the underlying fixed pattern sensor noise, mostly faint lines and hot/cold pixels that are present in every exposure.

A 32-second exposure time, in the context of normal astrophotography, is still comparatively short. When the exposure time extends beyond a few minutes, thermally generated noise (which includes both random and constant contributions) accumulates electrons and becomes increasingly significant, exceeding the noise of short exposures by some margin. Prolonged exposure also heats the sensor, which further increases the noise generation over the exposure period.

### Image Sorting

Just before looking at the treatment of particular scenes, it is important to consider another aspect of the averaging process; image sorting. While the averaging process diminishes the contribution of any particular image defect, it still impacts image quality. Before averaging aligned images, it is important to weed out the obvious rejects. These may be due to poor focus or tracking, both of which affect star shapes. In the case of simple landscape images, it may be a passing aircraft or satellite. With more images, typical of deep-sky imaging and with more advanced software these nuisances are selectively ignored at a pixel level and do not always require the rejection of the entire image. In the advanced imaging processing applications, there is usually an image-grading tool that analyzes each image and ranks them according to various quality criteria, or an inspection process that allows visual assessment by switching between successive superimposed images.

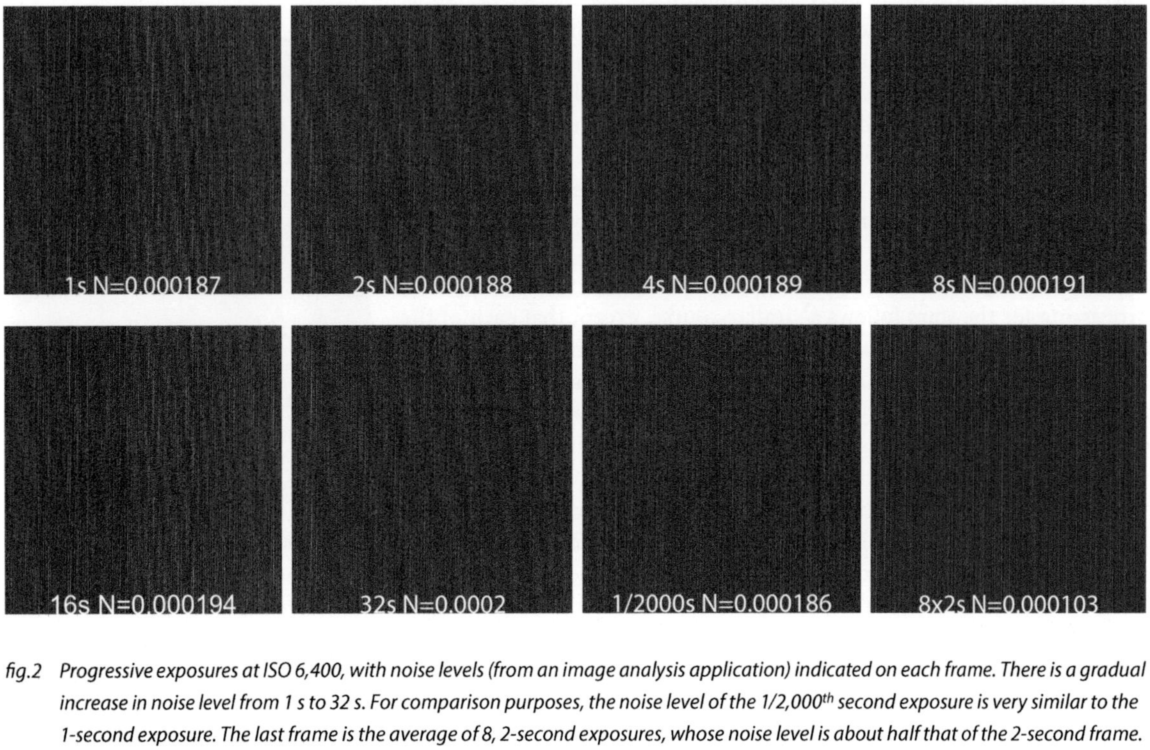


*fig.2 Progressive exposures at ISO 6,400, with noise levels (from an image analysis application) indicated on each frame. There is a gradual increase in noise level from 1 s to 32 s. For comparison purposes, the noise level of the 1/2,000$^{th}$ second exposure is very similar to the 1-second exposure. The last frame is the average of 8, 2-second exposures, whose noise level is about half that of the 2-second frame.*

## Composite Stacking

During a landscape assignment, it is a good idea to take additional exposures in short succession. This allows one to carefully align and blend them, to reduce the noise level, similar to the test exposures in fig.2. These steps require a little explanation. In deep-sky astrophotography, one uses sophisticated tools that automate the stacking process to a large extent, using dozens of exposures to radically improve image quality, as shown in the image comparison between fig.3 and fig.4. In the case of a vista that includes static and moving elements the land and sky need separate processing and, with fewer exposures, it allows us to use popular digital photo editing applications.

Starting with the static landscape portion, both Affinity and Adobe photo editors have stacking options. Here, the aim is to average the exposures and then mask off the sky portion for later blending with an improved sky image. In the case of the sky image, we first need to carefully align the images on the star positions before averaging the exposures and then overlay the landscape. These techniques involve some knowledge of layers and blending modes.

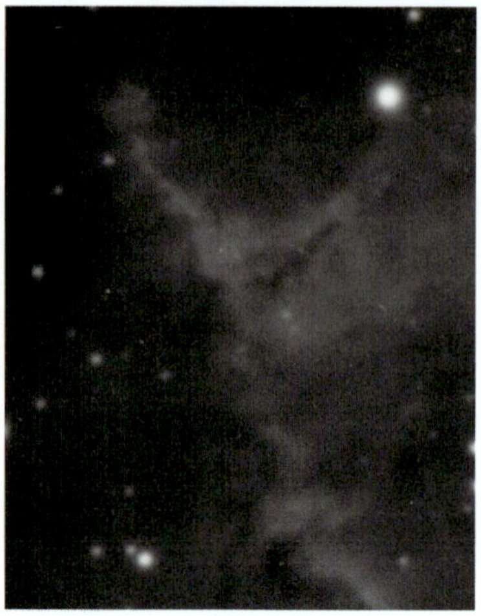

*fig.3 CMOS-based DSLR images often exhibit obvious color and luminance noise. It is not much better for a dedicated cooled CCD camera either; here, a single 20-minute frame from a cooled (-20 °C) monochrome CCD shows image noise and little detail after stretching.*

### Landscape Processing

The processing for the landscape content may be very simple if a single frame has sufficient quality. If it is noisy, averaging multiple frames will improve things. In this case, without changing their alignment, stack all the images. Affinity Photo has a stacking option which does just this and one of the Photoshop scripts does something similar. In both cases, disable any automatic alignment feature. Alternatively, one can manually load each image as a separate layer and change the opacity of each layer, starting from the bottom, in decreasing amounts, of 1/1 (100%), 1/2 (50%), 1/3 (33%), 1/4 and so on. Once that is done, flatten the image and tune the color and tonality to your liking, before carefully selecting the landscape portion using a mask. Modern applications have several intelligent mask tools that make this less arduous than the lasso tool and a steady hand! This selection excludes the sky and will be composited with a processed sky image in due course. It often helps to feather the selection by 1 or 2 pixels to create a more natural-looking join with the sky backdrop layer.

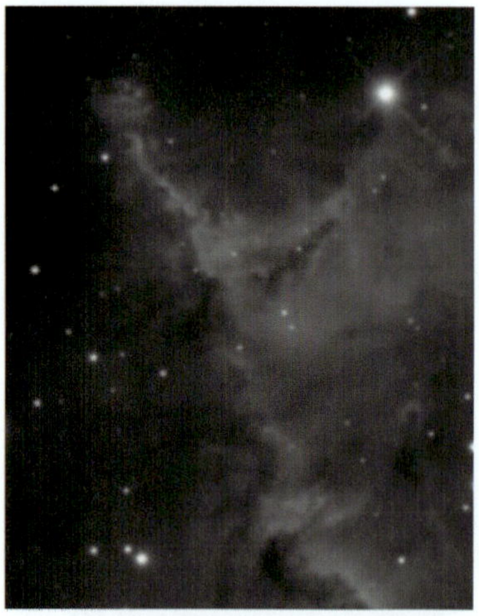

*fig.4 The average of 50 exposures (16 hours in total) shows much better detail, smoothness and lower image noise. Neither this image, nor the one in fig.3, have been otherwise processed other than by applying the same simple stretch.*

### Star Processing

Processing for the stars follows a similar course, except that each of the images require aligning to each other first. The automatic alignment feature is not always reliable, in which case, I align each layer manually by temporarily changing the layer blending mode to difference and using the cursor keys. This works for a short overall exposure period, as long as the stars' arcs appear as short straight lines. Not all versions of Photoshop have the ability to do an automatic mean of a stack. If so, one has to manually adjust the layer opacities as before. The end result will be a smoother, less noisy version of a single exposure that can withstand more stretching before image noise becomes objectionable. The landscape outline will be smudged, however, as each frame will have a different alignment. Once manipulated to your liking, lasso a boundary area between the land and sky and do a content-aware fill to cover any potential gaps, so that when you place the land image in a layer above the stars, the two blend naturally.

# Milky Way Vista

*Nothing beats a truly dark sky for capturing the grandeur of the Milky Way.*

***Equipment:***
*Canon EOS 7D, Samyang 14 mm f/2.8*
*Tripod with ball and socket head*
*Spare battery*
*Intervalometer*
*Small torch*

***Software:***
*Affinity Photo (or Photoshop)*
*PhotoPills app*

***Exposure:***
*15-seconds, ISO 1600, x 7*

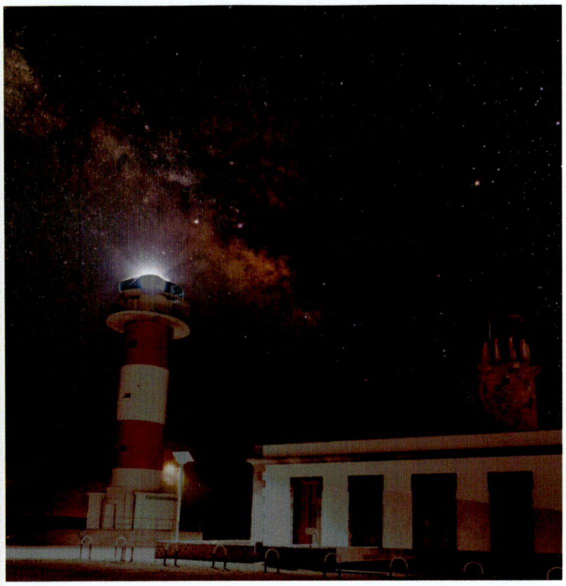

*© 2019 Andrew Duffield*

This assignment is beguilingly simple. Just take one camera, tripod and wide-angle lens and away you go. I wish it were so. For me, it was the most frustrating and brought me face-to-face with the plague of light pollution. It is a salutary lesson, as it is almost impossible to work around this reality of urban life using the means that are commonplace with what many expect to be more challenging forms of astrophotography. South East England is not known for dark skies and locating a suitable exciting venue is especially tricky. My first attempts to capture the Milky Way revealed a dense band of stars without any of the characteristic dark clouds or coloration, washed out by light pollution.

I decided to try again during a summer vacation in the Dorset countryside. The results from this were only slightly better, on account of another light pollutant, the Moon. I had not appreciated just how bright it is in an otherwise dark sky. It was a sobering lesson; my wife had arranged the holiday to help me out but we had not considered the phase of the Moon (she assures me I had an otherwise great time). After some careful tuning in Affinity Photo, I squeezed out the image in fig.1, using a tripod-mounted Fuji X-T2 with a 14-mm f/2.8 lens, blending a dozen aligned short exposures.

### Third Time Lucky

By now, I had a healthy respect for those who travel, or who are fortunate enough to image in really ideal conditions. I had a bad case of dark-site envy. With the submission date looming, I asked my friend Andrew Duffield to help me out, with his impending astro-tourism visit to the Canary Islands. The islands' remoteness makes them ideal for astrophotography and there are professional observatories on Tenerife, La Palma and Gran Canaria. The healthy tourism industry on the islands ensures frequent flights and inexpensive deals.

## Acquisition

Unlike my earlier effort, Andy had done his homework and chose to visit La Palma during a New Moon, in which the Moon does not appear at all. La Palma offers several wonderful imaging opportunities. Even in the tourist areas, the night sky is superb but the scenery can be improved upon. To give additional interest, he drove to the Fuencaliente lighthouse on the southernmost tip of the island. Andy used an iOS application called PhotoPills to predict the perfect spot and time for imaging for later on. (An alternative app is The Photographer's Ephemeris.)

*fig.1 The Milky Way, imaged from Dorset (UK) in the presence of lunar light pollution, using a Fuji X-T2.*

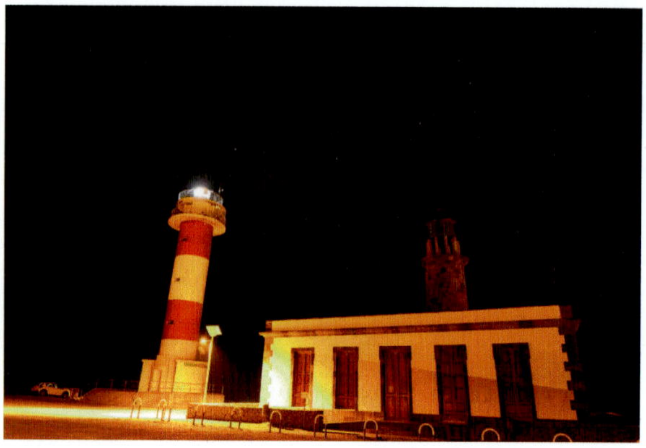

*fig.2 A more exotic, southerly latitude, offers the promise of something more interesting. This is the out-of-camera image.*

The Milky Way was very obvious and, after choosing a position to put it behind the lighthouse, he set up a tripod on firm ground. Andy knew this was possible as predictably, there were similar images on the Internet. After mounting his Canon EOS 7D he fitted a Samyang 14 mm f/2.8 lens. The Milky Way has good and not-so-good seasons, however, and the summer months are often the best pictorially. At the same time, this means short and late nights (he assures me this was not a problem; the hotel bar was still open on his return at 3 AM).

Focusing a wide-angle lens through the viewfinder is almost impossible on a DSLR. In this case, Andy manually focused, viewing the stars on a portable 7-inch LCD monitor, connected to the Canon's HDMI port and with its Live View set at 10x magnification.

Using the equation from *Getting Started*, the maximum recommended exposure duration in seconds for an APS-C camera is 330/14, about 22 seconds. After a few test exposures, Andy settled on 15 seconds at f/2.8 at ISO 1600. At this exposure, the star trailing was insignificant and the buildings were just within the dynamic range on the rear LCD. When DSLRs display a histogram, they base it on a JPEG conversion and the reality may be a little different. After traveling this distance, to be on the safe side, he captured additional exposures at the same and different exposures in short succession, saving them all as RAW files.

## Image Processing

The unprocessed image in fig.2 is not too promising. In this case, the upward tilt causes obvious converging verticals and the sky is indistinct against the bright buildings. It was obvious that perspective correction, color balance and tonal enhancements were needed for the sky and buildings. The image processing workflow considers the sky and foreground as separate images, processes each to taste and then combines them, with the help of a mask. This is a common practice for many landscape photographers. As this is astrophotography, it adds a few more challenges, in so much that the sky moves slightly between exposures and these long sky exposures are noisy and require noise-amplifying stretches to reveal their glory. There is no reason why one cannot make things simple and edit a single image but the image will be a little noisy. In this case, I had half a dozen images and Affinity Photo's Stacking option was too good an opportunity to ignore. This aligns and combines images to reduce image noise. I used two stacks; one aligned on the buildings and the other on the stars, which were processed separately, combined with a mask and then edited to correct the converging verticals.

### Foreground Processing

Starting with the buildings, I clicked New Stack in Affinity Photo and added the Canon's CR2 (RAW) files directly into the dialog and simply stacked them, without trying to align them. After about 20 seconds, it displayed a smooth version of the file, combined using the mean (average) function, designated with a μ by the layer name. Clicking this symbol brings up additional combination modes, including median (which also reduces noise) and standard deviation (which is good at showing any frame misalignments). Try as I might, I could not color balance the white walls with an adjustment layer. This foreground is lit by artificial lights and upon examining the individual color channels, I noticed the blue channel was blank in the mid-tones and shadows. The trick here was to donate some data from the other channels, accomplished by adding a Color Mixer adjustment layer and adding 20% of the

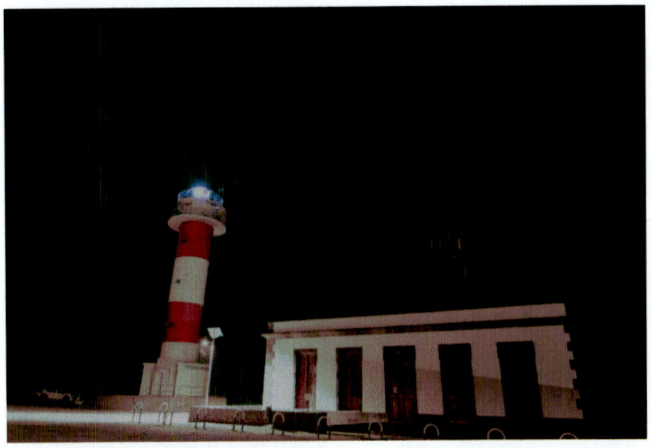

*fig.3 The completed foreground image, processed from a stack of exposures, creates a dark murky green sky. This is useful, as it is easily selected to create a mask.*

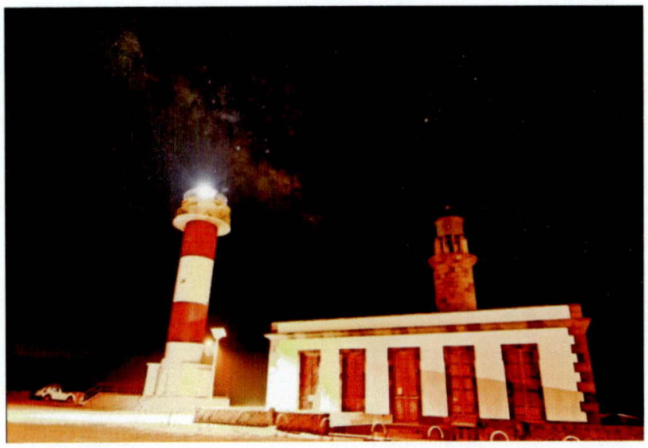

*fig.4 This is the completed sky image, the result of processing a stack of images, aligned on the stars, to show the Milky Way in all its glory.*

red and green channels to the blue. With some data to manipulate, I was then able to use a Levels adjustment layer, setting the Gamma individually for each of the RGB channels to improve the color balance. The foreground supports the sky image but it was too bright. In the same adjustment layer, I lowered the overall Gamma and throttled back the highlight output level to 80%. There is mixed lighting in this image and a perfectly neutral daylight rendering would look odd. I used a Color Balance adjustment layer to make some further adjustments to the image to make a believable rendering of the building. I then saved and closed the file.

## Sky Processing

I created a new image stack, using the 6 of the 7 RAW files that had similar exposures. This time I needed to align them manually to the stars and I used the layer stack in fig.6 to align each layer to a locked master layer.

To align these, I changed the stacking mode to Standard Deviation and locked one image and then hid all the other files in the stack. One at a time, I made them visible and shifted and rotated them, aligning the stars to my master image. To make things easier, I aligned a star in one corner and then set that as the Transformation Origin and rotated the image into place. When the two images align, the stars disappeared. With all the separate files aligned and locked, I revealed them all and changed the stacking mode to Median. The result had sharp stars with a hazy building outline. I created a pixel layer, using the Merge Visible command and knowing that image stretches would make the stars rather too obvious, I reduced their size by selecting highlights, expanded the selection by 1 pixel and applied a Minimum Blur active filter, also set at 1 pixel. To emphasize the sky, I stretched it using a Gamma boost in a Levels Adjustment layer. At the same time, I balanced the color by applying different levels to the Master, Red, Green and Blue selections. I judged this by eye and confirmed the RGB values in the info palette were similar for the sky background. With the band of the Milky Way more visible, it was evident that this lens had some light fall-off towards the corners. I lightened the image extremities with a soft, oval Vignette adjustment layer followed by a color saturation boost of all colors except green, using an hue/saturation/lightness (HSL) adjustment layer. I did not want to color the darkest tones and to restrict its application, I clicked the cog icon in the tool and pulled the Underlying Composition Ranges line to zero for tones from 0–25%.

The sky details were still subtle and I boosted its mid-band contrast with a Curves adjustment layer and an S-curve. This gave the required image balance and all that was needed was to emphasize the structures. Before that, however, I applied a Denoise live filter layer to the image, knowing that all forms of sharpening can accentuate image noise. To emphasize the dust lanes, I used Affinity's High Pass filter. At first, this creates an eerie monochrome file, showing vague outlines of objects. I increased the Radius slider until I could see the dust cloud outlines and then evaluated the Hard Light and Soft Light blending modes to accentuate the detail. To conclude, I created a pixel layer by merging the visible layers and copied it to the clipboard.

## Merging

Opening up the foreground image, I merged the visible layers and pasted the sky image on top and hid it. Selecting the foreground pixel image, I used the Selection Brush to select the green sky, starting with a big brush and then more carefully using smaller radiuses and in both Add and Subtract modes to include the small

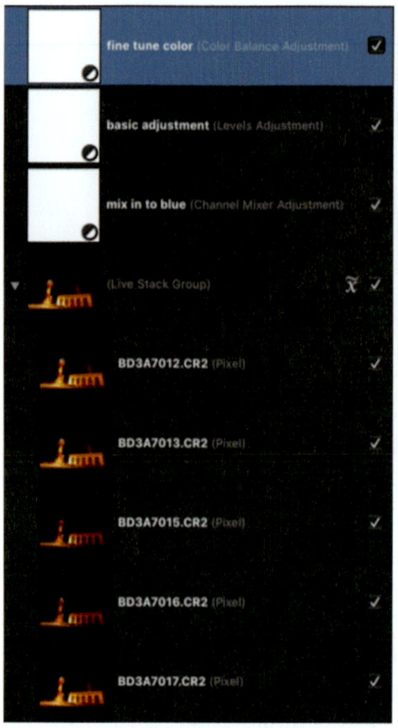


*fig.5* *The image layer stack usefully shows the history of the processing steps, any of which can be edited at will and at any time. I have given names to each, so you can follow the process. Here, you can see the image stack and three adjustment layers to yield a more realistic color balance and a subtle tonality to the foreground architecture.*

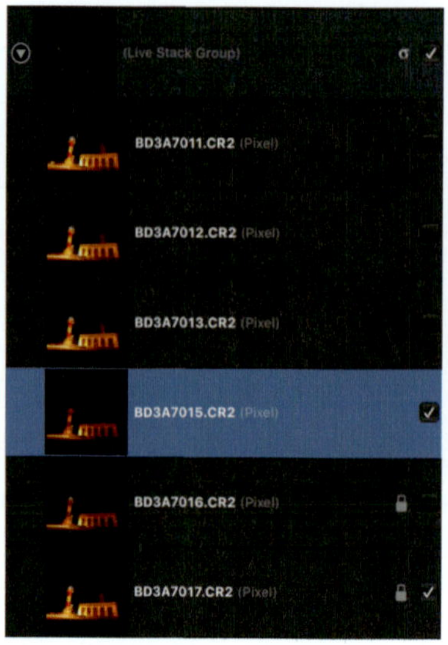


*fig.6* *This shows the Live Stack feature in Affinity Photo. Here, I have locked the bottom layer, aligned the layer above it and am now ready to align the third one up. After changing the stacking mode to Standard Deviation (note the Greek letter), I dragged the layer about with the mouse and cursor keys, centering on one star and then setting that as the transform rotation center and rotating the layer to align on the other stars. Once aligned, it is locked and hidden and I align the next image.*

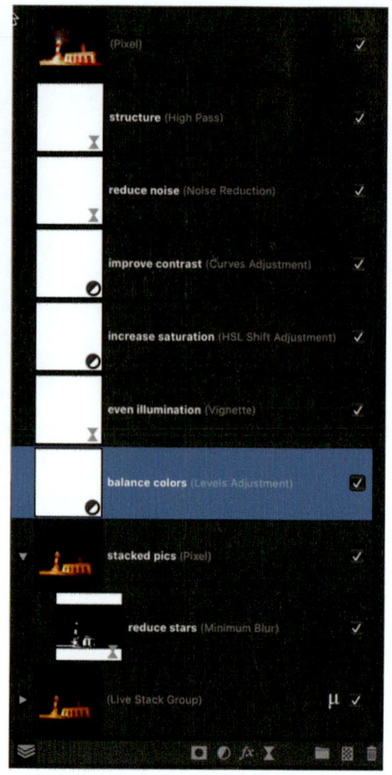


*fig.7* *This is the layer stack for the sky image showing the various adjustment and filter layers. Pixel layers are completed at various stages, as some tools will not work on a non-pixel layer. It is also convenient to have a pixel layer for copying onto another document.*

details at the top of the lighthouse. I refined the selection smoothing and feathering the edges and saved it as a mask, by clicking the Mask Layer button. I dragged the mask on top of the sky pixel layer. In this way, it combines the two images. The two images and the mask were not an exact fit and I used the Clone Stamp to extend the sky background to overlap the mask a little and some direct editing of the mask layer. The mask is made visible by Opt-clicking it, allowing me to tidy up some features. I then clicked on a pixel layer and then painted on the sky mask to blend the area around the floodlight with the sky area.

Saving the file, I merged the visible layers and corrected some of the convergence with a perspective adjustment. It is a matter of taste, but I am not keen on the distortion that occurs with full correction. Looking at the image, the old lighthouse was merging with the sky and I lightened it slightly, using the dodging tool.

## Conclusion

This assignment introduces us to some good practice during image acquisition, including the need for a rigid support, confining ourselves to the maximum exposure duration and an alternative method to focus accurately on stars with a wide-angle lens. The image processing was an interesting mixture of traditional digital photography techniques, using layers and masks, with the added (optional) step of combining multiple images of the same object to reduce image noise.

In this location, the Milky Way was in plain sight on account of the wonderfully dark skies in La Palma. This made image manipulation much easier, as it was not necessary to over-manipulate the image to remove light pollution or stretch weak signals. In poorer conditions, it would likely require a light-pollution filter in the lens throat and combining more exposures, taken in quick succession, to further reduce image noise.

# Wide-Field Deep-Sky Imaging

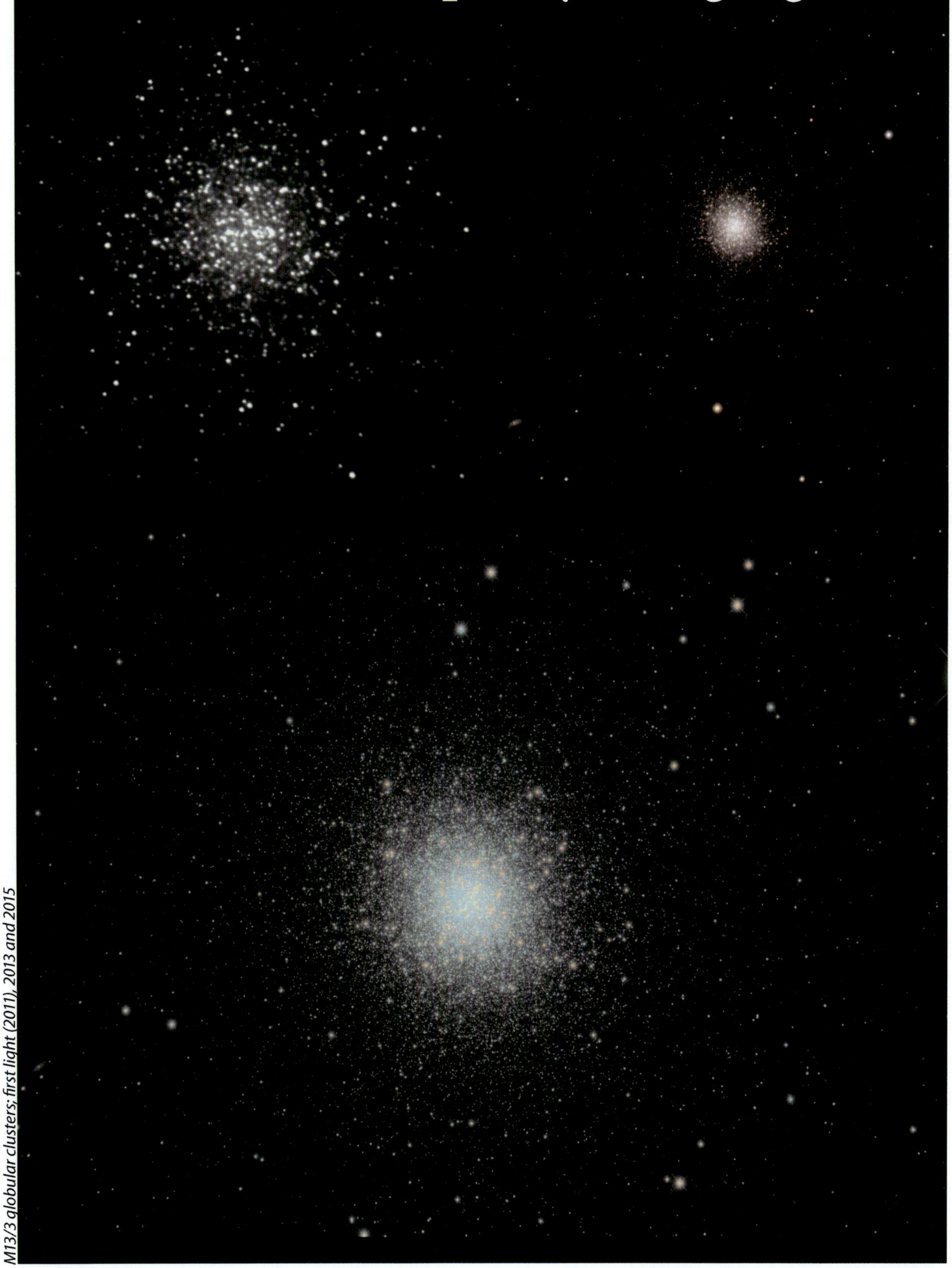

*M13/3 globular clusters; first light (2011), 2013 and 2015*

# The Limits of Amateur Astrophotography

*There are many practical realities that limit our ability to produce high-quality images, regardless of equipment or prowess.*

On the previous page are three images, taken two years apart. These show how improved technique, equipment and considerable patience dramatically impact image quality, in the face of the challenges imposed by the subject matter and my location. It is worth realizing though, that there is an upper limit to any improvement. What follows is a pragmatic walk through some of the practical limitations of telescopes, sensors, mounts and location, to avoid unrealistic expectations and unnecessary outlay. This admittedly covers a wide range of topics, many of which are picked up again in later chapters for an in-depth discussion.

## Optical Resolution

The first apparent challenge is the ability to take images of very small objects. We have to have a big telescope, right? Most of my most successful images are taken with a focal length of 600 mm or less. Advertising and consumer pressure tempt us to over-indulge in telescope purchases for astrophotography. There are many optical and physical properties that distinguish a "good" telescope from a "bad" one and, just like with any other pursuit, knowing what is important is the key to making the correct purchasing decision. The needs of the visual observer and astrophotographer are different too since the human eye has a higher resolution than a conventional sensor (though with less sensitivity). Expensive apochromatic (APO) refractors focus all wavelengths of light at the same point, a quality valued by visual users or those imaging with a color camera. It has less significance if the camera sensor is monochrome and one exposes, focuses and processes individual color channels through red, green and blue filters and combine the channels into a color image later on. What is more important is the ability for the telescope to focus uniformly across a flat sensor surface, something we take for granted in camera lenses but is a unique challenge with refractor and reflector telescopes.

Astrophotography shares the same physics as any other kind of photography; the final image quality has many factors and the overall performance is a combination of all the degradations in the imaging chain. Long before digital cameras were popular, premium optics, from companies such as Leica and Carl Zeiss, had more resolution than could be recorded on color film. The digital sensors that we use in practice still have less resolution than fine grain monochrome film (though they have far higher efficiency). In modern astrophotography, however, the sensor's role in the final resolution is more complex to assess and especially in the case of color cameras, which combine neighboring pixels to form color.

### Resolution and Sharpness

In photography, many amateurs and not a few professionals confuse resolution and sharpness. They are not completely unrelated, but in an image, they convey very different visual attributes. In simple terms, resolution is the ability to discern two close objects as separate entities. Photographic resolution tests often use alternate black and white lines in various sizes, stated in line pairs per mm (lp/mm) for convenience. Astronomers, not surprisingly, are more interested in closely-spaced points of light and it is more meaningful to quote angular resolution in arc seconds or radians. Image sharpness, on the other hand, has no agreed measure but is our perception of contrast between adjacent light and dark areas, especially in the transition area. Post-exposure image manipulation cannot restore lost image resolution, but image sharpness can be increased later on, using various sharpening tools. (Mild sharpening of an image may improve the actual perceived resolution of some coarser image details but often at the same time bludgeons delicate detail and accentuates noise.)

One key question is how much resolution do we really need? In many images, the stars appear randomly sprinkled throughout an image with plenty of space between them and we do not necessarily require a high resolution to see them, only contrast. We do not require a high-resolution photograph to appreciate clouds in the sky or in the same sense, images of nebulae and galaxies with diffuse object boundaries. In addition, many popular nebulae also span a wide angle and do not require a high optical magnification and hence angular resolution. In these cases, the optical resolution is usually better than the resolution of the sensor. In addition, the long exposure times required for dim deep-sky objects, coupled with the necessary image processing, accentuate the light scatter along the optical path, causing bright stars to bloat.

On the other hand, a high angular resolution is required to distinguish individual stars in globular clusters and double stars. In the 19th century, leading physicists discovered that optical resolution is ultimately limited by the diameter of the physical aperture, by a phenomenon known as diffraction. You will come across advertising statements such as "diffraction-limited optics." This just means that the imperfections in the mirror or glass have less effect than that of the diffraction from its aperture. There are a whole bunch of equations that predict optical resolution and the interesting feature of them is that while the image resolution improves with the size of the aperture, it is independent of the focal length or magnification. This fact drives many to get "aperture fever" and spend vast sums on telescopes that they can barely lift. There is a cure for this though; cruel reality.

Although stars have a finite size, they are so distant that they should focus to an infinitely small spot in an image. Even with perfect optics, diffraction causes stars to appear as diffuse blobs. The brightest part of the blob is at its center and a measure of its "blobbiness" is its diameter (measured in arc seconds, at the point where its intensity is half its peak value, as shown in fig.1). In later chapters, we will see that when this diameter is at a minimum, the object is in focus. A similar measure is also used to evaluate optical resolution.

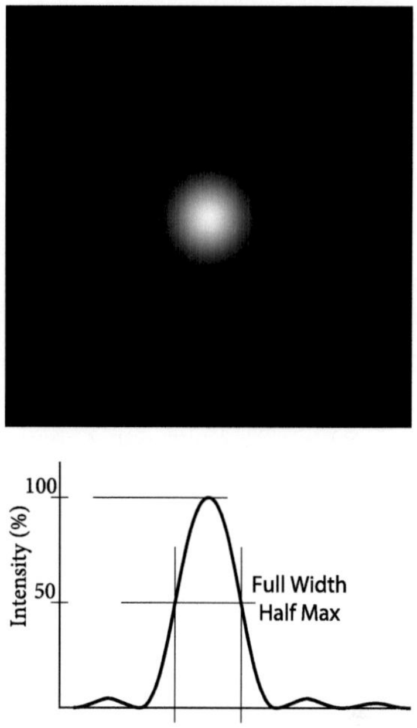


*fig.1* *A focused star, imaged through an idealized aperture creates a diffuse blob, with a central peak and increasingly pale concentric rings. This is known as the Airy Disc. Focusing programs calculate the Full Width Half Max (FWHM) or a similar measure, the Half Flux Diameter (HFD) of imaged stars at different focus points to find the minimum value, which corresponds to the optimum focus position.*

## The Cruel Reality of Earth-bound Astrophotography

A crucial part of the optical "system" is our atmosphere. It is anything but perfect and light has to travel through about 20 miles of it (looking up) and about double that, closer to the horizon, before it reaches our telescope and sensor. The air is not stable and its refractive index varies in space and time, causing a star to twinkle and shimmer, an effect we astrophotographers call astronomical seeing, or seeing, for short. Astronomical seeing is an empirical measure of the optical stability of our atmosphere. Turbulence causes rapid localized changes in air density and parallel light beams are deviated through refraction. At any one time, the light beams pass through small adjacent air pockets (a few centimeters across) with different refractive indices. This occurs mostly in the denser air near the ground and even from tiny convection currents within the telescope tube.

At high magnifications, typical of planetary video imaging, the individual frames jump about the screen, some are badly blurred and others remarkably sharp corresponding to when the light path has a variable and consistent refractive index. During longer exposures, the photons from these sharp, blurred and displaced images accumulate onto the sensor, creating a smeared star image. Seeing degrades angular resolution and it is measured in the same units (arc seconds). For a prime site, a seeing condition of 0.5 arc seconds is possible but values in the range of 1.5–3.5 are more typical. It is very likely that more often than not, the prevailing seeing conditions will rob the resolution of any modest telescope. The table in fig.2 shows the theoretical limits of visible light resolution for several common amateur telescope sizes in relation to typical seeing conditions. It is quite sobering to realize the limitation imposed by typical seeing conditions through the atmosphere is equivalent to a telescope with an aperture in the region of just 3 inches (75 mm).

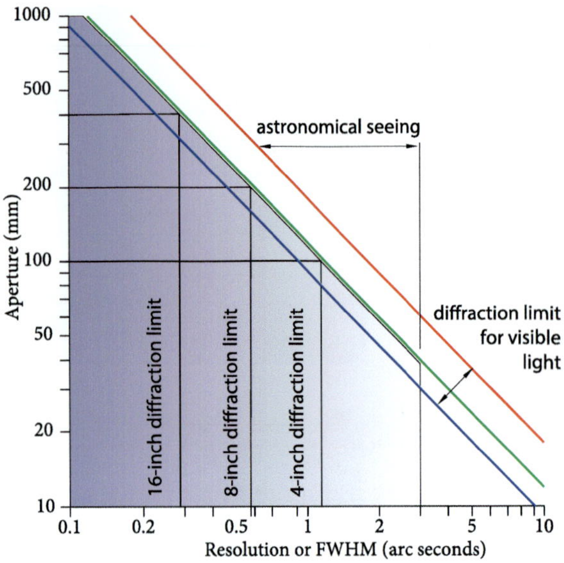


*fig.2 This diagram shows the diffraction-limited resolution for visible light, for any given aperture, in comparison to astronomical seeing. Here, seeing is likely to be in the range from about 0.5 to 3 arc seconds, robbing the resolution of large-aperture telescopes.*

There is not a whole lot one can do about this. Seeing conditions vary and are sensitive to changes in the dense atmosphere closest to Earth, and generally improve with altitude and proximity to large expanses of water (due to the moderating effect on thermal generation). The mountain observatories in Hawaii and the Canary Islands are good examples of prime locations. Seeing conditions also change with the season and the amount of daytime heating. The local site has an immediate bearing too; it is better to image in a cool open field than over an expanse of concrete that has received a day's sunshine. Astronomers choose remote sites not only to be anti-social; they just need to find high altitude, clear skies, low light pollution and low air turbulence. Knowing and predicting the prevailing conditions is a key part of our day-to-day activity. In some countries especially, each imaging opportunity is a precious one.

## Light Pollution

Light pollution is a perpetual issue for many of us. Urban living conspires with health and safety requirements to encourage local authorities to install street lighting. Even those in the uninhabited deserts suffer for a period each month as the Moon illuminates the sky. Each of the forms affect our imaging in a different way as do the coping strategies. Even with the same amount of illumination, light pollution may vary night-to-night, depending on the amount of dust, water vapor and aerosols in the atmosphere at the time and after it has been raining, the air is often much cleaner and the effects of light pollution are slightly reduced on account of the better atmospheric transparency.

### Emission Lamps (Sodium and Mercury Vapor)

We become increasingly aware of light pollution as soon as we take up astrophotography. Light pollution masks faint objects and if your local street lighting uses the traditional sodium lamps, a color image of what appears to be a dark sky reveals a muddy orange color as the light bounces back to Earth from all the airborne contaminants. These emission lights have distinct colors and although there are several different types, their major output is about 590 and 600 nm. The less popular mercury vapor lamps emit over a broader spectrum, but with strong components at specific wavelengths in the violet, blue, green and yellow-orange colors.

The coping strategies in these conditions take advantage of the specificity of the light pollution colors. In essence, special thin-film optical filters mask the unwanted colors, made by man, and pass the others. There are several approaches that reduce the impact on our astrophotography:

- light-pollution filters
- separate RGB filters on a monochrome sensor
- narrowband filters
- image from a darker site

In the first instance, light-pollution filters use interference filters to precisely exclude the principal street lamp colors. There are many on the market and most exclude the two sodium and four mercury lamp emissions yet pass the common nebula emission colors, with remarkable efficiency. They are typically found in 1.25- and 2-inch sizes and increasingly, in smaller tailored packages, to insert inside a consumer digital camera. Many telescopes have an internal 48-mm thread in the focus tube or field flattener to accept a 2-inch filter. The established manufacturers are Hutech (IDAS), Astronomik, with new entries from SkyTech and STC.

These filters cannot work miracles but they do reduce the light pollution effect. This not only improves the color balance (and sky gradient) of an image, but significantly, reduces the random shot noise that accompanies all incident light into the sensor. It is very noticeable that those astrophotographers imaging under glorious dark skies achieve spectacular results over one or two nights. To match that, I may image at every opportunity over a month to match their subtle detail and depth.

For those of you using a dedicated monochrome camera, one of the benefits arises from the use of discrete red, green and blue filters. The better filter sets use thin-film technology and the red and green filter responses are uniquely designed to exclude sodium yellow. (In a conventional digital camera, the RGB filter mosaic in front of the sensor passes yellow light.)

In the last case, it is fortunate that the principal nebula emissions in the blue-green and deep regions do not correspond to those from emission lamps. Special narrowband filters have incredibly selective transmission that excludes all light other than a specific nebula color. This removes most of the light pollution and its accompanying shot noise. As a result, many inner-city astrophotographers specialize in narrowband imaging.

### LED Lighting

Modern LED lighting is a concern to astrophotographers. While the lamp design and efficiency is considerably better and the beam pattern is confined downwards, the backscatter from the ground (especially when wet) is significant. The more worrying issue is the nature of the light itself. A simple LED emits light is a narrow range of colors and different doping of the silicon substrate changes the color. Modern LEDs cover a wide range of colors from deep blue to infrared. White LEDs are a combination of red, green and blue emitters. As such, the light output covers a broader spectrum and is more difficult to filter out. While physics determines the sodium and mercury light outputs, there is more variety in LED spectrum between manufacturers. Coping strategies are less successful; while some recent light-pollution filters, for example the IDAS D2 LPS filter, have been designed to remove the intense blue spectrum associated with LEDs, they still pass the lower intensity green and red wavelengths. This intense blue light pollution will pass through the blue filter of an RGB filter set but again, narrowband filters will continue to work well.

### Lunar Light Pollution

The Moon is a regular nuisance and for about 10 days each month, the reflected light from its surface is sufficient to floodlight the sky and degrade image backgrounds. Moonlight has a broad spectrum, peaking in the orange-red wavelengths. The obvious coping strategy is to avoid imaging during this period, or when the Moon is below the horizon. All is not lost; one may consider imaging with narrowband filters when the Moon is up. The light pollution fades as you aim further from the Moon and another alternative is to image in the opposite direction, in the two northerly quadrants.

### Big is Beautiful

Why then do so many sources recommend buying the largest affordable aperture and that "Aperture is King"? We know that larger apertures technically have the potential for better resolution, but above all, they simply capture more light. For visual use, the extra aperture is the difference between seeing a dim galaxy or not. For imagers, the extra light intensity delivers an opportunity for shorter exposure times or more light captured over a fixed period, which reaps benefits in sleep deprivation and lower image noise.

Not all telescope designs are equal; there are some subtle differences between the optical performance of the various telescope architectures; the more complex have additional losses in transmission, reflection and diffraction at each optical boundary. Of the telescope designs, for any given aperture, refractors are the simplest optically and have the highest image contrast, followed by Newtonian reflectors and then folded designs which, for the moment, we will collectively call Schmidt-Cassegrain Telescopes or SCTs. We will confine ourselves to refractors, as there is less to go wrong with one, they are more portable, robust to handling and are simpler to set up.

## Imaging Resolution

Apart from the limitations of the optical path, vibration, flexure, focus shifts, tracking accuracy and atmospheric effects also contribute to the blurring of the eventual star image and we have not yet considered the sensor resolution. If it was going to be easy, it would not be nearly as rewarding, or half as much fun!

### Sensor Resolution

The term sensor describes the light-sensitive device that resides in a camera. Digital sensors are complex and they are discussed in depth later on. For now, a single light-sensitive element on the sensor, a photosite, corresponds to a pixel in a monochrome image. A sensor converts photons into electrons and then into an electrical signal. This signal is amplified, sampled and stored as a digital value. An imaging sensor has a grid of photosites of fixed pitch, typically in the range of 3–7 microns. The photosites simply accumulate electrons, triggered by incident photons and largely irrespective of wavelength. To make a color "pixel" in an image file

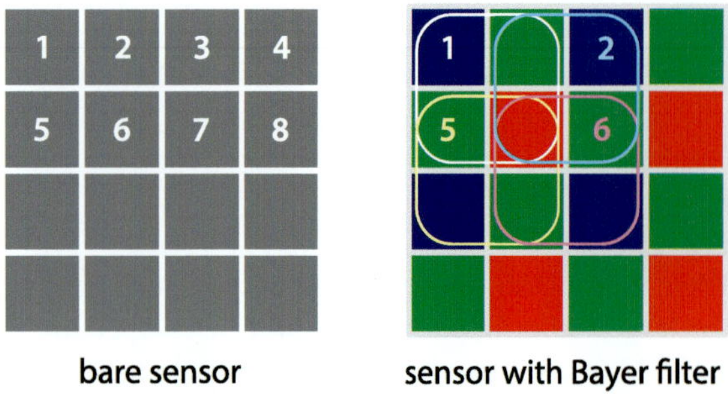


*fig.3 In a bare sensor, each photosite corresponds to an image pixel. In a color sensor (fitted with a Bayer Filter or similar) a color "pixel" is formed by combining 4 adjacent photosites, with 2 green, 1 blue and 1 red filtered photosites.*

requires a combination of exposures taken through red, green and blue filters. In many dedicated cameras, this is achieved by combining coincident pixels, from 3 separate exposures, each through a large color filter placed in front of a monochrome sensor. Color cameras, called One Shot Color or OSC, "see" color by combining adjacent pixels from a sensor fitted with a color filter array (fig.3). Astrophotographers use both approaches, each with benefits and drawbacks and these are also discussed in *Dedicated Camera Systems*.

The word pixel is familiar and although technically ambiguous, we will use it rather than "photosite" or "photodiode" when describing the tiny light-sensitive elements within a sensor. The pixel spacing on the sensor has a significant impact on image resolution. Up to now, the discussion has revolved around angular resolution. To consider the physical relationship between angular and linear resolution on an imaging sensor we need to know the focal length of the optics. A little math here is unavoidable but for a telescope with a focal length $f_L$ the angle subtended by 1 pixel, in arc seconds is given by the following simplified equation from basic trigonometry ($f_L$ in mm):

$$\textit{arc seconds/pixel} = 206.\frac{\textit{pixel spacing (microns)}}{f_L}$$

This important measure is called the imaging scale and in practical terms, it takes more than 2 pixels to resolve a pair of stars. A number of sources suggest about 3.3 adjacent pixels are required to guarantee the resolution of two points and hence the angular resolution of a sensor is 3.3x its pixel scale (in arc second/pixel). As we can see from the equation above, the pixel scale changes with the focal length of the optics.

This is very relevant and it is interesting to compare this to the diffraction limit of a telescope by calculating the equivalent pixel pitch to match the resolution of a typical refractor. In the case of a large refractor, with a focal length of 924 mm, an aperture of 132 mm and a sensor pixel pitch of 5.4 microns, it has a diffraction-limited resolution of approximately 1.0 arc second but the sensor's angular resolution is worse, at 4.0 arc seconds (note, this is the angular resolution, not the pixel scale). For the sensor to match the diffraction-limited performance of the optics, it would require a pixel pitch of 1.4 microns. That might look quite damning but there is another consideration, the effect of astronomical seeing: A sensor resolution of 4.0 arc seconds is only marginally worse than typical seeing conditions of say 2.5 arc seconds in a suburban setting. The overall system resolution is a combination of all the above and in this case, is approximately 5 arc seconds.

Putting our feet back on the ground, there are a couple of important observations here; the system resolution is a combination of the individual values and is always less than the weakest link in the imaging chain. If a, b and c are the individual optical, sensor and seeing resolutions, the system resolution is given by:

$$\textit{system resolution} = \sqrt{(a^2 + b^2 + c^2)}$$

Unfortunately, this is not the end of the story since the resolution is also further degraded by poor star tracking, caused by any number of tiny mechanical imperfections, or something as simple as a gust of wind.

To sum up, in this example, the telescope's optical diffraction has little influence on the final resolution and the sensor pitch is the weakest link. The seeing and sensor resolution are similar though, and while sensors with a finer pitch can be used (albeit with other issues), in astrophotography the most difficult thing to change is one's environment. All these factors are weighed up in the balance between resolution, image requirements, field of view, signal strength, cost and portability. Some texts suggest having a sensor whose arc seconds/pixel value is about 1/3rd of the limiting condition. This guideline, however, does not account for the practical and economic downside of owning multiple cameras.

### Tracking and Telescope Mounts

During every exposure, the telescope needs to follow the circular movement of the heavens to a high accuracy. The practical measures we take to ensure accurate tracking during long exposures dominate many discussions. There are lots of factors that affect tracking but, in general, sharp images require good tracking throughout each exposure. Some of these challenges arise from minute mechanical imperfections in the telescope mount, others from alignment or atmospheric refraction. The smaller, less expensive lighter mounts use simpler mechanisms and have wider tolerances on critical sub-components. The smaller dimensions make those tolerances all the more critical. It is easy, however, to become obsessive about tracking performance, polar alignment and so forth. I have seen colleagues beating themselves up, believing the need to target sub 0.5 arc second tracking accuracy. (This is equivalent to the angle subtended by a piece of videotape, edge on, from 5 m!) It is all about trade-offs. The aim is to ensure that the tracking does not introduce any significant image degradation and to keep in mind all the other compounding factors that affect resolution. In our earlier example, the static resolution was about 5 arc seconds, to degrade that by ~20% over the duration of an exposure only requires an RMS tracking accuracy of about 1.6 arc seconds. With shorter focal lengths, in the range of 300–600 mm, the tracking requirement is relaxed even further, as the imaging sensor's pixel pitch becomes the overwhelming limitation on image resolution. Imaging with longer focal lengths is more demanding upon technique and with this in mind, the assignments throughout this book use short and medium focal lengths.

## Dynamic Range

Astrophotography is particularly demanding on the dynamic range of a sensor. Classical monochrome photographers who use the Zone System are painfully aware of the need to modify film development and exposure to ensure that both shadow and highlight details are captured at the same time. With luck, they can capture a dynamic range of 10 stops, or a range of 1024:1. Modern digital sensors have improved over the years and perform similarly. In an astrophotograph, there may be bright stars, very dim clouds of ionized gas and somewhere in the middle, brighter regions in the core of a galaxy or nebula. Its dynamic range can easily exceed 20 stops or ~1,000,000:1. The Orion Nebula is a popular subject that exceeds the abilities of most sensors. Extensive manipulation is required to boost the dim clouds, maintain good contrast in the mid-range and at the same time emphasize the star color, without turning them into white blobs. For this to be a success, the original image data needs to not only record the bright stars without clipping but also to capture the dim elements with sufficient tonal resolution, so that they can both withstand subsequent image stretching without degradation. To make the image in fig.4, I blended a number of exposures, using different filters and with varying exposure lengths. Only the brightest parts of the short exposures were used.

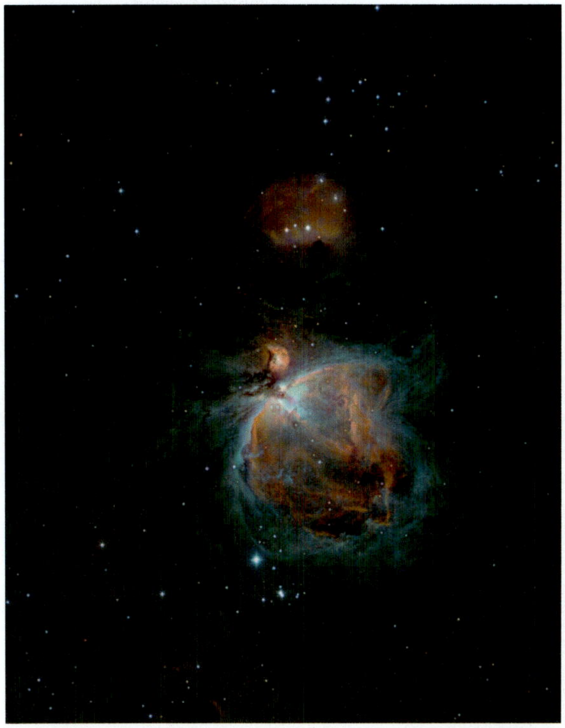

*fig.4 This very popular imaging target is also very difficult to do well, on account of its extreme dynamic range. With many subjects, a collection of similar exposures is sufficient. With this, several sets taken with very different exposure lengths were required to capture the faint details and bright interior and additionally required 64-bit image processing.*

There are a number of terms that loosely describe brightness in many texts, namely, luminosity, flux and magnitude. Luminosity relates to the total light energy output from a star; flux is a surface intensity, which, like an incident light reading in photography, falls off with distance. The brightness or magnitude of a star is its apparent intensity from an observed position. The magnitude of a star or galaxy in relation to the sky background and the sensitivity of the sensor are the key factors that affect the required exposure. Confusingly, this also relates to size and in the case of visually large galaxies, with a high magnitude, they can still be dim. Most planetarium programs indicate the magnitude information for any given galaxy and most stars using a simple measure called the "apparent" magnitude.

### Apparent Visual Magnitude

This is the most useful measure as it defines the luminosity of a star as it appears to an observer on Earth. Just as

| visibility | apparent magnitude | # objects brighter | example / notes |
|---|---|---|---|
| human eye urban sky | -1 | 1 | Sirius (-1.5) |
| | 0 | 4 | Vega |
| | 1 | 15 | Saturn (1.5) |
| | 2 | 50 | Jupiter (-2.9 to -1.6) |
| | 3 | <200 | Andromeda Galaxy (3.4) |
| human eye dark sky | 4 | 500 | Orion Nebula (M42) |
| | 5 | 1,600 | Uranus (5.5-6.0) |
| | 6 | 4,800 | Eagle Nebula (M16) |
| binoculars with 50-mm aperture | 7 | 14,000 | Bode's Nebula (M81) |
| | 8 | 42,000 | Crab Nebula (M1) |
| | 9 | 121,000 | M43 Nebula in Orion |
| typical visual 8-cm aperture | 10 | 340,000 | NGC4244 Galaxy |
| | 11 | - | Little Dumbbell (M76) |
| typical visual 15-cm aperture | 12 | - | beyond Messier Catalog |
| | 13 | - | Quasar 3C 273 |
| typical visual 30-cm aperture | 14 | - | Galaxy PGC 21789 nr. Pollux |
| | 15 | 20,000,000 | IC 4617 Galaxy nr. M13 |
| 10-cm refractor, CCD 10x 30 seconds suburban sky | 16 | - | faint star in image with simple stacking, about 20,000 times more sensitive than by eye alone |
| 10-cm refractor, CCD 10x 300 seconds suburban sky | 18 | - | faint star in image with simple stacking, about 1,000,000 times more sensitive than by eye alone |
| suburban sky | 20 | - | typical background magnitude in suburb |
| Hubble Space Telescope | 31 | - | galaxies 13.3 billion light-years distant |

*fig.5 This table indicates the objects we can see with the unaided eye, in different environments and with progressive levels of optical and electronic assistance, all the way to a space-borne telescope.*

with light measurements in photography, astronomical magnitudes are a logarithmic measure, which provides a convenient numerical index. Astronomy magnitudes employ a scale where an increase of one unit decreases the intensity by 2.5x, and five units by $2.5^5$ or 100x. The brightest star (apart from our own sun) is Sirius at -1.47 and the faintest object observable from the Hubble Space Telescope is about +31, or about 2.4 x $10^{13}$ dimmer.

Using logarithmic figures enables simpler math; in astronomy, any pair of similarly sized objects with a similar difference in magnitude value have the same brightness ratio. Similarly, if the magnitude limit for visual observation is magnitude 4 and a telescope boosts that by a factor, expressed in magnitude terms, say 5, the new magnitude limit is 9. There is a catch though, that we mentioned earlier; a visually large object, such as a galaxy, will not appear as intense as a star of the same magnitude, as the same light output is spread over a larger field of view. An example is the galaxy M33, a large galaxy but low image intensity.

The table in fig.5 sets out the apparent magnitude scale and some example objects with the number of stars that reach that magnitude. At the same time, it indicates the limitations imposed by the sensitivity of the human eye under typical light pollution as well as the exponential number of stars at lower magnitudes. Further down the table, at the lowest magnitudes, the practical benefit of using a telescope for visual use can be seen, and that improves even further when a modest exposure onto a sensor replaces the human eye. At the bottom of the table, the limit imposed by light pollution is removed by space-borne telescopes, whose sensors can see up to the limits of their electronic noise. It is interesting to note that film only captures about 5% of the incident light, compared to digital sensors that register 50–80% of the photons.

### The Challenges of Dynamic Range

In the early days of digital imaging, many wedding photographers preferred the results from color negative film, as it captured a larger brightness range than a digital camera. In effect, what they were saying was that film could distinguish a higher ratio of light levels between highlights and shadows. Comparing analog and digital systems, however, is not easy; there is more at play here than just the difference in light levels; there is tonal resolution too, and this is where film and digital sensors are very different.

In one respect, the dynamic range of a sensor is a simple measure of the ratio of the largest signal to the lowest, expressed either as a ratio or as a logarithmic value, like dB, which will be familiar to engineers.

In photography, dynamic range is also related to bit depth, which is measured by the number of binary digits output from the sensor's Analog to Digital Converter (ADC). A 16-bit ADC has $2^{16}$ voltage levels, over 65,000:1. In practice, this is not the whole story; sensors convert photons into electrons and there is rarely a 1:1 relationship between the number of stored electrons and the output value for a pixel. The other fly in the ointment is that sensors are linear devices and we are accustomed to working in logarithmic units; in a digital image, there are fewer signal levels per magnitude at the dark end than at the bright end of the exposure scale. Astrophotography sensors commonly have 12-, 14- or 16-bit ADCs. One has to read the sensor datasheets very carefully, since those sensors with 12- and 14-bit ADCs scale the value up to 16 bits and in reality only record 4,096 or 16,394 individual light levels respectively.

### Sensor Dynamic Range

Just as with system resolution, there is more to dynamic range than one single measure. We are familiar with bit depth; consumer cameras record JPEG files and these are made with three 8-bit values each for red, green and blue. That is just 256 levels per color channel, even though the sensor may have more detailed information, typically 16–32x or better. Color fidelity in astrophotography is not a primary concern, but low noise and fine tonal resolution in the shadow areas are essential qualities in the original downloaded file. An 8-bit value does not cut it in astrophotography.

There is a limit to how many electrons each photosite can store and many sensors have less tonal resolution than their ADC. A popular astro CCD sensor, the KAF8300M only requires 25,000 electrons to saturate (max-out) a photosite, corresponding to an output of 65,000 or so. In reality, it only has 25,000 states, equivalent to just over 14 bits. This number of electrons is known as the full well capacity or depth and varies between sensor models. It gets worse; random electrons introduced during the sensor readout process effectively reduce the effective dynamic range further. This concept is a little difficult to appreciate but sensor read noise reduces a sensor's dynamic range.

The effective dynamic range of any sensor is its full well capacity divided by its read noise (in electrons). In the case of a KAF8300M sensor, the read noise is about 8 electrons, yielding a dynamic range of 3,125:1 or just under 12-bit. The 16-bit ADC in the sensor has a finer resolution than the signal to ensure that it does not introduce further sampling (quantization) noise. Some of the latest CMOS astro cameras only have 12-bit ADCs, and their full well capacity of 20,000 exceeds this by some margin. They have the unusual property of their effective read noise decreasing with increased amplification of the analog signal in the sensor electronics (prior to sampling). This is due to the quantization noise becoming increasingly less significant than the effect of a single electron charge.

We also need to go back and consider the linear nature of imaging sensors: Starting at a data level of 1, successive doubling of light intensity (think aperture stops on a camera lens) produces pixel values in the sequence 1, 2, 4, 8, 16, 32, 128, 256, 512 and so on. At the dark end, there are no intermediate values and the tonal resolution is 1 stop. At the other extreme, there are 256 values between 256 and 512 giving a tonal resolution of 1/256$^{th}$ stop. Fortunately, in conventional photography, the human eye can discriminate smaller density changes in print highlights than it can in shadow areas, by a factor of about 5x and thankfully the highlights are the key to any photograph.

The challenge of astrophotography is that much of the interesting information is not only invisible to the visual observer but has to be boosted out of the shadows through extreme image manipulation. The astrophotographer needs all the tonal resolution they can muster in the shadow regions. The numbers tell a worrying story too: how can smooth images be stretched out of a signal with only 3,125 data levels? The answer is that it is highly unlikely from a single image exposure. We are jumping ahead of ourselves a little but it is good to know that the astrophotographer has two ways to significantly improve the dynamic range of an image. It requires a detailed discussion, but for now, here is a short preview.

### Improving Dynamic Range

The wonderful thing about noise is that it is mostly random and ironically, it can help us achieve better tonal resolution. If the average light level wants to be 1,001.67 units, successive exposures from the sensor will produce integer values of either 1,001 or 1,002 in the main or occasionally numbers further out, caused by the injection of a little noise. Assuming an even noise distribution, in this example, the values of 1,002 or higher will occur twice as often as the value 1,001 or lower in subsequent exposures. If many (aligned) images have their pixel values averaged, using a higher bit-depth file format, the math can achieve an intermediate level and close to the true noiseless value between 1,001 and 1,002. The averaging process and noise in effect improves the tonal resolution of the sensor.

There is another profound effect of averaging multiple exposures of the same object. With an increasing number of averaged exposures of the same subject, the random noise in the final image is reduced and the Signal to Noise Ratio (SNR) is improved. Each time the number of exposures is quadrupled, the SNR doubles. For example, if 16 samples are averaged from a KAF8300M sensor, the read noise of 8 electrons is reduced to about 2 electrons and the dynamic range is closer to 25,000 / 2 = 12,500:1 (equivalent to a bit depth just under 14 bits). In astrophotography, it is standard practice to combine (integrate/stack) multiple images to improve the image SNR and boost tonal resolution. This is the enabler to improve the appearance of faint nebulosity, extend galaxy peripheries and remove one-off events such as cosmic ray hits and plane trails.

The second trick is to cheat! For those of you who are familiar with advanced digital photography, you will likely know that a subject with a high dynamic range can be captured by combining long and short exposures of the same scene that have been individually optimized for their shadow and highlight regions. Photographers can use a fancy Photoshop plug-in to combine them but I'm not a fan of its unnatural smudgy ethereal look. In astrophotography, however, it is considerably easier to disguise the boundaries since there are fewer bright mid-tones in an image.

In practice, a simple combination of optimized exposure sequences for bright stars, bright nebulosity and dim nebulosity will improve the dynamic range of an image. The grouped images are aligned and averaged and then the (three) stacked images are aligned and selectively combined, for instance using Affinity Photo and a brightness mask. There are several ways to do this. One such is to assign each image to a layer and blend using a mask generated from the inverted image data. The result is a photo-fit of bright stars, bright nebulosity and dim nebulosity with a dynamic range many times greater than the imaging sensor. The masks are tuned and blurred to ensure smooth transitions between the image data in the three layers.

This trick is not always required, but there are a few objects, the aforementioned Orion Nebula is one, where it is a helpful technique to capture the entirety of its huge brightness range with finesse. Using a combination of exposure lengths, the dynamic range is extended by the ratio of the longest and shortest time, as much as 100x or about 5 magnitudes. The same is true for imaging the enormous Andromeda Galaxy (M31), although less obvious. The core of the galaxy overloads a typical sensor in under 2 minutes but the faint outer margins require 10 minutes or longer to bring out the details. It is fortunate that, unlike many conventional photographs, the subject is unchanging from night-to-night and we can patiently image over many weeks and at different settings. If we were to combine the 16-bit images in the processing software into another 16-bit image, we could never exceed 65,000 levels. The trick is to do the averaging and manipulation using 32-bit file formats, to make the most of the increased tonal resolution from averaging tens, if not hundreds, of separate images. Patience and a large disc drive are your companions.

### Exposure, Exposure, Exposure

The last reality check is the practical consideration of how much exposure is needed and how long it will take. Any single exposure must not over-expose crucial areas of the image and yet, we have just learned that averaging multiple exposures improves image quality. Just how many are required is dependent on the quality of the sensor, aperture, sky quality and the object brightness. You might notice a new item in this list, "sky quality." Without going into too much physics just now, the effect of light pollution adds noise to the image. Yes, it adds an orange-brown mush, but this is fairly easily subtracted from the image during image processing. What remains, however, is the "shot noise" that comes with it. In simple terms, photons, like raindrops, are random, and the absolute level of randomness increases with the average amount. Fortunately, the random bit increases at a slower rate, so that, each time the background light pollution quadruples in intensity, the associated shot noise only doubles. This is a law of nature and there is nothing we can do about it. In most urban and semi-urban environments, the shot noise from light pollution exceeds the sensor noise. The keen-witted of you will also realize that there must be shot noise from the image too, so even if a sensor itself were noiseless, the light from the object and light pollution will conspire to make the outcome noisy.

Casting our mind back a page or two, we recall that noise is random and that quadrupling the number of averaged exposures halves the relative noise contribution (doubles the SNR). This is also the trick to remove the effect of shot noise from the light pollution but the practical upshot is, an astrophotographer at a dark-field site may acquire enough exposures for an image during a single night, but back home, it takes several weeks to experience sufficient clear nights to achieve an equivalent quality.

It is possible to get an image with less, but in my country town, I now plan for 20–60 hours of exposure to achieve a high-quality result. I do not have equipment envy but I go green with those who post great images on the Internet, with 10 hours or less exposure using similar equipment from a dark-field site. If I achieve 10 good images in a year, I consider myself lucky. When my editor, unaccustomed to the demands of astrophotography, requested a couple of dozen images for consideration for the book cover, I had to explain we could not wait that long. Similarly, to write a book on the subject, with relevant up-to-date images, takes several seasons in which to acquire the image data. Writing is an activity for cloudy nights!

# Planning for Success

*Deep-sky imaging has no respect for sloppy technique and benefits from careful planning and choices.*

This and the previous chapter introduce you to the broader considerations that propel simple landscape images on to deep-sky astrophotography. They examine the practical aspects of imaging and introduce a healthy dose of realism, in an attempt to avoid unnecessary outlay and disappointment later on.

Apart from those few opportunist images, taken without specialist techniques, astrophotography benefits from preparation. It is cold, dark and the brain is dulled from a hard day at the office. (I'm not selling this hobby, am I?) In these circumstances, having a clear idea of what equipment to use, how to set it up quickly (and without hiccups) is exceedingly useful and makes the whole experience more enjoyable too. It is very easy to deviate into exciting details and lose sight of the bigger picture. What follows is a look-ahead mixture of practical considerations and some qualities that you look out for when you choose and set up equipment. The themes noted here, resonate through later chapters and become a guiding influence on system setup. Not everything is relevant when one starts out, but it does provide valuable insight into what is to come (or, put another way, what you are letting yourself in for). Understandably, some of these terms may be unfamiliar to you at present but they will be explained later. These essential requirements broadly group into three areas: planning, kit and imaging essentials.

Planning
- location
- safety
- power
- comfort
- weather and target planning
- timing

Equipment and Software (kit)
- familiarity
- mechanical integrity and stability
- tracking and alignment accuracy
- autoguiding
- dew control
- focus and focusers
- essential software

Imaging Essentials
- cleanliness
- sensor size
- pixel size
- sensitivity
- image noise reduction
- calibration and averaging
- optical correction
- setting up and record taking

### Location

It is most likely that the new astrophotographer will start with a portable setup and assemble it in their backyard each night, or travel to a nearby dark site for their photography. Light pollution and the prevailing atmospheric conditions for your location set the upper-performance limit for astrophotography. Although dark sites offer the best conditions, one must be willing, ready and able to set up at a remote dark site each time the weather looks good. On the other hand, the convenience of a backyard may outweigh greater light pollution, an interrupted horizon and a neighbor's insecurity lights. Early enthusiasm will wane and although we might start off with good intentions, the practicalities and effort of remote-site operation may be worthwhile for just a few guaranteed prolonged imaging runs in fantastic conditions or in a social context. In my case, I have a worthy dark site about 20 miles away on the coast, with low light pollution and a convenient grassy car park, set in the marshlands of East Essex. I have not tried it yet.

I started with a system mounted on a tripod and located in the middle of my lawn. With a little preparation, I was able to deploy this setup and be imaging within half an hour. The installation decision influences the equipment choice since telescopes, mounts, counterweights and tripods are neither light nor small. After all, a portable setup needs to be just that, portable, and it is surprising just how quickly the repeated assembly of a large articulated mass in cold dark damp conditions can reduce the appeal of this hobby! For example, my first acquisition was a used 8-inch Meade LX200 SCT, weighing in total at around 30 kg. The previous owner had bought it for his retirement but quickly

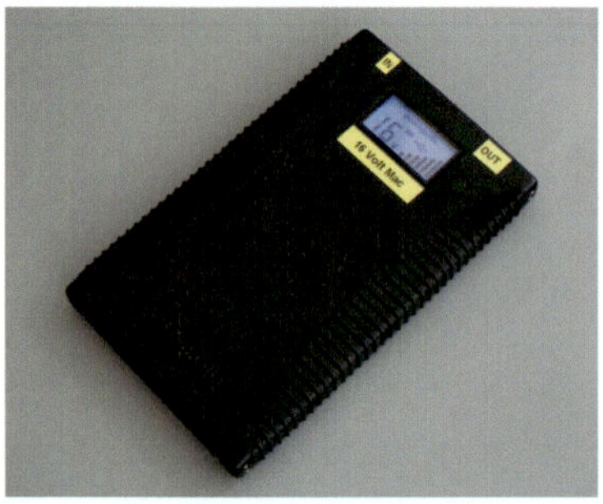

*fig.1 This programmable lithium polymer battery weighs just 600 grams and has multiple regulated output voltage options. It has slightly less effective capacity than the lead-acid gel battery in fig.2, but at 1/10th of the weight.*

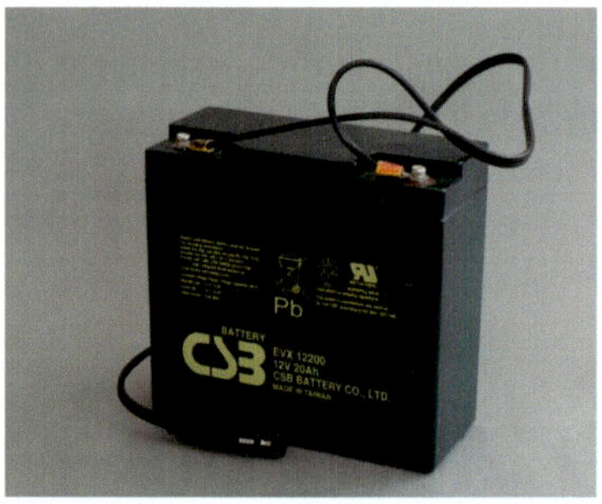


*fig.2 Although the capacity is rated at 20 AH, in reality, 50% discharge or more is to be avoided to preserve long-term capacity. Although 6.5 kg, it is 1/4 of the price of the LiPo battery in fig.1. I fitted it with a female in-line XLR connector, which I use as my standard power supply interface standard.*

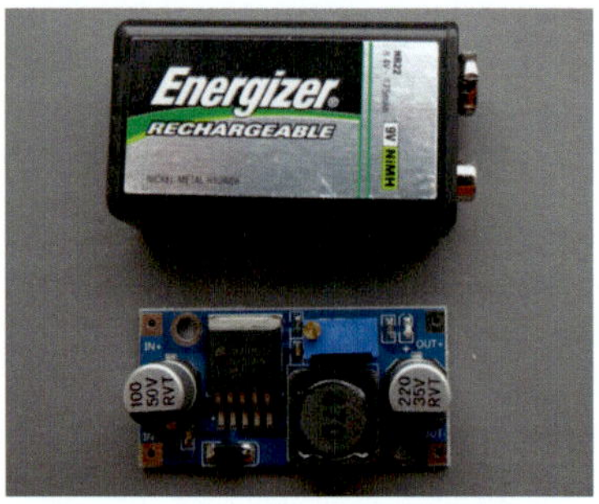


*fig.3 For low current applications, such as a substitute camera battery supply, these small DC-DC converters are one way to change voltage levels with minimum losses.*

changed to a lighter telescope. It is not just a case of lifting a large weight; equipment requires transport, carrying and assembly, often in the dark, without tripping, damage or injury. The box for the LX200 head filled the back of my car on its own. The same story plays out in the many adverts for large, used telescopes. In a permanent setting, however, larger and heavier mounts, scopes and installations are a one-time problem. These systems remain fully assembled and the cables routed permanently so that remote operation is safe and feasible. In a portable setup, the needs are different: there will be trade-offs in weight and rigidity and all the mechanical and electronic components must endure repeated assembly without failure or damage. In these situations, refractor telescopes are certainly more compact and robust during transport and normally do not require alignment before use.

## Safety

I must also highlight a few safety considerations; something that is sadly almost totally absent in other publications. The remote locations that astrophotography encourages may create some unique situations; lighting, personal security and heavy lifting are three obvious ones. Although mobile phones are commonplace, there may be no signal at a remote location and someone should know where you are. I carry several torches, including a wind-up one and prefer a vehicle with a flat load-space, as it allows heavy objects to be slid in and out without lifting them with a bent back. Dewy counterweights can easily slip through cold fingers and capped shoes are a sensible precaution.

## Power

Power has its own safety considerations. Most astronomy equipment requires 12–14 volts DC, but some devices, like most USB hubs, only require 5, for which I use a small 12 to 5 volt DC converter. (A few mounts also use 24 or 48 volts for which I use a larger, sealed DC-DC module.) Lead-acid cells are the most common source for mobile DC power. They conveniently store a high capacity (measured in amp-hours) but this lowers after each discharge/charge cycle, an effect which accelerates with the level of discharge. There are several lead-acid battery designs; those ones for hobby-use are best; they are designed to be more tolerant of deep discharge, whereas automotive battery versions are optimized to deliver bursts of high current but quickly lose charge capacity after repeated cycling. Gel-filled batteries or AGM designs are maintenance free and do not need to be kept upright. Large capacity lithium-ion batteries, up to about 85 Wh are also available, but at premium prices. These are considerably lighter and smaller than the lead-acid versions (figs.1, 2).

Power supply quality is important too and the DC supply to the camera should have as little electrical noise as possible. A simple solution is to use two batteries; one for the imaging and guiding cameras and the other for the "noisy" motor and switching functions such as dew heaters, focus control and the telescope mount. Modern batteries require a little care to prolong life and capacity; always follow the charging recommendations and use the recommended charger in each case.

For a domestic site, mains power is also an option but only with care. Any mains extension cable run out through the backyard or buried in the backyard should not only be armored, to protect from accidental rupture but employ an earth leakage current breaker (ELCB) at the power source. Just as significantly, many regulated DC power supplies, including some that are supplied by telescope mount OEMs, are only certified for indoor use. Dew and general dampness go hand-in-hand with astronomy and there is an obvious risk of electrocution or electrical failure with any model that exposes themselves to moisture. One alternative is to place the power supply in the house or in a suitable enclosure and use a length of heavy-duty speaker cable to carry DC, with minimal loss, to the telescope site. The same safety considerations apply to domestic plug-in-the-wall power adaptors and desktop computers sited in outdoor situations. They need protection from moisture and should be appropriately earthed. Spare mains sockets should be kept away from moisture and fitted with a plastic child safety cover for good measure. If there is any doubt, please ask a qualified electrician to check over your installation. I don't want you seeing "stars" with your eyes closed.

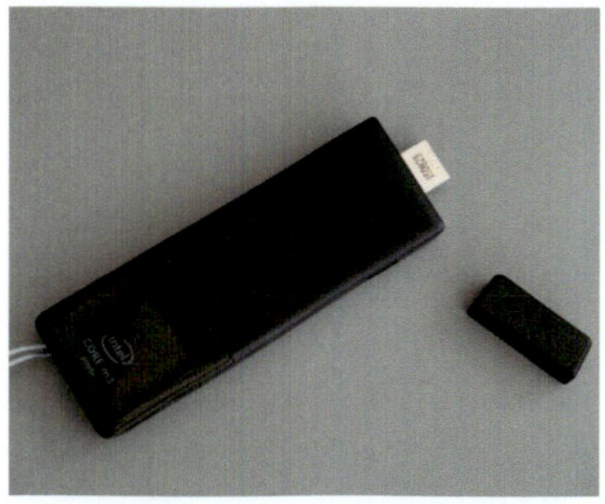

*fig.4 I decided early on to dedicate a computer for astronomy use, allowing it to run lean and mean. Astrophotography does not require a powerful computer and one alternative, worth considering, is to use a computing stick. This resides next to the telescope and is controlled remotely via WiFi, using remote desktop software on a PC, Mac or iPad. At the same time, this avoids the trip hazard from trailing cables around in the dark.*

In a portable setup, the laptop is often the computer of choice if you are close by the mount. It is useful if its battery life exceeds 5 hours. Aggressive power saving settings are great but remember to turn off all the advanced power settings sleep modes or your imaging run may unexpectedly terminate! A few models have hot-swappable batteries but most are internal. An alternative is to use a reserve high capacity lithium-ion battery. Some popular models are designed for laptop use and have a programmable output voltage and multiple adaptors to suit the laptop power sockets (fig.1). A third power option is to use an inverter. These convert 12 volt DC into AC mains, which then supplies the laptop's power adaptor. As before, please check the model is safe for outdoor operation, as most are not. Many modern power supplies are highly efficient switched-mode models but their output is "floating" and can create static discharges when they are plugged in live, into a grounded piece of equipment. These high voltages can easily damage sensitive electronics. To avoid this, connect all the equipment together (signal and power cables) before switching on. The various 0 volt or ground connections will keep things tied together and static voltage discharges will not occur. I am not a huge fan of sharing the backyard with the mosquitos and I prefer a dedicated Intel Stick or NUC computer under the mount (fig.4). I have a sealed DC-DC converter that provides a 5-volt power supply for the stick and I control the computer remotely, using Microsoft's free Remote Desktop software, from any computer or tablet via WiFi and the comfort of the car or home.

### Keeping Comfortable

In colder climates, the manual exertion during the setup phase may initially keep you warm but the body quickly cools down during the extended inactivity of image-capture. As with hill-walking, you need to layer up when you stop exerting yourself. I prefer to look ridiculous than sit shivering in the car. It is easy to forget that even at night, long attended sessions require food, drink and diversion. I use an iPod rather than the vehicle radio or mobile phone to preserve these important batteries. A number of laptops and tablets have touch-sensitive controls and will not work with conventional gloves. If you look around, gloves are now available with conductive fingertips. Extreme cold does have one advantage though; it keeps the insects at bay. Insects like astronomers,

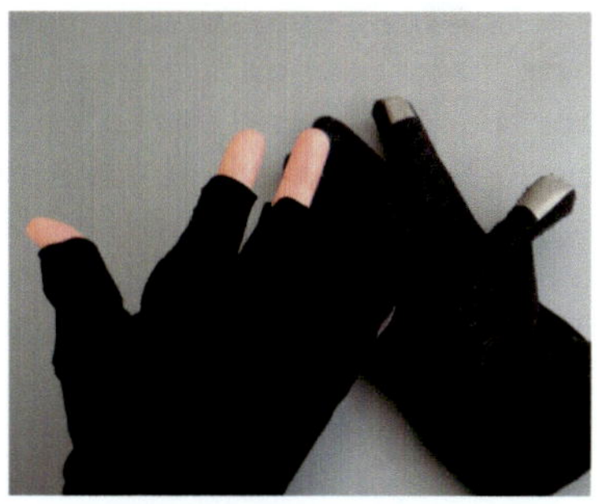

*fig.5 In cold conditions, gloves are an essential accessory. Choosing a pair that are warm and still have sufficient dexterity for keyboard use is challenging. These gloves work with touchpads and touchscreens. In really cold conditions, I use an over-mitt too. My wife offered to attach them with a string, to pass through my sleeves, to stop me losing them in the dark!*

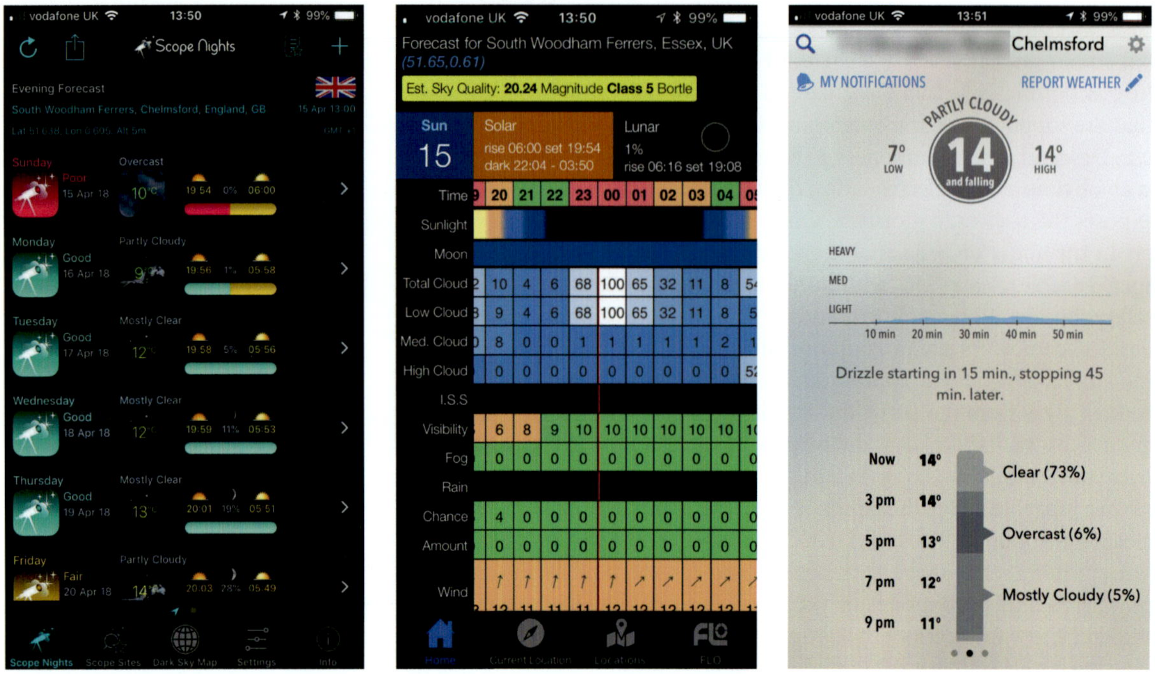


*fig.6 Weather Apps are commonplace. For the UK at least, it helps to have three on the go for a majority vote! From the left, these apps are Scope Nights, Clear Outside and Dark Sky. Dark Sky links to the iWatch and gives you a timely reminder if rain is on the way.*

so mosquito repellent and bite cream are a few more essentials to keep in mind.

### Weather and Target Planning

Not everyone images from the Atacama Desert. More often than not, the cloud is unbroken and each weather front merges into an unending overcast month and you have almost come to the point of giving up astronomy and taking up fishing. Then, one evening, surprise, it is clear. There is suddenly an opportunity for imaging ... of what exactly? Meteorology is an inexact science but weather systems are predictable to some extent (even in the UK). On-line weather reports and weather Apps are constantly being released and updated and a useful resource to anticipate clear skies. Forecasts are quite accurate, up to 48-hours ahead, but the timing may be out. It helps to recognize cloud sequences so that one can realign the forecast. This is not necessary with those applications, like Dark Sky, which constantly maps weather fronts, and gives you customized warnings for your location.

Next, we have to decide what to image. Many objects, especially those at lower declinations have a preferred season when they are visible for some of the night. Ideally, they rise to an altitude of about 30° at twilight, which maximizes imaging time. Those objects nearer the pole never set and those over 70° from the celestial horizon (for UK sites) are safe bets too. At a latitude of +50°, objects with a DEC between -10° to -90° will never have an altitude greater than 30° and will be a challenge to image. Planetarium programs usually have a "tonight's best" option, usually for short term visual purposes. Luckily, every object's visibility is predictable and it helps to identify those objects that will be optimally positioned each month beforehand. There are plenty of sources for information: magazines, books, planisphere, free planetarium programs and so on. These are typically very good for stars, constellations, planets and the most obvious deep-sky objects. If you wish to find something more unusual, you may need a specialized planning tool that has access to more exotic catalogs or use a printed resource such as Charles Bracken's *The Astrophotography Sky Atlas*, which has a comprehensive set of maps of objects, including interesting nebulae, for all seasons and latitudes.

### Timing Matters

Image acquisition is rarely a quick fix (well, other than short planetary videos). When you start out, it is often the case that the beginner tries to photograph several objects in one night. It is highly unlikely that they will be able to make sufficient exposures for a quality result and they quickly realize that it may take one or more nights to record enough exposures to overcome the various operational hiccups and the degrading effect of light pollution. There are a few bright objects that can be captured in a single night but to do justice to the majority, you will need more perseverance. On the same note, it is really frustrating to start a promising imaging sequence only to be obliged to break off before you have enough imaging data to make a quality image. To prevent this from happening too often, I make a quick evaluation of the likely overall exposure duration and available time. Once more, the Internet is a useful resource to plan exposures: a search for the deep-sky object often links to another's images that usefully indicate the equipment setup and exposure details, possibly with the same camera or sensor.

By now, it is starting to sink in that image quality benefits hugely from multiple exposures (also called "subs" or "lights" by astrophotographers), the more the merrier. Dim galaxies and nebulae benefit from at least 10 hours of data or more, especially when there is light pollution or using special filters. As you become more sophisticated, the more advanced acquisition programs remember what you were imaging, and how far through the intended exposure plan you were at the point when you shut down. They can start up and continue on another night with relative ease.

## Equipment and Software

### Familiarity

How well do you know your set-up? So many times, a frustrated imager posts a forum question that is explicitly answered in the manual (which of course they had not read), me included, guilty as charged. Equipment and software setups are complex and complete familiarity really helps with quick and reliable operation, especially at night, when we are cold and tired. Many problems occur with inconsistences in a portable setup or after a software update. Where possible, I rehearse assembly, balancing and alignment during the day to make night-time operation second nature. Some preparatory work can make this easier still; for instance, the balance points can be marked on the telescope dovetail plates for instant assembly and the polar scope reticle calibrated to the RA scale and centered during the day. I also identify USB cables with simple colored markers made from electrical tape around each end. It sometimes occurs that a new hookup into a different USB hub confuses the device configuration. The cable labeling not only allows me to assembly USB/USB hubs in a consistent manner but it is also handy for when a single device requires a USB reset, without disturbing the scope-end of things.

> USB connections and the device drivers can be problematic if they are plugged in a different sequence each time. This is especially true of USB-Serial adaptors, as COM port numbers are assigned dynamically. I set out a table of adaptors and COM port settings and then manually assign the port number in Window's device manager, under the advanced settings for the port. These settings then stick and do not change, allowing you to set up your devices once and forget about it.

Little things like systematic storage, say using clear and labeled plastic food storage boxes, protect small items from dew and dust and avoid rummaging in the dark. The smaller boxes can be organized into larger ones for imaging, observing and common accessories, which helps when loading up. I save the silica gel bags from various purchases and pop them in for good measure, or you can use a small paper or cloth bag filled with rice, which works similarly.

Understanding software operation is essential and easily underestimated. Even when you think everything is perfect, software can kick back; in some versions of Windows, if a USB camera plugs into a different port connector, the operating system demands the driver file is loaded again, causing unnecessary aggravation, especially at a remote site, without the device's CD or an Internet link.

I am not the first, nor will I be the last to waste imaging time due to poor preparation. The software startup sequence can be established during a dry run and the hardware drivers checked beforehand for reliable operation on each of the different USB ports, especially after any software upgrade or update. One simple way is to record the power-up sequence and having found a reliable order, repeat it each night. This increases the chances that the equipment connects flawlessly and follows a reliable alignment and calibration sequence, including polar alignment, focus, star alignment, object location, autoguider calibration and exposure determination. (Dedicated astro equipment often communicates with Windows via a universal "ASCOM" interface. The ASCOM suite of utilities also includes simulator versions of camera, focuser and mount drivers, that can be tested in the comfort of the home.) Astrophotography challenges our ability to remain clear-headed; few of us are at our best at night and I keep a checklist on my smartphone. That also includes the essential data for the telescope and imaging configurations that are required by the imaging software. For instance, I manually set all the COM ports for the various USB-serial modules, rather than work out the random assignment made by Windows.

### Mechanical Integrity and Stability

Long image exposures require a very rigid structure to get the best out of the telescope and mount. Every vibration, slippage, flex or movement, degrades the final image. Working from the ground up, the tripod or pier and the telescope mount must be as inert as possible and minimize vibration, slippage or flexure. This is principally a combination of mass and stiffness. A permanent pier, set in concrete is very rigid, especially if it is de-coupled from external sources of vibration. For portable setups, a rigid and secure tripod is essential, with feet that do not slip or slowly sink into the ground under load. Height is not important for imaging and one way to minimize leg flex is to extend them as little as possible (providing it is stable) and choose a tripod that has a spreader bar, for improved stability.

Once you move beyond shooting landscapes and brief exposures of the Moon, you will need a telescope mount. The mount has the lead role in the imaging system's performance. Even without tracking, the mount's many mechanical interfaces affect general stability. Mounts vary enormously in weight, stiffness, payload and price. When planning the equipment budget, the mount takes the top spot. Conventional wisdom suggests that for the best results, the weight of the scope, camera and hardware should not exceed 2/3rds of the mount's alleged payload capacity. It is also essential that the mount cannot move about on the tripod or pier. This requires tightening all four polar adjustment knobs on the mount and re-tightening any base fixings after polar alignment (which requires the mount to move).

Every piece of metal will flex under load and every mechanical interface is a source of movement too. One of the troublesome areas is the back half of the telescope. Mechanical flexure is not an issue for visual use but it is clearly important when a camera is substituted for an eyepiece. The focus mechanism, field-flattener and filter/camera system need to be secure and orthogonal to the optical axis. This may require a little ingenuity in some cases: for example, many standard telescopes are equipped for observing use, with 2-inch and 1.25-inch clamping systems for diagonals and eyepieces. The imaging adaptors often make use of these but the weight of heavier cameras causes these to flex or slip. The less expensive adaptors use a single screw fixing, which not only marks the tube but will almost certainly wobble on the other axis. The up-market versions use a brass ring, clamped at three points; these are not perfect but are an improvement. Any tilt or flex has a direct impact on the shape and focus of stars around the image periphery. It is always better to screw things together to ensure an aligned and stiff assembly and an Internet search will often locate a threaded adaptor to mate items together or a machining company who can make an affordable custom design. To prevent thin rings binding together, I apply a little lithium-based lubricant to the threads before assembly and to unscrew them, grip with a pair of rubber gloves. I devised my own threaded adaptor to couple a telescope to its adjustable field flattener. The idea has caught on and the OEM now sell their own.

Mechanical play, sometimes known as backlash or hysteresis, is a consequence of mechanical tolerances. Each mechanical interface has a little gap or flexure that results in backlash between each gear, belt and bearing. Backlash is not an issue if things only move in one direction; it becomes a problem when a motor changes direction (or when the balance point flips over and the engagement forces reverse). It can be measured by noting the amount of reverse movement needed before a system starts to change direction. It is normal to have a small amount of play in any system but it is prudent to check your mount for severe issues, as consistent factory adjustment is not guaranteed, even in a premium product. Manufacturers have become wise to this and there is often an instruction or a helpful YouTube video on how to tune a mount. To avoid any later unpleasantness, check with your dealer before making any adjustments to your mount, as your actions may invalidate its warranty.

### Tracking and Alignment Accuracy

As a star's declination is fixed an equatorial mount only has to rotate around the right ascension (RA) axis and in a perfect world, a telescope mount tracks a star's movement around the sky using a single motor. Real life is not that accommodating and there are two principal issues, drift and periodic error, that compromise tracking performance and cause a star to wander during each exposure.

When a mount's RA motor axis is misaligned with the celestial pole the tracking issue "drift" occurs. As the center of the star's rotation and that of the telescope are not perfectly aligned, the stars slowly and steadily drift across the sensor during each exposure. The rate of drift is minimized by improving the mount's polar alignment. Alignment is covered in detail in later chapters but for now, we assume it can be improved through the use of a polar scope or computer assisted polar alignment. Modern techniques can get very close in under 10 minutes and there are countless utilities to help you make the right adjustments. At low altitudes, additional drift is caused by the rapidly increasing effect of atmospheric refraction. (When the sun sets, it actually is not where it appears to be, but its full width of 0.5 degrees beneath the horizon.)

Periodic Error (PE) is the consequence of the dimensional tolerances of the drive system in the mount. As the main right-ascension (abbreviated RA) motor turns, the imperfections of the mechanics cause the mount to run a little fast or slow at different angles and the image does not track perfectly. The error from each part of a rotating system repeats for each cycle, hence the name. These errors combine to form a complex repeating tracking error. During a long exposure, these cause a star's image to elongate, to an oval or worse. We measure PE in terms of angular error; typical peak-to-peak values range between 1–7 arc seconds for premium brands and between 15–50 arc seconds for less expensive systems. I owned one system, that cost over £1,000, which will remain nameless. Its PE was over 80 arc seconds peak-to-peak, as were its two replacements. It could not be tamed and I sent it back.

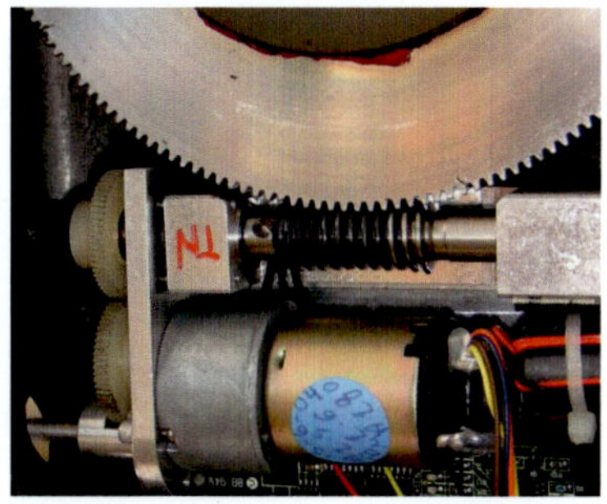

*fig.7 Many mounts use a worm gear assembly to move one or both axes. The worm, the part that looks like fusilli pasta, has small irregularities that cause the mount's movement to deviate from a perfect tracking rate. These repeat every cycle of the worm (normally about 5 to 10 minutes) and the repeating irregularity is hence called periodic error.*

Price and brand are not always a guarantee of good PE performance either; many forums bear testimony to the performance differences between identical models, which is an unavoidable outcome of the real world of production tolerances. In addition to physical tolerances, grit or debris trapped in the lubricant of the gear mechanism can create an abrupt tracking error, or more simply, by an incorrect adjustment of the assembly. Some of the more confident engineers in the community may thoroughly de-grease, clean and re-lube the gear system of a used or even a new mount and adjust the gears and drive, checking for play or binding. This is not a task for the faint hearted and should only be considered after the warranty situation, ease of access and a risk assessment is completed! After some tense moments, I successfully replaced, aligned and cleaned the gears in my used Meade LX200, to huge benefit and later on, cleaned and lubricated the drive on a Paramount MX too.

I am fortunate to have a permanent installation with a mount that has good PE. Its accompanying software has a feature that measures the error and then applies it, in reverse, to correct it out. This Periodic Error Correction (or PEC) halves the overall error but even so, it is sometimes larger than the telescope resolution, sensor resolution and astronomical seeing combined. For that reason, and even though it is accurately aligned, I additionally guide this mount to remove any residual tracking errors, to ensure small round star images. In this well-behaved system, the error changes slowly and I only make corrections every 5 seconds or so.

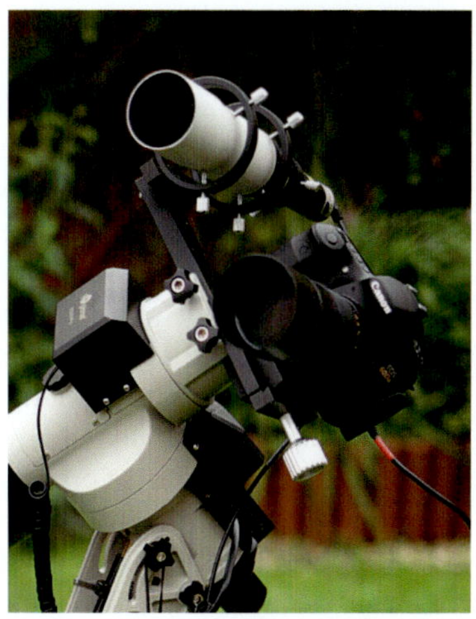

*fig.8 With longer focal lengths or exposures, it is often necessary to correct for alignment and mechanical tolerance issues by autoguiding. This uses a separate autoguider system, here fitted alongside the imaging camera. A small guide scope is fitted with a small sensitive CCD camera. A PC repeatedly checks the location of a solitary star and issues small corrections to both the motors on both axes.*

By way of comparison, my semi-portable system uses a mount that has a larger PE and does not have any correction feature. It is less likely to be accurately polar aligned and in this setup I make corrections every 1–2.5 seconds and typically restrict imaging to shorter focal length telescopes, whose lower magnification is less sensitive to tracking errors.

There is an insatiable interest in perfecting polar alignment and reducing periodic error on the Internet forums. These go to the extremes of replacing or upgrading gears, bearings, complete overhauls of new equipment and extensive use of numerous polar alignment routines. It seems to be a goal in itself to achieve perfect alignment and run long, unguided exposures without star elongation. There is no such thing as perfect polar alignment. It varies depending on what you are trying to achieve. I apply the 80–20 rule to these goals and use autoguiding to correct for both drift, flexure and periodic error during exposure. To put things into perspective, if a single tripod leg sinks into soft ground by just 1 mm, the resulting polar alignment error dwarfs any benefit from advanced polar alignment.

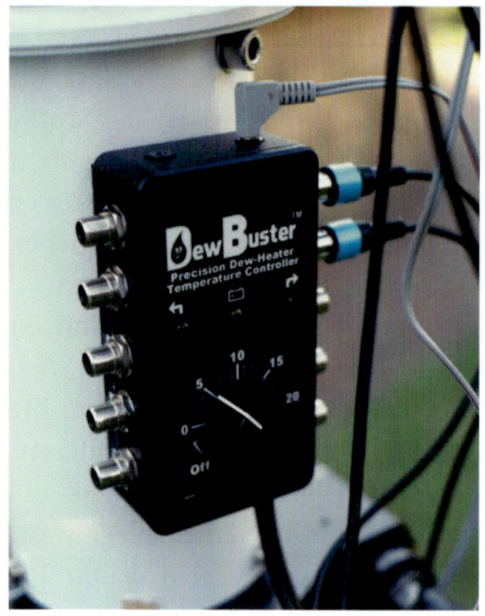


*fig.9* *Dew is particularly an issue in Spring and Autumn or when there are large variations in ambient temperature. This control unit has multiple pulsed 12-volt outputs and has a feature that regulates the output to maintain a constant temperature difference between the telescope tube and the air. Less expensive alternatives re-purpose a LED light or motor controller to regulate the 12-volt output. These can be effective but one has to watch out for electrical interference from the switching circuit. Linear regulators (using voltage regulator chips) are another alternative, they have low electrical noise but waste more energy as heat.*

### Autoguiding

Practically and within reason, autoguiding corrects for periodic error and drift by repeatedly checking the position of a single star on a small sensor and making small adjustments to the mount motor(s) to keep it centered at the same position. This is easier said than done and this process has its own chapter and within some assignments too. At this stage, it is important to realize that autoguiding is not a cure-all; it requires a reasonable level of mount performance and polar alignment accuracy to give it a chance of working. Guiding can be performed on one axis (the RA axis) or both axes. The former system is found on simpler mounts that only have a motorized RA axis. Some of the simplest trackers do not have any facility for guiding.

### Dew Control

The need for dew control varies by region, season and by telescope type too. As the air temperature falls at night its relative humidity increases. At the dew-point temperature, air is fully saturated and cannot retain all its water vapor. Below the dew point, excess water vapor condenses onto cold surfaces. Those surfaces which are exposed are worse affected and optical surfaces will not perform with condensation on them. Long dew shields (lens-hoods) certainly help but do not entirely eliminate the problem. In reflecting telescopes the primary optics are generally well protected at the bottom of the telescope tube. In the case of truss design reflector, a cloth shroud will similarly shield the primary mirror but the secondary mirror will require mild heating to prevent dew from condensing.

Those designs with glass optics at the front will likely need gentle warming too; just enough to prevent condensation forming. (Aggressive heating is counter-productive as it will create your very own poor seeing conditions within the telescope's optical path.) This is conveniently achieved by passing current through an electrically resistive band (dew heater tape) which wraps around the telescope close to the exposed optical surface. Most dew heater tapes conveniently plug into a 12-volt power source for full power or a pulsed supply for finer regulation. Camera users, who otherwise do not need to take heavy 12-volt power supplies into the field make do with a disposable hand warmers, fixed around the camera lens with an elastic band, or one of the latest 5-volt based dew heater tapes that plug into a portable USB charger. With dew control, prevention is always better than cure and it is best to turn on a dew heater before you remove the lens cap. It takes a lot more heat to remove condensation and there is a risk of image blurring from eddy currents at the air/glass boundary, until the glass cools down again to ambient conditions.

### Focus and Focusers

Several refractor telescopes from competing companies share the same optical cells, sourced from the same Far-Eastern supplier. Their major differences lie in the other aspects of the telescope. Putting aside optical quality for the moment, these include the finish, internal baffling, sundry accessories and most importantly, the focus mechanism. Precise focus is critical and it took me some time to realize just how important it is for high-quality imaging. A poor focuser design is a source of intense frustration. In the more demanding setups, the precise focus position may require updating during an imaging session to account for thermal contraction of optics and tubes, as well as any differences in focus position introduced by inserting a filter in front of the sensor.

Focus mechanisms must be a dilemma for telescope manufacturers; the needs of the visual user are less demanding and it makes no sense to design-in cost, for a performance that is superfluous to many customers. In astrophotography, the combined mass of the imaging equipment can place a considerable load on a focus drawtube and more importantly, the need for maintaining precise focus and stability. A good focus mechanism is strong, does not slip or flex in any direction under load and, when the focus position is changed, the image does not shift laterally.

Even with a substantial mount, touching the focus knob on a telescope will cause the image to jump about, making any focus assessment tricky. For that reason, when astrophotographers get serious, they focus via an electronically driven motor attached to the focus knob (fig.10). The better designs use a gearbox reducer and a stepper motor, which enable fine control and absolute positioning, usually via a serial or USB interface. Several companies specialize in upgrade focus mechanisms, with and without motor actuation. Some telescope manufacturers offer third-party focusers as a point-of-sale option for imaging use, at a price.

You might be wondering just how essential is accurate focus for high-quality imaging? A star is smaller and brighter when it is in focus since all the photons fall on the fewest number of pixels (this also maximizes the signal to noise ratio). It was a revelation, however, to realize just how sensitive the focus accuracy has to be. As an example, a short telescope of just 400-mm focal length and with a 100-mm aperture has a diffraction-limited image star diameter of 2.7 microns (less than a typical pixel). A focus error of 0.1 mm blurs this to a circular patch 25 microns wide, about 10x wider and covering 25 pixels of a typical sensor! Correspondingly, the photon intensity per pixel will be, even accounting for average seeing conditions, about 25x less. With the luxury of a motor controlled focuser, it should be possible to move the mechanism in steps of about 0.004 mm.

While not mandatory, in retrospect, motor-controlled focusing has a significant impact on image quality, over that of a better telescope or camera and especially when focusing is automated using computer analysis.

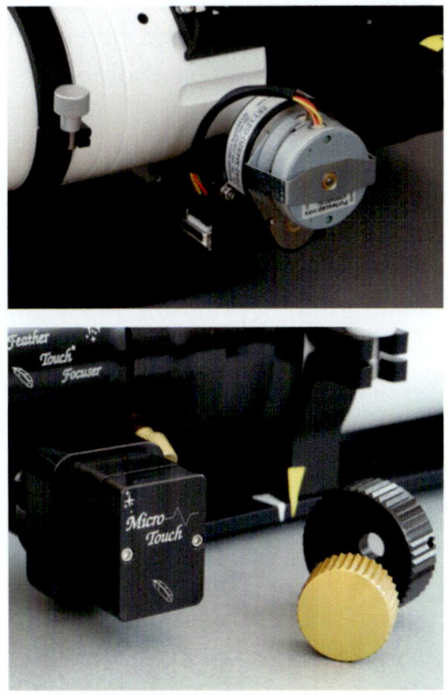


*fig.10 The back end of a telescope is just as important as the front. Here are two examples of electronic focus motors with linked gearboxes. Both systems have separate control modules with a USB interface and ASCOM control, allowing remote autofocus. The highest quality focus systems cost as much as a good refractor but, as with many things, give a diminishing return on expenditure. Finding the happy medium relies on a bit of research and luck.*

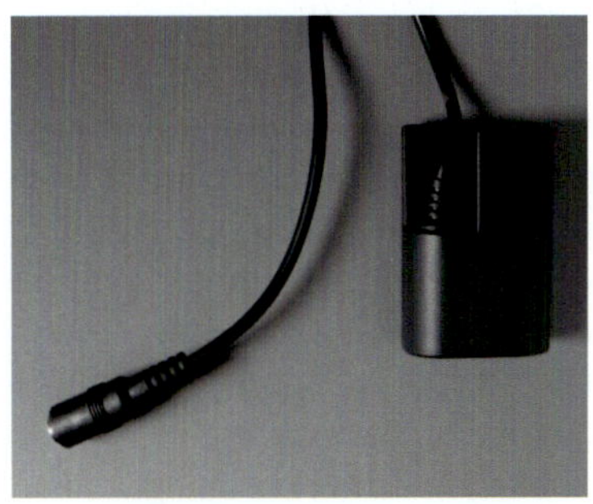

*fig.11 A Canon EOS battery eliminator. The Canon EOS 60Da comes with one and a PSU. One may also purchase the adaptor from the Far East too. In practice, I found the DC supply voltage into the device is usually less than the indicated voltage on the camera battery.*

### Essential Software

Software is clearly a necessity for digital astrophotography and at the same time it is the most difficult subject to give specific guidance on. Whatever type of astrophotography you embark on, image-processing software is a must. Introducing a telescope mount and computer-controlled equipment into the equation requires some form of planetarium and mount control

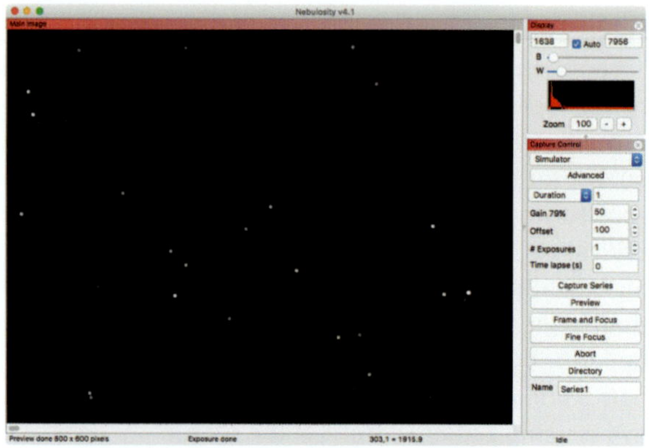

*fig.11 Nebulosity 4 is an often overlooked value-for-money data acquisition and image processing application. It has both Mac OSX and Windows versions. I started with this application, which prides itself on keeping things simple.*

with image capture and potentially autoguiding applications running on a computer.

This represents a fraction of what is available: It is easy to collect multiple software applications and utilities like confetti as many are free, or reasonably priced, like the thousands of apps for mobile phones. The trick is to be selective; in addition to the essentials, an autoguiding program, polar alignment utility and an autofocus program are the most useful for imaging.

One of the dilemmas with software selection is there are no consistent feature boundaries between software genres and many applications' functions overlap with others. The variety and permutations are almost endless. One way to look at this problem is to turn it around and consider the necessary functions rather than applications. Where the functions reside is a matter of product selection and preference. If we adopt the same strategy as this book, we start with the barest essentials and work our way up to more sophisticated applications.

Irrespective of how one captures images, or the equipment you use to do so, the images themselves will need extensive manipulation. The dominant player in the digital photography market is Adobe Photoshop®. The latest version is only available for rent and although its pricing is easily justified by professional users, the cost-savvy user may look elsewhere. Prior versions may run on your PC or Mac (10.14 or earlier) but Adobe's update policy for their Camera Raw plug-in encourages new camera model buyers to upgrade. Serif's Affinity Photo is a better multi-platform alternative, with a one-time cost of about £40 and offers several unique features that are of interest to the astrophotographer and is also compatible with many Photoshop plug-ins.

There is another consideration too; digital image processing of deep-sky objects is very different from normal digital imaging and apart from landscape astrophotography, it requires several special tools to produce a quality result. A typical imaging session produces multiple exposures either in color or as separate monochrome files. Processing these images is a long journey; the key image processing steps are calibration, alignment, stacking and a group of functions that collectively enhance the image. Some users persist with Photoshop. Although I own the CS6 suite, I get better results using applications designed for astrophotography or which can manipulate 32-bit images across their entire toolset. Calibration, alignment and stacking functions are included in several image-capture programs or available as separate utilities, several of which are free. The initial image enhancements are applied to very dim images and benefit from 32-bit image manipulation to avoid unsightly artifacts. Image enhancement is a huge subject and worthy of its own book and it is important to realize that there is no one way to process an image: each application has a range of tools for reducing noise, sharpening, enhancing faint detail, improving color saturation and so on, some of which will be more or less effective for your particular image than a similar function in another application. It is an area for patient experimentation on a cloudy day.

A dedicated image-capture program is desirable for camera control, image download and storage of high bit-depth files. In the beginning, it is also possible to get by, without specific image capture software, by using a standard camera with a simple cable, timer release or in some, their built-in intervalometer. Image-capture programs allow remote operation and have drivers for most dedicated astronomy cameras, webcams, and increasingly leading brands of DSLR too. Many photographic cameras are enabled for tethered operation through their USB port. This facilitates remote control and various options such as mirror lock-up and complex exposure sequences. In addition to the key functions to control exposure and download images to disk, capture programs display a stretched preview and provide essential data to assess focus and exposure, including maximum pixel values, noise levels, star sizes and image histograms. As you advance, they can also set and report the sensor temperature and filter wheel position.

An added benefit of using an image capture program is that they can save images in the FITS file format, which is commonly used by dedicated image processing applications and includes all manner of supplementary data in its header, which becomes increasingly useful later on.

Deep sky objects are rarely visible through a viewfinder and it requires a combination of computerized mount control and some form of planetarium (either as part of the mount system or an external application) to quickly find and align on a subject. The main function for a planetarium is to display a chart of the sky for planning purposes, for any geographic time and place, with the essential information for thousands of objects. There are many applications for computers, tablets and smartphones. They commonly have additional functions to point and synchronize the telescope mount to an object and show the field of view for an eyepiece or camera. These programs, especially those with high-resolution graphics, use a good share of the microprocessor time and memory as they calculate and update thousands of star positions each second on screen. Once a portable system is imaging, there is little need for the planetarium and the application merely wastes computer power. There are many program styles; some are educational and graphics intensive, others more scientific in approach, with simpler graphics, aimed at practical astronomers. Several excellent programs are free but the more sophisticated can exceed $100. Most use or can access the same star catalogs for reference but the controls, ease of use and interfaces vary widely. I often use an iPad application to plan my imaging sessions; the intuitive screen controls and built-in GPS are equally convenient in the daytime or at night and on occasion, the direct control of the telescope too, via WiFi.

Mount control comes in many forms too. A telescope mount houses a motor-control board that connects to an external handset or computer, which has the essential functions of alignment, object selection, slewing and tracking. After physical polar alignment, which, say, points the RA axis within 5 arc minutes of the pole, the mount needs to know exactly where the stars are. It already roughly knows, within the field of view of a finder scope, but it needs precise registration to improve pointing accuracy. The basic alignment function has two steps; slewing the mount to where it thinks a prominent star is, followed by manually centering it with the handset controls and "syncing" the mount. Basic registration or alignment calibrates both motor positions for a single star and in a perfect world, every other star falls into place. Sadly, mechanics and physics play their part and to compensate, the more advanced alignment functions repeat the registration process with several bright stars in different parts of the sky. A hand controller typically uses three stars to align itself but the sophisticated computer programs use more to build a more accurate "pointing model."

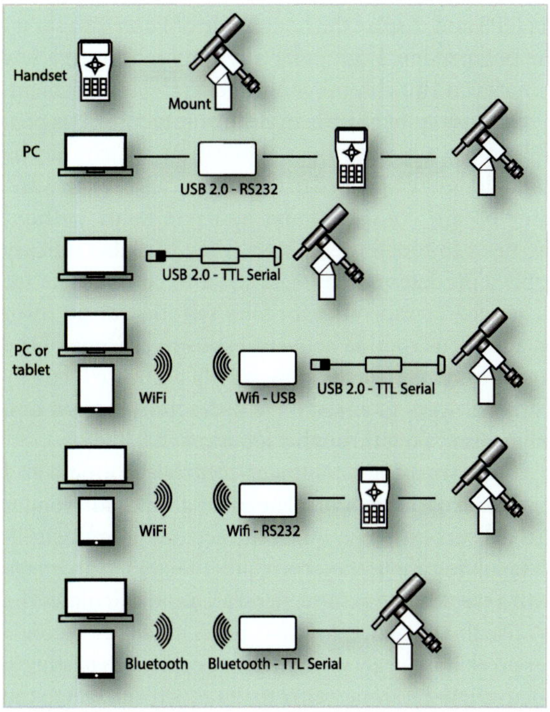


*fig.12 In the case of the popular SkyWatcher SynScan system, there are numerous ways to connect and control their telescope mounts, through serial, USB, WiFi and Bluetooth interfaces.*

It is easy to go over-board; ultimately, we just need to place the object we wish to image onto the sensor and then rotate and fine-tune the camera position to give the most pleasing composition.

In the case of a mount with an external computer interface, either directly into the mount or via the handset, things become more complicated. For example, the popular Sky-Watcher® EQ and early Avalon mounts have two software interface protocols; the one between the supplied SynScan™ handset and the mount and a basic one with an external computer. The physical connection possibilities are numerous and are shown in fig.12. In each case, a computer, tablet or mobile device, controls the mount using a simple command protocol. This program, in turn, accepts commands from the planetarium and autoguiding applications that are linked to it through a software driver. This example serves as an illustration of just how complicated these systems can become. To improve the chances of success, start off simply and check things work before introducing additional hardware and software complexity.

There are several utility functions that are particularly helpful in a portable setup; polar scope alignment, autofocus and plate-solving functions. I use a small iPad application to determine my location and time via GPS and display the hour angle of Polaris to set up the polar scope. I can polar align my mount in a few minutes and quickly move onto general star alignment.

Of those systems with an electronic focuser, it is possible to use a remote focus application, which may be standalone or embedded in the image-capture software. They obtain consistent and accurate focus without the need to touch the focuser. They can also quickly apply a predetermined focus offset to compensate for a change in temperature or filter selection or perform an autofocus routine between exposures. Some of the recent developments achieve highly accurate focus by using software to analyze the diffraction pattern of a bright star taken through a focus mask.

If you are using a computer-controlled camera and dual-axis robotic mount, plate solving is a small wonder that delivers extremely quick and accurate alignment. A plate solving application correlates the stars in an image with a star catalog of the general area and calculates the precise image center, rotation and scale in arc seconds per pixel. It is hugely satisfying on a cold night to simply polar-align the scope, retire to the couch and watch the telescope automatically slew, expose and self calibrate its alignment within a few pixels. The satisfaction grows when a slew to the object places it dead center in the image, ready for focus confirmation, autoguider calibration and a night of imaging. It is quite surprising just how useful plate solving is in general astrophotography and with the many free applications, it is no longer an indulgence. With the introduction of ASTAP, Linux, Mac and PC platforms are all supported. Ten years ago, PC's were the only pragmatic imaging platform. In recent years, Linux (Indie) platforms have overtaken Mac as a viable alternative.

| focal length [mm] | aperture [mm] | optical FWHM [μm] | +seeing FWHM [μm] | pixel pitch [μm] |
|---|---|---|---|---|
| 10 | 3 | 2.2 | 2.2 | 1.1 |
| 20 | 8 | 1.7 | 1.7 | 0.8 |
| 50 | 12 | 2.8 | 2.8 | 1.4 |
| 100 | 25 | 2.7 | 2.8 | 1.4 |
| 350 | 70 | 3.4 | 4.8 | 2.4 |
| 700 | 110 | 4.3 | 8.0 | 4.0 |
| 1,000 | 130 | 5.1 | 11.0 | 5.5 |

*fig.13 This table compares a selection of typical camera lens and telescope configurations and a theoretical sensor pixel size, to avoid losing further resolution. The optical FWHM is the diffraction-limited resolution, calculated for green light. The +seeing FWHM is the combination of the optical resolution and a seeing condition of 2.0 arc seconds. For short focal lengths, optical aberrations limit the achievable resolution and at long focal lengths, astronomical seeing dominates. In practice other effects rob further resolution and larger pixel sizes will suffice. It makes interesting reading when it is compared with the common sensor pixel sizes in fig.14.*

## Imaging Essentials

Without digital sensors, astrophotography would be a very different affair. In a single decade they have changed the popularity and quality of astrophotography beyond recognition. They deserve a detailed analysis of their workings and shortcomings but just now we consider only the simpler specifications and requirements. In the beginning, one is likely to use an existing DSLR or mirrorless camera and potentially upgrade to a purpose-built cooled camera system later on. Camera choice is ever increasing. More models appear each year, in ever increasing frequency, to tempt the customer but in reality the differences are minor. As far as the context of this chapter is concerned, whichever camera you choose, there are a number of practical considerations in their use and choice.

### Cleanliness

In the days of film-based photography and darkrooms, with a little care, one could avoid the need to "spot" a darkroom print. Those precautions are still valid today and in conventional digital photography, we need to be equally careful to avoid dirt on the sensor and optical surfaces. Astrophotography is even more demanding; a dim dust spot on the original image becomes a black hole after image manipulation. The calibration techniques (described later on) that compensate for variations across the sensor also hides the effect of dust spots. These techniques work up to a point but are ineffective if the dust pattern changes between imaging and calibration.

The shadow of a dust speck becomes smaller and more pronounced the closer it is to the sensor and indeed its diameter can infer its position along the optical path. Cleaning dichroic filters and sensors is not without risk; they are delicate devices that can be ruined from the wrong cleaning technique. Dichroic filters can be ruined by a fingerprint. SLR cameras have a mode that exposes their sensor for cleaning but

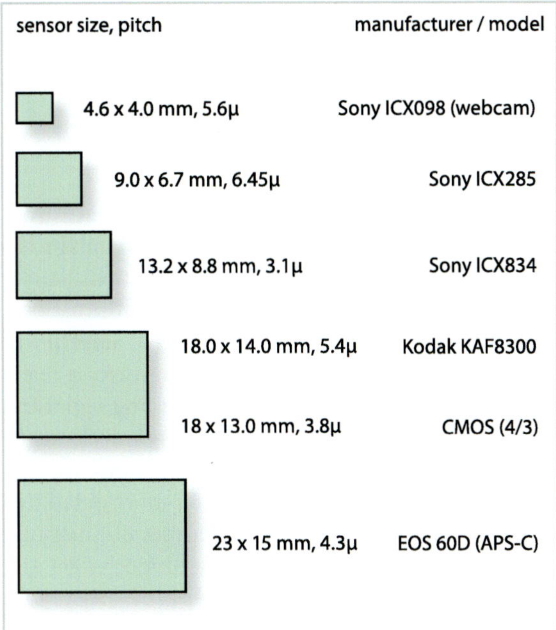


*fig.14 A number of popular CCD and CMOS sensors used for astrophotography and modest telescopes. The sensor size is reproduced at 1:1 and the pixel size is indicated for each. Few sensors have small enough pixels to match the diffraction-limited resolution of traditional camera lens optics.*

a dedicated astrophotography sensor is often inaccessible and requires disassembly to gain access, which may invalidate its warranty. A few specs are normal but excessive amounts within a sealed camera unit may require a service by the manufacturer. If you do choose to clean your own, proceed with care, check the available resources on how to proceed and observe the normal precautions of a clean environment, natural fiber clothing and static electricity protection. (Some sensors are fitted in a sealed cavity, filled with dry argon gas and cannot be opened without causing potential condensation issues later on.) A sensor brush is useful to remove dust from the sensor and filter. Exposed optics need a quality cleaning fluid and a soft cloth.

A few sensible precautions reduce the chance of dust build-up. In a portable setup, the camera/filter/telescope are frequently disassembled for transport. Any optical surfaces should only be exposed in a dust-free environment and capped as quickly as possible. It also helps to cap the telescope at both ends and store small items in a clean container with a bag of desiccant. Many cameras have threaded adaptors and protective caps. I find push-on plastic caps are often lacerated by the end threads. I use threaded dust caps for C and T-mounts to avoid this potential source of contamination. Screw-on covers for T-threads are quite common and an Internet search will locate a source for the less common C-mount caps.

### Sensor Size

One of the surprises to many is the need for a wide field of view to image large nebulae and galaxies. The ratio of sensor size to focal length determines the field of view and although full size (35 mm) digital cameras are now common, many telescopes are unable to produce an image circle large enough to cover a full frame sensor or do so with significant shading (vignetting) at the corners. Most telescopes, however, do have an image circle that covers the smaller and more popular APS-C or Four Thirds sensor size. Many of the popular deep sky objects (for example, the Messier objects) will only occupy a part of the sensor area and an image will require cropping for aesthetic reasons. Conversely, very large objects can be imaged by photographing and assembling a mosaic of several partially overlapping exposures, but realistically, this is a considerable challenge for a portable setup. If you wish to try out different things and if funds allow, two telescope and reducer sizes will broaden the range of possible deep-sky subjects, providing the telescope's image circle covers the sensor. This is where a telephoto lens on an SLR comes in handy or adapted to fit a dedicated camera, for those few wide-field shots.

### Pixel Size

As mentioned earlier, the sensor pixel size (pitch) affects image resolution and can limit the overall optical resolution of the combined optical system and prevailing seeing conditions. If the pitch is too large, the image is said to be under-sampled and the resolution is compromised, too small (over-sampled) and fewer photons will land on each pixel for a given exposure, with the outcome that pixel variation is more noticeable in the final image. Large megapixel counts sell consumer cameras but in the case of astrophotography, less is often more, as the pitch has to decrease to accommodate more pixels within the same sensor area. Most of my images are taken with an ancient 8 MP CCD camera with a 5.4 μm pitch.

In general terms, for a sensor to resolve the light falling on it, with small optics the sensor's pitch should be about 1/2 that of the telescope's diffraction limit and for larger telescopes, about 1/3 of the typical seeing conditions. The table in fig.13 suggests the theoretical pixel pitch to deliver the full resolution of several focal lengths and aperture combinations. In practice, using a sensor with a larger pitch size (or "binning" a monochrome sensor with a small pitch) can improve image noise levels but at the same time will lower resolution. CMOS sensor development is growing and new models are increasingly being deployed into astro cameras.

### Sensor Sensitivity

Both CCD and CMOS sensors translate the number of electrons at each photosite into a digital value. The sensitivity or quantum efficiency of a sensor is the ratio of incident photons converted into electrons. This changes between models and varies with wavelength too. High efficiencies are obviously desirable for visible wavelengths and especially, in the case of deep-sky objects, deep red wavelengths associated with Hydrogen-alpha (Hα) emission nebulae. Typical peak values for sensitive sensors are in the range of 50–80%. Many popular SLRs have built-in infrared cut-off filters that reduce their Hα sensitivity to about 20%, although these filters can be replaced or modified by third-party specialist companies to improve sensitivity. The benefit of high efficiency is two-fold; shorter exposures and better image noise from a given imaging session.

### Color and Monochrome

Color images require color filtration in one form or another since photosites are monochromatic. Color cameras achieve this with a fixed RGB color filter mosaic (Bayer array) precisely aligned in front of the sensor elements. An unfiltered sensor requires three separate exposures, taken through colored filters, one at a time, to form a color image. A common method is to fix the filters into a carousel or filter wheel directly in front of the sensor, a kind of cosmic disco light. Later, combining the three separate images during image processing forms a color image. In the latter case, separate filtration provides additional flexibility when we replace the common red, green and blue filters with specific narrowband filters, tuned to the common nebula emission wavelengths.

External filters may also reduce the effect of light pollution. Light pollution comes from all light sources, the most prevalent of which, for the moment, is from low-pressure sodium and mercury-vapor street lamps. Luckily, these sources emit distinct wavelengths that do not correspond to the common nebula emissions but do, unfortunately, affect images taken with sensors fitted with Bayer arrays. Although light-pollution filters block the majority of these unwanted wavelengths, they are not a perfect solution and require careful color correction during image processing. Separate RGB filters and narrowband filters also minimize the effect of common light pollution wavelengths by using carefully selected transmission characteristics. In particular, the red and green filter pass-bands do not overlap and exclude the principle yellow sodium street-lamp wavelength.

### Image Noise Reduction

Sensors are not perfect and teasing faint details from an image's background places extreme demands on the initial file quality. Of the various imperfections, sensor noise is increasingly influential in low light situations. If an exposure has a relatively high level of noise, it will show up in the final manipulated image. Noise refers to any additional unwanted image-forming signal and is typically made up of constant and random elements. It is worth digressing a little to discuss noise mechanisms, even though it is explained in more detail in a later chapter. There are several sources of noise in a sensor system, each of which has a unique effect and remedy. The two most significant types of sensor noise are thermal (dark) noise and read noise.

Thermal noise adds random electrons to each sensor photosite. The average count increases with time and temperature, as does its randomness. In many sensors, the electron rate per second doubles for each 6–8 °C increase. To combat this, many dedicated sensors have cooling systems that lower the temperature by 20–40 °C. This significantly reduces the average and random level of non image-forming electrons. Consumer cameras do not have cooled sensors (yet) and their electronics warm up with extended use. While the average level is removed by calibrating each image before image processing, this does not remove the associated random noise.

Read noise originates in the interface electronics and is present in every exposure, including zero-length exposures. Again, there is a random and constant element. The random noise level is a critical specification for astronomy cameras and as discussed in the prior chapter, high levels have the effect of lowering the dynamic range. The constant element is called bias noise and like dark noise, can be calibrated away. Even if one was to have a perfect sensor, there is a third noise source; the randomness of the image forming photons themselves, or shot noise, such that an image of a perfectly evenly illuminated object has different pixel values.

### Calibration and Averaging

Calibration is a process that analyzes the consistent noise elements (like hot pixels), generated by the sensor itself, and then removes them from each image file. These calibrated image files still contain random noise in them though. The key to removing any random noise is to take multiple exposures and combine them. Combining (averaging) the results from multiple events reduces their randomness. It is standard practice to acquire and combine multiple exposures and, for each doubling of the exposure count, the noise level of the averaged images reduces by 40%. As soon as one starts taking images of dim objects, requiring exposures at extreme ISO sensitivity or exposure duration, it requires special measures to improve the image quality. To reduce all forms of image noise, the astrophotographer carries out a series of calibration and averaging processes that require additional images:

- multiple dark-frames, with the same temperature and duration as the image frames.
- multiple zero-length exposures (bias frames) with the lens cap on
- multiple exposures of a uniformly lit target (flat frames)

We do not need to know the intricacies of image calibration right now, suffice to say that we need it for high-quality deep-sky images and should consider it during the planning stage.

### Interface and Software Support

Modern astrophotography cameras transmit images via a USB interface and come with software drivers. In addition, there may be a standardized driver for a camera and additionally, dedicated driver support within the imaging software. ASCOM.org defines a set of standardized device commands for mounts, focusers, filter wheels, observatories and so on. These allow application developers to design their control and acquisition programs without knowing how every potential piece of equipment works. There is no equivalent in Mac OSX but there is no reason why it could not be done. Developers are, however, actively working on Linux systems. Consumer cameras are a mixed bag and some are capable of remote operation or image download through a USB interface. Although low-cost T-adaptors exist for most camera bayonet mounts, remote triggers and compatible remote control software do not. In the case of my Fuji X cameras, although tethering software exists, it is designed for studio use and is not ideal for long-exposure astrophotography.

## Optical Correction

The optical design of a telescope is quite different from the telephoto lenses we fit onto photographic cameras. A standard refractor telescope has just two or three glass elements at the front end and, irrespective of its optical prowess, focuses light onto a curved plane. When stars are in focus in the middle of a flat sensor, the stars near the edge are elongated and fuzzy. A long focal length or small central sensor will be less sensitive to these issues but shorter scopes and or larger sensors will show obvious focus issues and distortion at the image periphery.

For imaging purposes, you need a special lens module, called a field-flattener, which is fitted just in front of the camera. Field-flatteners typically have two or three glass elements and have a generic optical design compatible with a range of focal lengths and apertures. Some advanced telescopes (astrographs) incorporate these additional elements within the telescope body and are specifically optimized for astrophotography. (A telephoto lens for a digital camera already has additional elements near the camera-body end to do the same thing.)

Some field-flatteners also shorten the effective focal length. These are known as "reducers" and give a wider field of view as well as improve the focal ratio. The most common have a power of 0.7–0.8x. There are two other common sizes, 0.63x and 0.33x, especially suited for the longer focal lengths found on Schmidt-Cassegrain Telescopes (SCTs). I incorrectly assumed the touted benefits of apochromatic and triplet refractor designs implied excellent imaging quality too. This is only true of astrographs but whose precisely aligned 4- or 5-element optical designs command premium prices.

In the scope of this chapter, the telescope choice is the sum of its parts: the principal characteristics, its aperture and focal length need to match the imaging field of view and the focuser design needs to be robust enough for imaging. As far as the optics are concerned, there are some differences between designs. Apochromatic performance will reduce color fringing on color sensors (though it is of lesser significance when a monochrome sensor has separate color filters placed in front) or by careful image processing. Some optical configurations are naturally better suited for imaging but in the case of many, a good telescope for visual use is not necessarily the last word in imaging.

When using a field-flattener, its distance to the sensor is critical for optimum performance and it takes some experimentation to find the best position. The reward is sharp round stars across the entire image. The camera-end of a field-flattener is often fitted with a T-thread that assumes the sensor spacing is 55 mm from the lens flange. There are many compatible camera adaptors, which automatically maintain this sensor spacing but in the case of filter wheels and astronomical cameras, some careful measurement and juggling with spacers is required to find the sweet spot.

In the case of most Newtonian reflector telescope designs, the principal optical flaw is coma. As the name suggests, stars assume the appearance of little comets. The issue is more apparent at the image periphery and with faster focal ratios. This is not a flaw in the optical elements, as dispersion is for glass optics; it is physically not possible to design a parabolic mirror that eliminates the problem. Again, larger sensors are most affected and in a similar fashion to the issues with simple refractors, a coma corrector lens reduces the effect when placed in front of the sensor. Some coma correctors are optimized for imaging and have additional lenses which improve the field flatness at the same time.

The Newtonian reflector is the granddaddy and there are many derivative designs, with folded light-paths and additional lenses at the top end or sometimes in the focus tube, to trade-off optical defects (aberrations). Some of these designs replace the parabolic and plane mirrors in a Newtonian with hyperbolic or spherical mirrors. Each is a compromise between economy and performance. As manufacturing methods improve, designs that were previously reserved for professional observatories have become increasingly affordable to the enthusiast. At one time it was the APO refractor, then the Schmidt Cassegrain, the Ritchey Chrétien design, with its pair of hyperbolic mirrors and now, refractors with synthetic fluorite glass allowing doublets to perform like triplets. Some other designs have wonderful double-barrelled names and a price tag to match. Suffice to say, depending on the configuration, your purchase may require an additional corrector lens to ensure pinpoint stars on four-thirds and larger sized sensors.

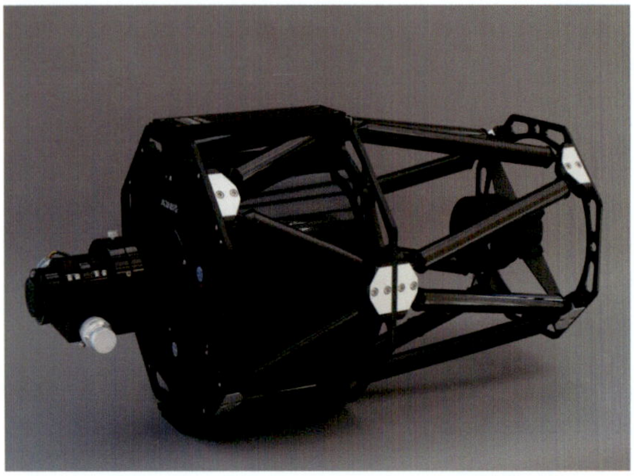

*fig.15 While refractors are convenient, robust and ideal for the newcomer, for the same money, you can buy a reflector telescope with a much larger aperture. As an example only, this 10-inch RCT costs less than a modest 5-inch refractor but it does require a more substantial mount.*

## Setting Up and Record Taking

The setup process for a portable system is extensive. In my location, imaging opportunities are sometimes weeks apart, if not months, and it is easy to forget an essential step that causes a problem later on. It might sound mundane and boring but good record keeping can save considerable frustration later on. In one of my systems, the software connection order was critical. Using a simple checklist is a simple solution or, for more advanced users, set this all down into a small program (scripting language) that runs like a macro in Windows. Automation is fantastic until something goes wrong, so it is best to consider this only after one is satisfied that their system is reliable during manual operation.

Although most imaging software includes essential information in the image file header, a few choice notes help with future setups. Several imaging applications have a facility to save equipment and setup "configurations"; storing and retrieving all the hardware and software settings in one click, as well as key operating parameters. In addition to manual note-taking, most programs and device drivers have an option to record a log-file. These are stored as simple text files and are a useful resource for debugging problems and as a record of activity. Many software providers ask for a copy of the log file before they will actively debug code. It is not uncommon to return to a favorite subject and keeping a record of how you acquired an image is very useful. This book's resource website includes an imaging log file template for you to adopt and modify.

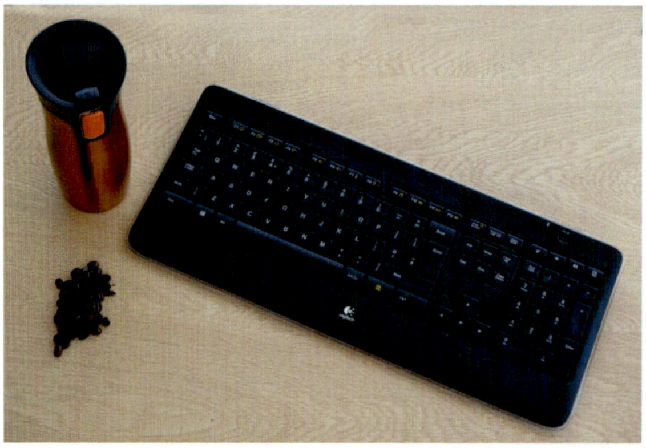

*fig.16 Happiness is a clear sky, backlit keyboard, an insulated spill-proof flask and plenty of coffee. Some music helps too.*

# Simple Telescope Mounts

*Choosing and using a simple tracking mount.*

We are at a crossroads; up to this point, we have been able to take images with the minimum of specialist equipment or software. From now on, the images track the stars across the sky throughout each exposure. At first, we can still use conventional camera lenses, but it will soon become necessary, as we want to image smaller objects, to consider using small telescopes. When starting out, it is very easy to believe that the telescope is the major purchase. It rarely is and in some ways, it is less important than the camera and it certainly has less impact on the final result than the telescope mount.

A good mount is expensive. They represent the most significant investment in any budget. There are two classes: single- and dual-axis mounts, referring to the number of motorized mechanisms. Single-axis mounts have a motorized right ascension axis, dual ones have an additional motor-driven declination axis too. (In this book, I do not consider "Alt/Az" mounts as practical imaging platforms as although they may be able to track a star within the image, the image slowly rotates during any exposure.) Since the wheeling stars appear to move in perfect circles, at constant declination, it is easy to assume that one only requires one motor axis to track this movement. Well, yes and no. Unfortunately, atmospheric refraction and the effect of any small misalignment of the mount's orientation will cause every star image to have motion blur. Perfect tracking generally requires continual small adjustments, to both axes, to keep the stars focused onto the same pixels throughout each exposure. This is usually accomplished with an autoguider system that, every second or so, commands the mount to make small motor movements, some of which may be less than a $1/2{,}000^{\text{th}}$ degree.

We look at single-axis mounts and how they are best used and then, in later chapters, move on to dual-axis models and autoguiding for more demanding assignments. Single-axis mounts are aimed at first-time users, working on strict budgets and for those who require extreme portability. Since they only move in one axis, their performance is practically limited by the accuracy of its polar alignment and the mount's periodic error. Single-axis mounts fall into several camps: swing-arm devices, miniature motorized rotators and more sophisticated single-axis equatorial mounts. As they only move in the right ascension (RA) axis, they cannot correct for any tracking errors on the other axis. The principal error is declination (DEC) drift, caused when the rotation axis of the mount does not aim perfectly at the celestial pole (in other words, the stars and the camera's rotation are not concentric). In the Northern Hemisphere, Polaris is an approximation, but for best results, or longer lenses, it needs to be more precise.

For example, and without going into the underlying geometry, if you take a typical sensor with a pixel size of 5 microns and with a 300-mm f/4 lens, each pixel subtends 3.5 arc seconds (an arc second is $1/3{,}600^{\text{th}}$ degree). If, for example, you aimed the rotation axis perfectly at Polaris, it would be 45 arc minutes out of alignment to the North Celestial Pole. This misalignment causes up to 24 arc seconds of drift over a 2-minute exposure. In this example, each star image becomes a streak, a few pixels wide and about 7 pixels long. It is possible to disguise a drift issue with an extreme wide angle lens, but it becomes increasingly objectionable as the image magnification increases with focal length and the exposure time lengthens.

At the other extreme, I have a reflector telescope with a 2,000-mm focal length. Each camera pixel subtends 0.5 arc seconds and I try and keep the average tracking error within that for maximum image resolution. Even on an expensive mount, aligned within 0.5 arc minutes and costing many thousands, I have to use autoguiding to guarantee precise tracking during each exposure. It is not trivial to achieve precise alignment and in practice, the example above illustrates that it is the mount's mechanical accuracy and one's ability to accurately align, that sets the performance ceiling, placing an upper limit on the focal length and the exposure duration. For that reason, our starter budget places us in in camera-lens territory and with the better performing mounts, short focal-length refractors, providing the exposure is no more than a couple of minutes.

It is also worthy to note that autoguiding is not a panacea. If the mount mechanism has rapidly changing tracking error, backlash, stiction or contamination, autoguiding can only act as a sticking plaster and in some cases, make matters worse.

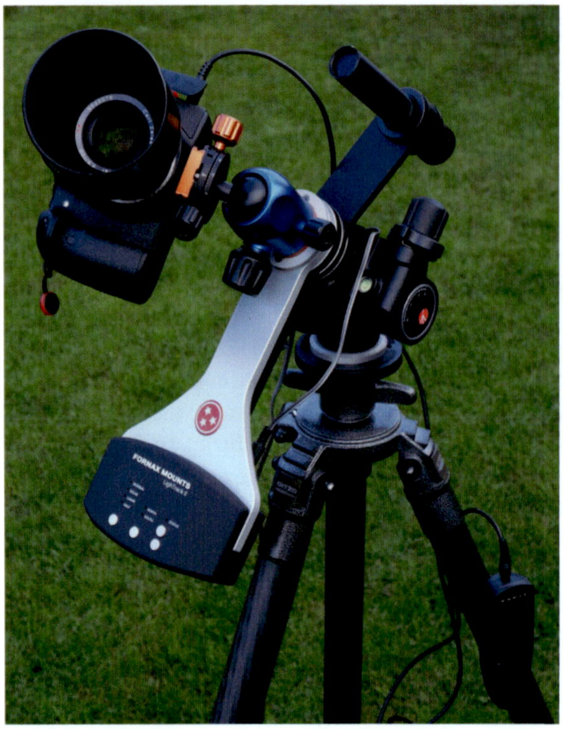


*fig.1 The Fornax LighTrack II® is a deceptively simple piece of engineering, small, light and with extremely low periodic error. Here, the wedge is a modified Manfrotto geared head.*

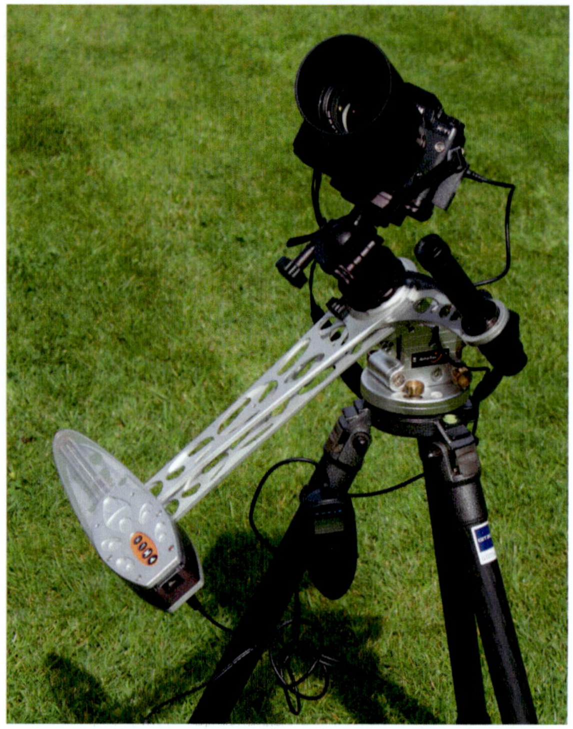

*fig.2 The AstroTrac uses the same principle as the Fornax unit, but uses a motor and lead-screw. Each is good for about 1.5 hours imaging before it requires re-setting and re-targeting.*

## Single-Axis Mounts

### Swing-Arm Designs

These are epitomized by the models from AstroTrac, Fornax and the diminutive wind-up Omegon Mini-Track. These units comprise a scissor-like arrangement with a drive unit at the end (figs.1, 2). The clever aspect of this architecture is that the angular tracking error of a mount is proportional to the movement error divided by its distance from the pivot. By creating an extended arm, these mounts exhibit extremely low RA tracking errors that rival dual-axis mounts costing thousands. The AstroTrac and Fornax mounts also feature an RA guider input, which enables an external autoguider system to feed tracking corrections to the motor. The Fornax unit uses an unusual direct-drive system (a scaled down version of that used in professional mounts) and the AstroTrac uses a long lead screw, a modern equivalent to the many home-made tracking mounts affectionately termed "barn-door trackers." With these mounts, their reported tracking performance is in the range of 1–5 arc seconds peak to peak and in practice, image quality is limited by the accuracy of the polar alignment. Their main design limitation is that the scissor movement is limited to about 90 minutes, after which the mechanism has to be re-set and the camera re-aligned on the target before the next set of exposures are taken. Both consume little power and will run off 8 AA-cells.

### Camera Tracking Mounts

These inexpensive units are simple tracking devices, with a motorized rotator housed in a plastic box of about 100 x 140 mm, into which a polar scope is fitted for alignment purposes. These units are light, at about 800 g and many run on just 3 volts, conveniently supplied by a few rechargeable cells. They are designed for a lightweight camera with a short lens and a brief exposure. SkyWatcher, iOptron, Vixen produce similar models, with simplicity and portability top of the agenda. They are built around a traditional worm-drive mechanism but the 2-inch worm gear diameter makes it challenging to achieve accurate tracking but at the same time, these units do not have a restricted rotation period. A guideline maximum exposure time for an APS-C camera is given by the equation ($f_L$ is in mm):

$$6{,}000 / f_L = \text{exposure (seconds)}$$

These units are fun, and ideal for teaching purposes. Their payload and tracking performance, however, limit their application and I prefer to purchase a mount that also works with a short refractor too.

### Single-Axis Equatorial Mounts

These devices resemble small equatorial telescope mounts but only have one motor in them. Most, if not all, use worm-gear mechanisms and their tracking ability is limited by the mechanical tolerance of the worm and diameter of the mating worm gear. These units and the camera tracking mounts have a larger periodic error than the swing-arm designs; potentially up to 10 x more. In some ways, they are re-packaged camera tracking mounts but they have more robust (and heavier) housings and offer an RA guider input, alternative guide rates and an integrated polar scope. These units still typically require an additional attachment to effect the DEC adjustment in the form of a tripod head or accessory plate that additionally facilitates counterbalancing. Some of these have a built-in wedge adjustment, others offer a proprietary accessory, to make polar alignment more convenient. The popular amateur mount manufacturers have similar models, including units from SkyWatcher, iOptron, Bresser and even Takahashi. For the most part, they are priced to be attractive to first-time buyers and are ideal for wide-field star-scapes. With their more significant periodic error and used without autoguiding, however, they are again best used with short exposures and focal lengths.

The guideline maximum exposure for these mounts is similar to that of the motorized rotators, assuming equally good polar alignment and without using an autoguider. Assuming a rigid assembly and a typical polar alignment error of 5 arc minutes, the following guideline gives an upper exposure duration that keeps star elongation to acceptable levels.

$500 / f_L$ = exposure (minutes)

For longer focal length lenses, an RA guider input is key. These three mount systems are comparatively light and it is easy to forget that their performance is heavily influenced by the stability of their support. Even with a pier-mounted system bolted into a concrete base, I never touch it while it is imaging. The suitability of the tripod and its feet placement are extremely important. My carbon fiber tripod has just three leg-sections and, as each joint is a source of flexure, I removed the center column and replaced it with a simple 5/8-inch threaded plate. The ability for digital cameras to use high ISO values and image stabilization has allowed tripod design to focus on lighter and more compact support. Those tripods that are designed to fold up small and have 4- or 5-section legs are less suitable for the long exposures required in astrophotography.

### You Get What You Pay For (mostly)

This is not the last word on telescope mounts. We have just scratched the surface by considering the most affordable units to kick off with. There are many differences between the entry-level and portable units and the more expensive, larger portable and static designs. The demands of higher magnification work are considered in later chapters. Ultimately, these differences cover many attributes, all of which affect their tracking ability:

- structural rigidity
- payload
- bearing quality
- periodic error
- durability
- gear backlash
- reliability
- durability
- consistency of manufacture
- polar alignment capability

## In Practice

For the first of the tracked images, I chose what appeared to be the best portable mount in terms of weight, payload, price and performance. The Fornax unit is a relative newcomer and after an industrial-looking initial version that proved the design concept, they released the Mark II, a more sophisticated offering. I specifically chose this unit since, on paper at least, it appears capable of more than just wide-angle shots and has more long-term value as a serious traveling companion.

This single-axis mount has a remarkably low periodic error and is very compact; under 30 cm in length and weighing just 1.3 kg. The test report accompanying my unit declared a periodic error of 1 arc second peak-to-peak. That is extremely good by any standard and all but makes its guider input redundant (except perhaps for deliberate micro-movements between frames, called dithering, that are used for statistically removing sensor hot pixels during image processing). With such low PE, this mount is ultimately limited by the accuracy of its alignment to the celestial pole. Angling the mount requires a mounting it to a wedge or a tripod head with pan and tilt micro-adjustments. There are countless utilities

| model | weight [kg] | load [kg] | Price [£] | PE [typical, arc secs] | Guide | Comment |
|---|---|---|---|---|---|---|
| **Single Axis** | Price and weight for mount mount only. | | | | | |
| AstroTrac | 1.1 | 15 | 450 | 5 | ST4 | 2-hour movement |
| Baader Nano | 0.35 | 2 | 250 | not stated | none | tiny! |
| SkyWatcher Star Adventurer | 1.2 | 5 | 200 | not stated | ST4 | extensive accessories |
| Vixen Polarie | 0.8 | 2 | 300 | not stated | none | compact |
| Fornax LighTrack II | 1.3 | 12 | 463 | 1 | ST4 | 2-hour movement |
| iOptron SkyTracker Pro | 0.7 | 3 | 300 | not stated | none | includes wedge |
| Bresser Tracking Mount | 1 | 2.5 | 300 | not stated | none | compact |
| **Dual Axis** | Goto GEM mounts, bundle price including tripod etc . | | | | | |
| iOptron CEM40/GEM45 | 7.2 | 18 | 1,500 | 7 | ST4+ | PPEC, ASCOM |
| SkyWatcher HEQ5 Pro | 10 | 11 | 800 | not stated | ST4+ | ASCOM |
| SkyWatcher AZ-EQ5 Pro | 7.7 | 15 | 990 | ±5 | ST4+ | PPEC, ASCOM |
| SkyWatcher NEQ6 Pro | 16 | 18 | 995 | not stated | ST4+ | ASCOM |
| Celestron AVX | 7.7 | 14 | 800 | not stated | ST4+ | ASCOM |

*fig.3 A comparison of starter telescope mounts, with guideline prices, payload and capabilities. Many manufacturers do not specify their periodic error and in practice, customer reports suggest a wide variation for the same model. I have included some dual-axis mounts for comparison. On the dual-axis mounts, while the periodic error is important, the quality of the declination mechanism impacts dual-axis guiding and shows up any mechanical backlash. This sets models apart, but is hard to specify. Those with simple washers, rather than high-quality bearings, are generally more likely to have stiction and those with multiple-spur gears, rather than belt- or direct-drives, are more prone to backlash.*

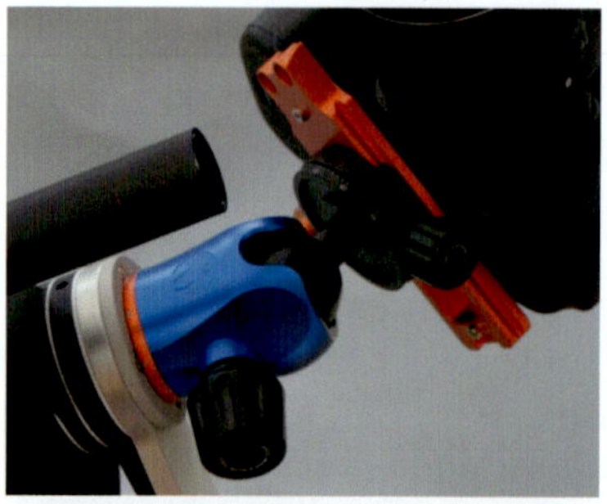

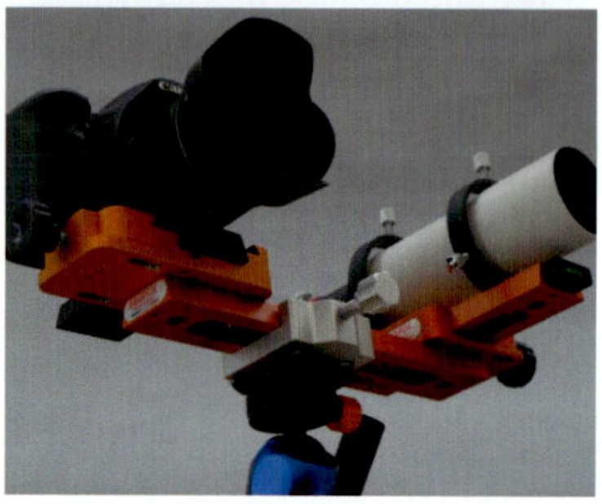

*fig.4 Single axis mounts work best if they are balanced, in so much that there is no moment about the rotation axis, which may cause tracking issues. As the imaging system becomes heavier, it is more critical to have accurate balance. A panoramic base on the head is essential for checking good balance in situ and, in the case of the swing-arm designs, resetting the camera position too. The trick is to position the center of gravity of the imaging system on the rotation axis of the mount. A selection of sliding bars and clamps can be used in a variety of ingenious ways to accomplish this, described in more detail in the chapter* Setting Up a Simple Mount System.

to aid polar alignment, with and without a computer, including miniature camera systems and calibrated polar scopes. The next chapter looks at the common methods, in considerable detail, as part of the hardware setup.

While it is possible to use a standard pan and tilt head, in practice they are difficult to make small adjustments with, compared to a dedicated pan and tilt wedge. You also have to consider that, while the price of these single-axis mounts is attractive, when you factor in the cost of a wedge, tripod and a counterweight system, the system price may exceed that of an entry-level motorized equatorial mount system. The inexpensive SkyWatcher equatorial wedge is a popular choice and will work with most of these single-track units. I bought a used Manfrotto 410 geared head for under £100. This has both geared and quick adjustments, which make it ideal for all forms of photography. This head has pan, tilt and level axes that use a common mechanism. It really only needs pan and tilt and it is possible to dismantle the three moving pieces and re-assemble, excluding the middle section. The result is lighter, stronger and better balanced over the tripod center. Details of the modification are in the chapter *Cloudy Projects*.

I also chose a strong ball and socket head, to attach the camera to the mount and set the declination. This head features a panoramic adjustment on the bottom and top plate, with an Arca-Swiss compatible clamp system. This bottom adjustment is very useful; all of the swing-arm designs have a limited tracking period of about 2 hours. After that time, the mount is reset and started again. At this point, the camera needs to be re-aligned to the target. With this head (fig.4), one can simply rotate the head on its panoramic baseplate by about ~30 degrees rather than loosening the main ball and socket joint. The entire system, comprising the tripod, head, mount, ball and socket adaptor with camera weighs just 6 kg.

Our plan starts with a lightweight digital camera body fitted with a wide-angle zoom and then moves onto "standard" lenses, modest telephoto lenses and finally short refractor telescopes. As the optics increase in mass and focal length, the system becomes progressively lens heavy and it soon requires balancing to ensure the mechanism is not strained, or slips. This is the point where astrophotography reminds me of my childhood days and Meccano sets, playing with nuts, bolts and metal plates. Amateur setups typically couple the optical assembly to the mount via a dovetail clamp system, conforming to one of two sizes: the 44-mm Vixen system or the 4-inch Losmandy system for heavier assemblies. There are also numerous camera-mounting plates, some of which use smaller dovetail sizes, of which the Arca-Swiss size is most commonly adopted, at 38 mm wide. Some recent plates have both Vixen and Arca-Swiss dovetails.

A single-axis mount is balanced when there is no angular moment around the RA axis. This allows a light-duty motor to move a large mass. (Incidentally, the huge, balanced lock gates on the Panama canal weigh more than 600 tonnes and are moved with two engines with the same power as the original VW Beetle.) In practice, this means the camera does not swing around when the panoramic base of the ball and socket joint is loosened. With a simple imaging system, it may be sufficient to shift the center of gravity of the camera and lens over the ball and socket fixing (figs.4, 5). Later on, if an autoguider system is required, the camera and autoguider systems are both mounted side by side and slide along their dovetail clamps to ensure balancing on both axes (fig.4). In the more traditional equatorial mounts, it is common to see long counterweight bars. These are required since the optical system is mounted off-axis and requires counterbalancing. In the scissor design, the system is mounted on the axis and with a little ingenuity, it may be possible to avoid using an accessory counterweight, which just adds more weight and expense. The other fortunate aspect of the scissor design is its low power consumption; I usually carry a heavy 20 VA lead-acid cell for my equatorial mounts but this design consumes just 130 mA @ 12 volts and will happily run all night on 8 NiMH AA-cells, weighing about 250 g. I bought a battery holder and a DC plug (5.5 x 2.1 mm) on eBay for the price of a beer. While most astronomical equipment is nominally rated at 12 volts it often accepts a range of 11.5–15 volts. (If in doubt, consult the device's specification sheet.) While equipment and batteries are marked "12 volts," lead-acid cells vary from about 13.8–11.0 volts over a full discharge, depending on load current and temperature. In practice, 11.5 volts is a working minimum, since discharging a battery to that level reduces its life and is for some mounts a minimum requirement too, beneath which their motors stall.

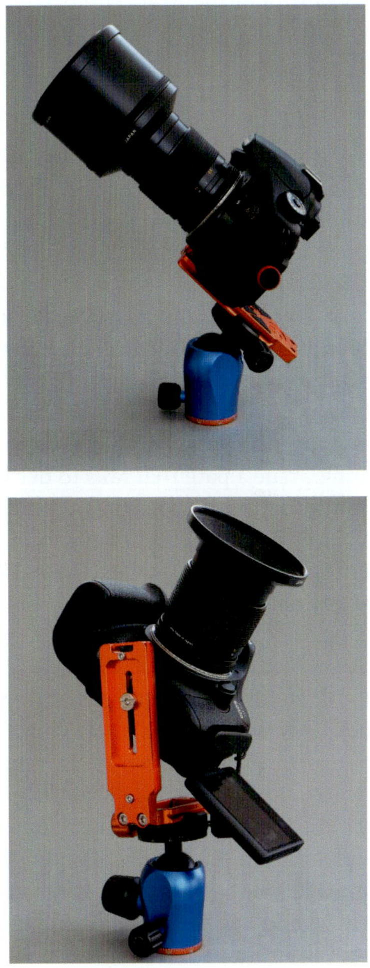

*fig.5* *Following on from fig.4, another way of conceptualizing balance, especially if the ball and socket joint does not have a low-friction panoramic base, is to ensure the camera and lens assemblies' center of gravity is over the ball of the head. When this is the case, the assembly will simply balance on the small head's baseplate and not topple over. Right-angled plates are available from several tripod head suppliers. These are made by 3 Legged Thing®. The top clamp is interchangeable, one of the versions has a panoramic adjustment, making mosaic images easier to accomplish.*

# Imaging Hardware

*Making initial hardware choices is tricky and requires anticipating the evolving demands of astrophotography, in the face of overwhelming consumer choice.*

At some point, we will move on from traditional camera lenses and bodies. The equipment emphasis so far has been on the tracking system and other than some initial considerations for scenic applications, not a lot has been said on the subject of optics or the best camera selection for astrophotography. Making this choice (or choices) is complicated and influenced by many things, both inside and outside of our control. The marketplace for affordable hardware is very exciting right now and it is easy to be distracted and pursue a path that fails to deliver on its promise. While there is an indisputable logic to hardware selection, it is not necessarily apparent in the beginning, unless someone who has "been there" draws one's attention to the various compromises. This chapter considers the imaging equipment selection for a range of applications, from astro-landscapes to moderate focal lengths. The more demanding requirements of deep-sky imaging, which will stretch and exceed the budget, are discussed in later chapters.

In an attempt to highlight some of the decision-making considerations, fig.1 shows the interplay of influences in the decision-making process. It shows a considerable tension between what one would like to and what one is able to image effectively, which is often dictated by our geographic location. Our location not only determines the visible targets during each season but, in addition, the weather and light pollution may limit our practical options. There is also a domino effect, where a more demanding subject shows up the weaknesses in your current system and triggers upgrade fever.

So, for example, it is not uncommon for an optimistic inner-city astrophotographer to start off with a color camera in a simple setup, only to be disenchanted by the effect of light pollution. They either abandon the hobby entirely or migrate to narrowband imaging, which then leads them to use a cooled monochrome astro camera with a filter wheel. The extended exposure times required for narrowband then, in turn, demand better tracking, a more expensive mount, all-night imaging and all that it entails. Those imaging in more favorable conditions are in the lucky position to have more imaging options and often with less expense and with shorter sessions. In these less demanding circumstances, it takes less extreme measures to make quality images.

This chapter takes a look at some of the various options and their pros and cons. Even if we exclude planetary imaging, earlier chapters have shown the amazing diversity of the imaging targets themselves and this, in turn, similarly drives very different system approaches. The marketplace is very buoyant with new entry-level products every year. While it is not feasible to make any particular recommendation that will stick for several years, it is possible to arm oneself with some pertinent questions that guide the decision-making process. With the many possible combinations, one way to pick our way through the maze is to consider the options for various fields of view, from wide landscape images through to deep sky imaging.

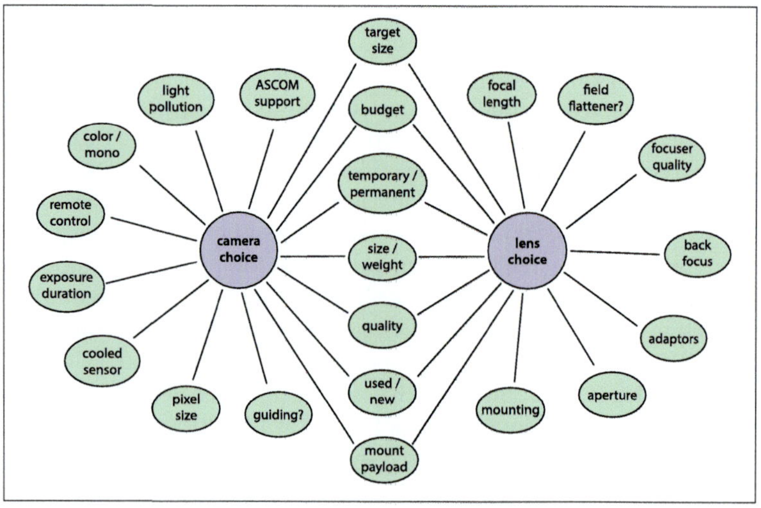


*fig.1 Choosing a camera and lens is a complex affair; there is no perfect combination and the outcome weighs up many considerations to make an informed decision. While smartphones pervade our lives, for the price of three monthly payments, a used DSLR delivers far, far better results in every respect.*

| classification | field of view | focal length |
|---|---|---|
| landscape | 70° – 23° | 8 – 50 |
| wide-field | 23° – 3.3° | 50 – 350 |
| deep sky | 3.3° – 0.6° | 350 – 2,000 |

*fig.2 Focal-length classifications are less well defined for astrophotography. This suggestion is carried through in the book, with the focal lengths referring to APS-C sized cameras.*

# Fields of View

The imaging field of view is a simple function of the sensor size and the focal length of the optics. It is expressed as the angle across the sensor's diagonal. While we are familiar with the concept of wide angle, standard and telephoto lenses in general photography, the classifications are not as well-defined for astrophotography. For the sake of argument, fig.2 proposes a broad definition.

## Landscape Systems

In the case of landscape images, in astrophotography terms, these are extreme wide-angle shots. Focal lengths start from distorted fish-eye lenses, through rectilinear wide-angle designs and on to standard focal lengths, up to say 50-mm focal length. That covers a vast range from a 180° field of view through to about 32°. These focal lengths assume conventional photographic camera optics. Over the years, optical designs have changed to meet the demands of their partnering camera systems, fueled by the need for lightness, zoom lenses and nimble autofocus. Before computers, optical design was a resource-intensive manual process. Now, advanced computer design and modern manufacturing techniques are able to construct highly complex and dynamic lens assemblies, with spherical and aspherical elements and all at an affordable cost.

Most traditional manual focus lenses have a fixed lens group whose glass elements have spherical surfaces and move together to focus with a simple helicoid mechanism. Early zoom lenses altered the spacing of selected lens elements to change the focal length, using complex mechanical cam systems. The first autofocus systems motorized traditional optics but it was soon apparent that high-speed autofocus required lower inertias. This pushed lens designers to achieve focus by only moving a few lens elements and keeping the others stationary (often accompanied by a slight change in effective focal length at the same time). In a similar fashion, early zoom lenses started off with a 2x zoom ratio and were an optical compromise. With aspherical elements, low dispersion glass and modern manufacturing techniques, the zoom and maximum aperture ratios increased and now, many of the premium models with modest zoom ratios are used interchangeably with fixed focal-length lenses without obvious detriment. In more recent times a more worrying trend is emerging, in which lens aberrations are removed in software, either during the JPEG conversion process, specialist RAW converters or utilities, such as those from DxO Labs. This allows the lens designer to take liberties with the optical design and assume they are fixed in the image processing software. Specialist astrophotography acquisition applications do not apply these corrections to camera RAW files and it may require more radical manipulations later on to remove things like chromatic aberration and other issues, which are usually at their worse in the image margins.

Extreme wide-angle lenses are very challenging to design and manufacture and are mostly fixed focal lengths. Zoom lenses start from about 8-mm focal length and are commonplace at a variety of price points, with significant trade-offs between versatility, quality, bulk and price. Modern zoom lenses are constructed with many lens elements and each air/glass surface causes a little light loss and increased potential for ghosting and loss of contrast. Multi-coating helps but cannot eliminate this problem. Even so, the initial quality gap to fixed focal length designs has significantly

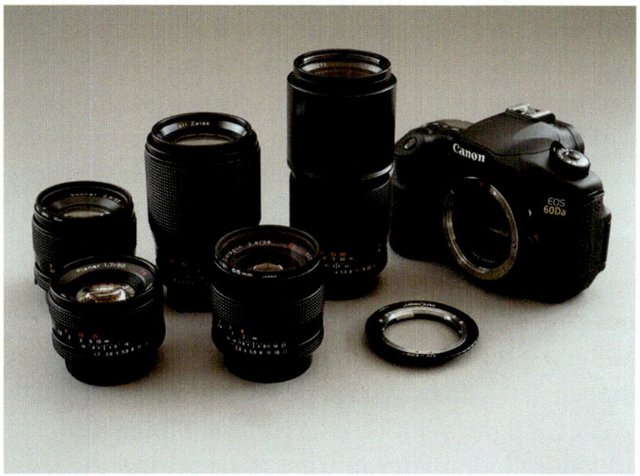


*fig.3 The availability of affordable lens adaptors for DSLRs and mirrorless cameras gives a new lease of life to classic prime optics. These are set for infinity focus but require manual stopping down prior to taking the image. In the case of the longer lenses, I had mine modified to focus past infinity, which guarantees infinity focus when the camera has its UV/IR blocking filter removed.*

reduced and the better wide-angle zoom lenses are a practical and flexible alternative for those starting off.

Fixed focal-length lenses, especially manual focus ones are simpler designs using fewer lens elements. Manual focus lenses are in decline and some seek the premium optics of the last generation and give them a new lease of life. They suit a certain style of photography but need to be chosen and used with care for astrophotography. Their optical design and technology is typically older and scaled to cover 35-mm film and as such, they are not necessarily optimized for sensors in general or the smaller field of view of an APS-C sized camera. I have a range of highly-regarded (in their day) Carl Zeiss Contax lenses, from 18–180 mm. Mounted to a digital body, they only start to compare to top-notch modern optics from about 50-mm focal lengths onwards, even on the smaller APS-C sensor. This is unfortunate as their mechanical construction is more robust and suits manual focusing (fig.3).

Lens choice is a complicated subject, heavily influenced by the systems one already owns as well the economics of specialist optics for a particular project. Optical stabilization is not required and is best disabled for astrophotography. Similarly, fast autofocus capabilities are irrelevant. Faster aperture ratios (f/4 and faster) are increasingly expensive and allow for shorter exposures in scenic astrophotography. This advantage diminishes with star trails (as you typically stop down to prevent background overexposure) and it is worth remembering that dichroic filters, which include narrowband and light-pollution filters, work best with slower f/ratios of f/4 and less. Those designs with fewer aperture blades have an irregular aperture shape, that in turn create a star-burst effect on bright stars (which becomes more obvious with image stretching). With the availability of inexpensive body adaptors, it is possible to carefully select old lens models from the Contax, Leica, Nikon and Pentax stables. If light pollution is an issue, it is possible to use a light-pollution filter in front of or behind the lens. These filters have a restricted range of sizes and large ones are particularly expensive, so are only a practical proposition for smaller filter-thread sizes. Fitting a light-pollution filter behind the lens is another option, made possible by clip-in DSLR body filters but has some limitations, including compatibility with lens mounts that protrude behind the body flange.

> Note: not all camera lens caps are created equal; some of those that clip into the filter thread leak light and/or are transparent to infrared, noticeable when one is creating long-duration dark frames for calibration purposes.

### Choosing a Consumer Digital Camera

Choosing a camera follows a completely different path to that of conventional photography. Keeping in mind that a body may be used with a telescope, although full-frame models may appeal to the landscape artist, they may be overkill for general astrophotography. Not only are full frame cameras expensive and heavier than their APS-C cousins but, more significantly, only a minority of telescopes have an image circle that covers a 35-mm frame. As a result, four-thirds or mid-range APS-C cameras are a prudent and inexpensive choice. These are now entirely CMOS-based and while their sensor's megapixel count is not a key consideration, its efficiency and noise performance are. Modern sensors also control amp glow better than those from the previous decade and while it is possible to remove the most obvious effect using image calibration, the associated random shot noise remains.

Standard cameras are not particularly efficient when imaging the expansive deep-red emission nebulae in the night sky. This is quite deliberate, as unfiltered sensors detect infrared light and are the basis of a video camera's "night-vision" mode. Photographic cameras have an infrared blocking filter sandwiched to the front of the sensor to prevent the near infrared light affecting the image. The implementation of that filter is critical and, as Leica found out with one of their early digital M models, if the filter is not aggressive enough, image coloration occurs. On the other hand, common DSLR brands play safe and in doing so, remove infrared light completely and at the same time reduce the transmission of the deep red wavelengths too. To overcome this, Canon and Nikon make dedicated astro cameras, in which their infrared blocking filter specifications are relaxed to pass deeper red wavelengths but are still usable as a standard camera. These versions are more expensive than their stablemates and some users choose a less expensive route, purchasing a used mid-range camera body and have it modified by a third party. This allows one to select an IR blocking filter specifically designed for astro work and transmit both hydrogen and sulfur transmissions, with even better efficiencies. Another option is to remove the IR blocking filter from the sensor. This permits the camera to make creative IR images (with a red filter) or astro pictures with a IR blocking filter (or light-pollution filter) for astro work.

I also do not like the idea of imaging with a heavy, premium camera out in the dew. They stress the focus mechanisms of the less expensive telescopes and may fail if they get wet. Where possible, I use a model from the mid/bottom of the range; these are lighter, a fraction of the cost and usually have a lower pixel count (whose pixels' spacing is better matched for astrophotography and typically lower pixel noise too). For instance, prior to finding an EOS 60Da in a sale, I bought a used EOS 1100D. It had better noise performance than its higher megapixels brothers and the current EOS 1300 offers no significant advantage. These lighter cameras often have lower battery consumption and these days, all have full remote capability via USB. If the camera is being used without PC control, a model with a fold-out LCD screen and a live-view feature, or similar, helps with backache during framing and focus.

| landscape | wide field | in deep |
|---|---|---|
| lightweight mount with motorized RA axis, and tripod | equatorial mount with computer control over both axes, sturdy tripod to match | equatorial mount with computer control over both axes, sturdy tripod to match |
| existing camera optics or small aperture (60–70 mm) refractor with field flattener | doublet refractor up to 80-mm diameter or 150-mm reflector with fine focus | triplet refractor up to 80-mm diameter or 150-mm reflector with motorized focus control |
| polar scope for alignment | guide scope and guider camera/module | guide scope and guider camera/module |
| telescope to camera adaptors (if required) including telescope to T-adaptor and T-adaptor for camera | field flattener/reducer or coma corrector to suit telescope | field flattener/reducer or coma corrector to suit telescope |
| DSLR or mirrorless camera with a removable lens | DSLR, controlled through USB, or low cost color CMOS astro camera | low-cost cooled mono astro CMOS sensor |
| remote release for camera, intervalometer, spare batteries or USB cable and PC for control | existing computer system, with USB/serial control of cameras and telescope mount | existing computer system, with USB/serial control of cameras, focuser and telescope mount |
| PC, smartphone or tablet with inexpensive planetarium software | PC controls telescope and guiding functions, freeware planetarium | PC controls telescope, focus, guiding functions and planetarium |
| light-pollution filter (either screw-in or clip-in) | simple dew heater, screw- or clip-in filters for camera body | dew heater controller, electronic filter wheel with color and narrowband filters |
| manual acquisition, freeware image stacking and manipulation using photo editing software | simple image capture software, with calibration, alignment, stacking and processing features | image sequence software, plate solving and dedicated image processing software |
| £1,000 | £2,000 | £3,000 |

*fig.4 The equipment choices change with the imaging challenge, with longer focal lengths and extended imaging time being the most demanding.*

In the beginning, it is likely that camera lenses will be fitted to their partnering DSLR or mirrorless body. Those lenses with mechanical focus rings can potentially be used with dedicated astro cameras for even better results. Several astro camera manufacturers sell an accessory bayonet adaptor that directly couples their camera body to Nikon, Canon and Olympus lenses. The shorter flange-to-sensor distance of a mirrorless camera permits many SLR lenses to be fitted to a mirrorless body, providing an adaptor is available. Conversely, the larger SLR lens' flange-to-sensor distance also permits an upgrade path from a cooled, color astro camera to a monochrome camera coupled to a slim filter wheel and perhaps an off-axis guider mirror that diverts a little peripheral light to a separate guide camera.

## Wide-Field Systems

Standard and telephoto camera lenses are still considered wide-field by astrophotographers and are commonly available up to about 300-mm focal length, bringing the field of view down to 4 degrees. The longest telephoto lenses overlap with the shortest refractor telescope designs, in the 300–400 mm range, with a field of view of around 3 degrees. The pricing overlaps too; 300- and 400-mm camera lenses are similarly priced to their refractor telescope counterparts. With the samples I have used, the SLR lenses do not perform as well as a corrected refractor, fitted with a field flattener. This is not because the SLR lenses are bad, it is just their design is optimized for a different purpose and focusing distance.

There is another issue that can arise with adopted camera lenses. In some cases, the lens adaptor does not allow the lens to reach infinity focusing (if it is too deep). In the case of traditional manual focus lenses, it is usually quite easy to modify them, to allow the focus ring to go beyond the infinity mark, to compensate for this. The only downside is that the traditional engraved scale will be slightly out for classical use.

## Refractor Telescopes

Small doublet and triplet refractors, commonly referred to as "grab and go scopes" together with matching field flatteners, are the practical scope for wide-field imaging. Refractor telescopes are deceptively simple designs, with 2 or 3 glass lens elements at the front end (fig.5). The two-element doublets are mostly referred to as achromatic and those with three elements (triplets) are often apochromatic, relating to their ability to focus all visible wavelengths of light at the same point. This is achieved by combining glass elements with different dispersion characteristics and strengths. The distinction blurs, however, with several doublets branded as "apo" that use exotic glasses and approach the performance of a triplet with only two elements. It is easy to get carried away with all the fancy glass names, just as it was when carbon fiber bicycles were first introduced. In both cases, the implementation is equally critical. Of particular note are fluorite glasses, known for their very low dispersion characteristics. This is not an easy material to work with and recently developed synthetic alternatives have stimulated a surge in affordable, high-quality portable refractors with focal lengths between 350–500 mm. Many share the same optical cell and differentiate only by tube construction, accessories and most importantly, the focus mechanism. These meet the demands of the increasing number of lightweight tracking mount users.

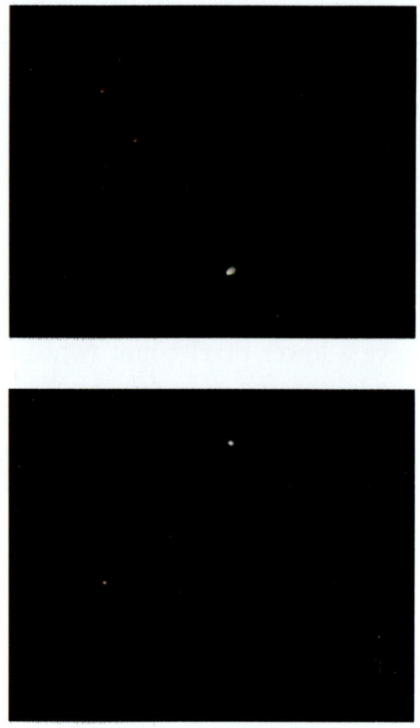

*fig.6 A doublet or triplet refractor focuses onto a curved field. If you do not add a flattener, the stars at the corners will be out of focus and elongated. These corner-cropped images show the difference. The one at the top is without a flattener.*

## Field Flatteners/Reducers

A two- or three-element refractor focuses stars onto a curved plane. A camera image may show sharp stars in the middle but increasingly blurred and elongated stars towards the edges (fig.6). It may be just possible to image with a 0.5-inch sensor, but with 4/3 and APS-C sensor sizes and larger, it requires a field flattener to improve star focus and shape all the way into the image corners. Field flatteners are weak negative lenses, designed to be inserted in front of the sensor and at a particular spacing. They work by transforming a curved focus plane into a flat one. They are not necessary for visual use, as our eyes accommodate the range of focus when we look through an eyepiece, whereas a sensor cannot. When choosing a refractor,

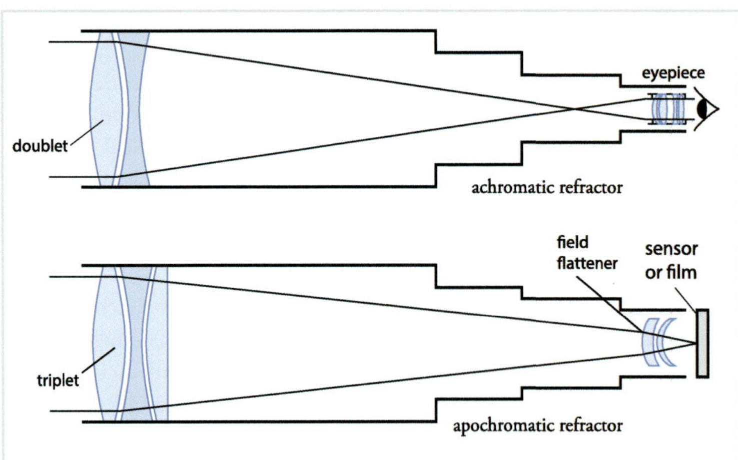


*fig.5 The most likely first telescope will be a two- or three-element refractor, with an aperture of between 60–90 mm and fitted with a field flattener or reducer.*

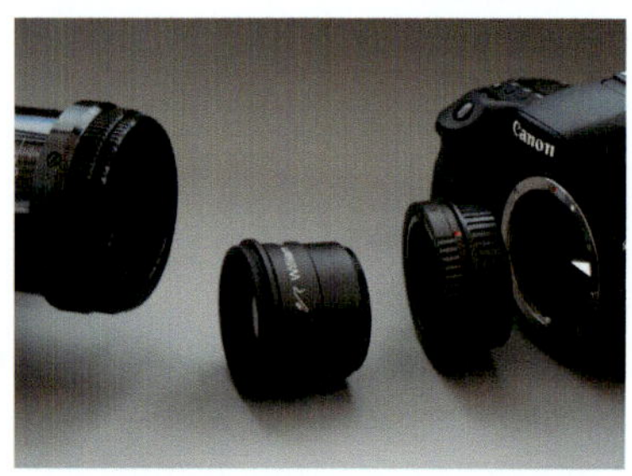

*fig.7 Screw-fitting reducers and flatteners are preferable to push-fit, as they make a rigid assembly. The common T-thread standard (42 mm) is a little restrictive for short focal length scopes with a large imaging field and several manufacturers are now standardizing on 48 mm.*

it is easy to forget the cost of the accessory flattener. Many early versions couple with the focus mechanism using the common 2- or 1.25-inch eyepiece coupling. This is not ideal, especially those that have a single clamping screw. Any flexure from the weight of the optics and camera body will cause a tilt in the focal plane and uneven focus across the sensor frame. Those with expanding rubber glands, designed to enhance centering are unfortunately tricky to ensure alignment and the compliant coupling is less rigid. A three-screw brass ring clamp is ok.

Where possible, choose a system whose field flatteners or reducers screw into the focus mechanism and camera adaptor (fig.7). In addition to a field flattener, many telescopes also have an accessory reducer. Like a field flattener, these are made from a few lens elements but uniquely flatten the image plane and at the same time, effectively reduce the focal length of the telescope. A 0.8x reducer will change a 400-mm f/5.6 refractor to a 320-mm f/4.4. It is not mandatory to use the telescope manufacturer's model since there is a degree of interchangeability between reducers for similar telescopes. Reducers concentrate the image over a smaller imaging circle, usually sufficient for APS-C sensors and with an increased image intensity. Reflector telescopes are not exempt from the need for optical correction for imaging purposes and may require field flatteners or coma reducers to improve the quality of the captured image over the entire field.

The use of refractor telescopes and field flatteners does not rule out using DSLRs. Most flattener and reducer designs have a standardized back-focus that allows for common T-thread adaptors (figs.8, 10) to place the camera sensor at the right distance (55 mm) from the field flattener. A little too close and the focus plane curves one way, a little too far away and it curves the other; a millimeter can make all the difference. There are a few esoteric designs that have unique spacings and these usually have proprietary adaptors for Canon, T2 and Nikon couplings. Somewhat ironically, dedicated astro cameras have no such standardization and it requires close scrutiny of their specification sheets to find the sensor distance from the camera's mounting flange. It is then a matter of accounting for filters

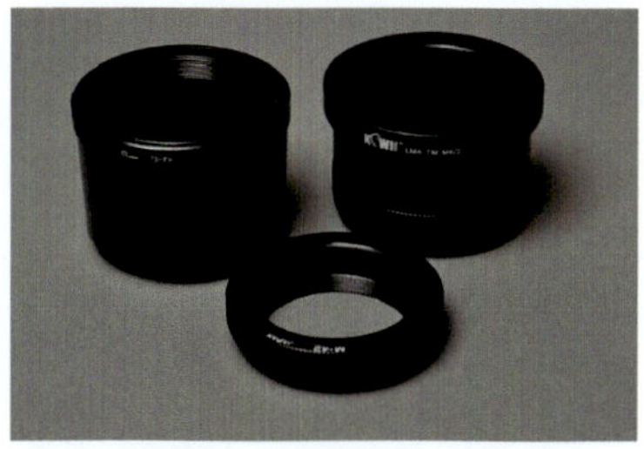

*fig.8 There are T-thread adaptors for most cameras, which couple the 42 x 0.75-mm thread to the camera bayonet and, at the right spacing. A new, larger size, at 48 mm to support full-frame cameras and larger fields of view is increasingly becoming popular.*

Astronomers often use an alternative nomenclature for describing telescopes; whereas a photographer would say a 500-mm f/5.6, an astronomer refers to the aperture *diameter* and focal ratio, rather than the focal length; in this example, a 90-mm f/5.6.

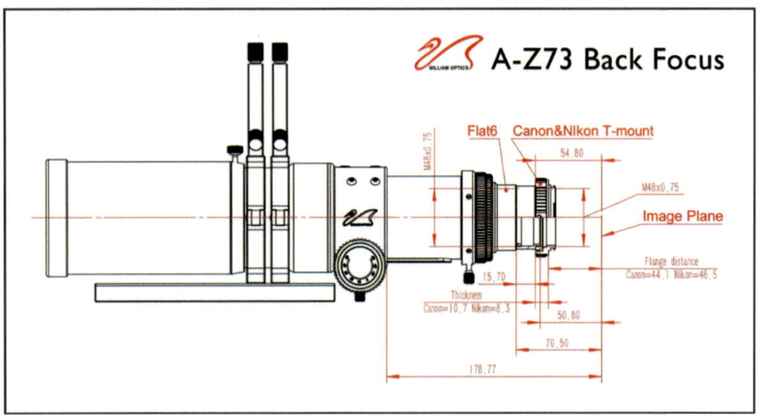


*fig.10 A field flattener is designed to have an optimum spacing to the image plane. Here, this William Optics' Zenithstar 73 drawing (reproduced with permission) indicates the requirement and conveniently, has Nikon and Canon adaptors that suit. If one is using a dedicated camera, it will require a little more work to determine the coupling and spacing to the sensor.*

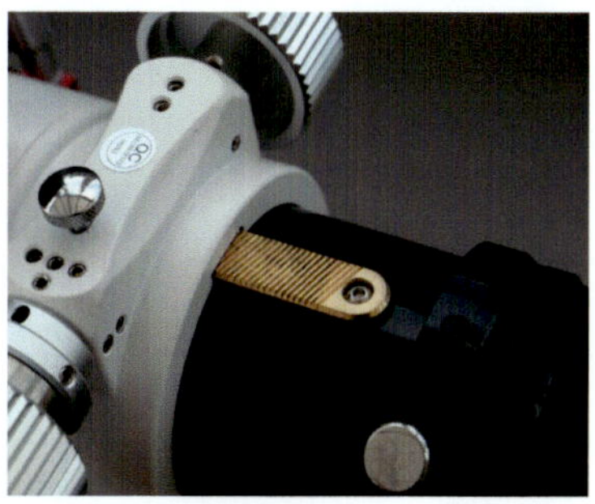

*fig.9 The quality of the focus mechanism sets manufacturers apart. A good quality rack and pinion mechanism should not slip but may have backlash. It only takes a small flexure or lateral movement during an exposure to ruin the result. It is worth reminding ourselves that an arc second is only 1/3600th degree.*

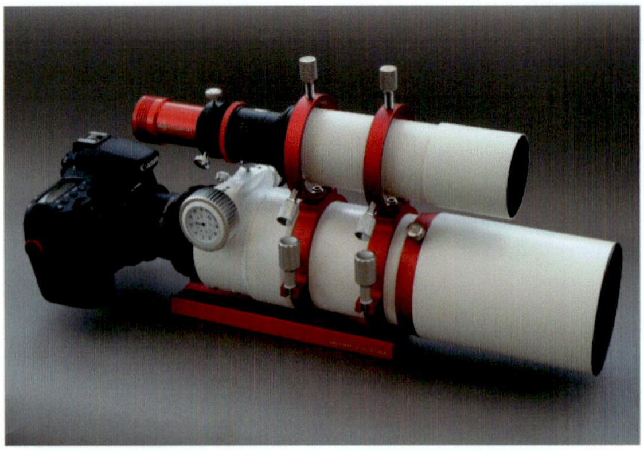

*fig.11* *This William Optics Zenithstar 73-mm f/5.9 refractor sets a practical upper limit for a simple tracking mount. Fore-aft balancing a small refractor is always difficult and the focus mechanism is placed on top so that the mounting plate can be positioned aft. The 50-mm guide scope is fitted with a CMOS guide camera.*

in the optical path, filter wheel and mounting adaptor depth and shuffling extension tubes and plastic shims to make the spacing just right. The nominal setting is rarely the best and a little experimentation may improve focus consistency across the frame.

When choosing a scope, the design and implementation of the focuser are just as important as the optics. I once bought a model with a digital readout, but it wagged around like a puppy tail. It went back. The better refractor packages largely abandon the Crayford focus mechanism and use a rack and pinion mechanism (fig.9). The Crayford focus mechanism was introduced as a low-cost alternative to the rack and pinion focuser for amateur use. It has the benefit of no toothed gears, with a smooth roller compressed against a milled flat on the focus tube. The friction between the metal surfaces moves the focus mechanism. It has the benefit of low backlash but with too little tension, it slips under load, for instance when a telescope is pointing upward with a heavy camera on the end. A rack and pinion uses teeth on the focus spindle and along the focus tube. Unlike the smooth mating surfaces in the Crayford, this mechanism cannot slip in the same way. While it is possible to have backlash between the teeth, it is not a significant issue for imaging, especially when a mechanism is under motor control. In these cases, the focusing process and compensation methods effectively remove backlash arising from a direction reversal.

With the growth of the amateur market, there is a huge array of affordable refractor models with apertures from 60–90 mm. There is little to distinguish them and in reality, several share similar optical cells. I have had several William Optics refractors over the years with apertures of 81, 71 and 51 mm, all with excellent wide-field imaging performance.

## Deep-Sky Systems

Optical systems diversify in the 350–1,000 mm focal lengths (covering 3.3–1.2 degrees), with a variety of refractor and reflector designs. For the purposes of this book, I bought the slightly longer 73-mm aperture William Optics model. With a 430-mm focal length and 3 kg, I judged this to be at the limit of a simple tracking mount and perfect for later on, as its guide scope rings will additionally hold a small 50-mm guide scope to facilitate extended deep-sky imaging on a dual-axis mount (fig.11). Above 1,000 mm, modest aperture refractors become prohibitive and the less expensive reflector designs take over with a different set of compromises (fig.12). With a 1,000-mm focal length, at just under 1° field of view, the Moon is still completely within the frame of an APS-C camera.

For many of us, focal lengths over 1,000 mm have diminishing returns in regard to image resolution. On a typical sensor, the imaging scale is around 0.5–1.0-arc second/pixel and this is frequently less than the prevailing atmospheric seeing value of 1.5–2.5 arc seconds. The larger refractors and their reflector cousins, are heavy and together with the higher effective image magnification, they require a mount with good tracking ability and capable of a larger payload. The tube weight of my 924-mm focal length refractor is 9 kg and with mounting hardware and camera, over 12 kg. If we are keeping to a budget, a pragmatic focal length limit is about 700 mm, beyond which the overall load may exceed the capability of a budget mount and trigger an expensive upgrade.

### Moving on from DSLRs

Staying on the subject of upgrades, longer focal-length instruments with their smaller aperture ratios and fainter objects, especially those imaged in narrowband, require extended exposure times in the region of 300–1,200 seconds. Such exposure times push DSLR or mirror-less camera performance to their practical limit and are often the trigger to consider a dedicated astrophotography camera. Armed with some great pictures from your DSLR, it is time to make your pitch for a budget extension. A few years ago, most of the models were based around a handful of CCD sensors made by Kodak and Sony. While implementations varied, due to the fact that the peripheral electronics had an effect on performance, they all had many things in common, including USB 2.0 interfacing, cooling capability and power requirements. These cameras were usually offered in a number of configurations: as a bare cooled camera, a combined camera and filter wheel or an integrated camera, filter wheel and off-axis guider in one housing.

In recent years, there has been a surge in new camera designs using full frame and APS-C sized CMOS sensors, primarily driven by two Chinese companies, ZWO and QHY. Their attractive price-point and the intense consumer interest have not gone unnoticed and the rest of the industry is in catch-up mode. At present, the CMOS market is very dynamic, with products experiencing rapid evolution. CMOS and CCD cameras have different performance characteristics and the chapter *Dedicated Camera Systems* looks at the pros and cons of sensors in detail. At this point, however, it is better to keep things simple.

| design | advantages | disadvantages |
|---|---|---|
| refractor | robust to handling, quick to reach temperature, highest imaging contrast, sealed tube has less tube-current problems, relatively compact, easy to care for, focus point allows easy camera coupling, with care, glass optics, unlike mirrors are virtually maintenance free, with a range of focal lengths from very short to long | color fringes (apochromatics less-so), field curvature at image periphery, much higher cost for aperture compared with a reflector design, front optics prone to condensation, practical size limit, longer models can clash with tripod legs at high altitudes, range of eyepiece heights for viewing |
| Newtonian reflector | lowest price for aperture, no color fringing, condensation is less likely, big sizes, open tube has less cool down time than sealed reflector designs, fixed mirror cannot shift image, smaller central obstruction than folded designs | off-axis camera, mirrors tarnish, short back focus, lower contrast than refractor, coma at image periphery, require adjusting and less robust than refractor designs, short focal lengths less common, larger mirrors take longer to cool down, tube currents, diffraction spikes, poor wind resistance and require more substantial mount |

*fig.12 This table lists the principal pros and cons of refractor and reflector designs. Many start with smaller refractors and when aperture fever takes hold, add a reflector to their system since they offer more aperture for the same money.*

There are a number of key specifications that discriminate between models, including pixel size, full well depth and their various noise characteristics. A general comparison is difficult and changes with time too, as both CCD and CMOS products are still improving with each new development. It is further complicated by the fact that CMOS sensors change their principal characteristics at different gain settings. Several CMOS camera adverts make the most of this, citing the peak attributes but failing to mention that they do not occur at the same time!

The price point of CMOS-based astro cameras and their availability in color versions make them a natural step up from consumer digital cameras. Compared to CCD models, CMOS cameras have fast image download times and some users make use of this to take many short exposures in any given period (some using USB 3.0, to reduce image download times) and then ruthlessly discard any imperfect images. Shorter exposures, reduce the need for accurate tracking and defeat the worst excesses of poor seeing, to improve image resolution. The down-side is that it consumes a lot of file space and the large file count takes proportionally longer for initial image processing.

The principal performance advantage of a dedicated astro camera over a consumer digital camera is not the sensor itself (after all, astro cameras use consumer camera sensors) but the benefit of cooling the sensor to sub-freezing temperatures. This is because extended exposures proportionally accumulate more thermally generated electrons and, to make matters worse, the sensor warms up during the exposure, which increases the rate of electron generation. These electrons are also referred to as dark current. With the accumulation of more "dark" electrons during the exposure, the random "dark" noise increases too and quickly becomes the dominant noise source in the image. Fortunately, if a typical sensor is cooled by 5–8 °C, it halves the dark current. Many astro cameras have two-stage electronic coolers and reduce the sensor's temperature by 35–40 °C below ambient, potentially reducing dark current by about 200x and the associated troublesome random noise by 14x. The burgeoning marketplace makes choosing a camera model increasingly difficult. Every individual has different requirements and as such, the outcome of any logical selection process will be equally unique. What follows is a simple guide to narrow the search, starting with a number of key decisions. More detailed analysis of sensor specifications enjoy the luxury of their own chapter later on.

### Sensor Size?

The most common sensor size is APS-C, about 17 x 25 mm and makes good use of the imaging circle of most telescope optics to provide fairly wide coverage. The slightly smaller 4/3 sensors work with 1.25-inch external filters, with minimal vignetting, providing they are positioned close enough to the sensor. For those who do not need the field of view, say for imaging small galaxies, a 1-inch sensor is less expensive and will happily work with 1.25-inch filters without any vignetting. If wide-field imaging is your thing and your optics have sufficient field of view, a full frame sensor (24 x 36 mm or similar) will make the most of it. The camera and larger filters are significantly more expensive though.

| sensor | pixel \ $f_L$ | 300 | 400 | 500 | 600 | 700 | 800 | 900 | 1,000 |
|---|---|---|---|---|---|---|---|---|---|
| IMX183 | 2.5 | 1.7 | 1.3 | 1.0 | 0.9 | 0.7 | 0.6 | 0.6 | 0.5 |
| ICX834 | 3.0 | 2.1 | 1.5 | 1.2 | 1.0 | 0.9 | 0.8 | 0.7 | 0.6 |
| ICX814 | 3.5 | 2.4 | 1.8 | 1.4 | 1.2 | 1.0 | 0.9 | 0.8 | 0.6 |
| MN34230 | 4.0 | 2.8 | 2.1 | 1.7 | 1.4 | 1.2 | 1.0 | 0.9 | 0.8 |
| IMX294 | 4.5 | 3.1 | 2.3 | 1.9 | 1.5 | 1.3 | 1.2 | 1.0 | 0.9 |
| IMX071/094 | 5.0 | 3.4 | 2.6 | 2.1 | 1.7 | 1.5 | 1.3 | 1.1 | 1.0 |
| KAF8300 | 5.5 | 3.8 | 2.8 | 2.3 | 1.9 | 1.6 | 1.4 | 1.3 | 1.1 |
| IMX174/128 | 6.0 | 4.1 | 3.1 | 2.5 | 2.1 | 1.8 | 1.5 | 1.4 | 1.2 |
| ICX825 | 6.5 | 4.5 | 3.4 | 2.7 | 2.2 | 1.9 | 1.7 | 1.5 | 1.3 |
| | 7.0 | 4.8 | 3.6 | 2.9 | 2.4 | 2.1 | 1.8 | 1.6 | 1.4 |
| KAI04022 | 7.5 | 5.2 | 3.9 | 3.1 | 2.6 | 2.2 | 1.9 | 1.7 | 1.5 |
| | 8.0 | 5.5 | 4.1 | 3.3 | 2.8 | 2.4 | 2.1 | 1.8 | 1.7 |
| | 8.5 | 5.8 | 4.4 | 3.5 | 2.9 | 2.5 | 2.2 | 1.9 | 1.8 |
| KAI11002 | 9.0 | 6.2 | 4.6 | 3.7 | 3.1 | 2.7 | 2.3 | 2.1 | 1.9 |

*fig.13 Pixel scale, the angle subtended by one pixel (in arc seconds) changes with focal length and pixel size. This conservative guideline highlights a range of pixel sizes (microns) and focal lengths, based on average seeing conditions. CMOS and CCD sensors are identified on the left, adjacent to their approximate pixel size (the Canon EOS 60Da is 4.3 microns). The guideline is less effective for wide angle lenses or lenses with an aperture of 60 mm or less.*

### Pixel Size?

The digital photography industry is obsessed with pixels. A higher pixel count within any fixed area obviously requires smaller pixels, and, as we have seen in the earlier chapters, may have little effect on image resolution due to the overriding limit set by optical diffraction and seeing conditions. If we consider one extreme, with a long telescope and small pixels, the image of a star will cover many pixels to form a diffuse blob and the pixel intensity will be less for a given exposure as a result. It may also highlight optical defects too. At the other extreme, with a short focal length and large pixels, the light from a star may only fall on a few pixels and will saturate them more easily. In the case of a color sensor, a star's light has to hit red, green and blue filtered photosites equally for accurate color reproduction.

So, what is an ideal pixel size? For a small refractor, one published recommendation, to minimize resolution loss, is to choose a pixel pitch at about half that of the telescope's diffraction limit (full-width half-max value), and with larger apertures, when the resolution is limited by atmospheric seeing, it is to image with a pixel size equivalent to 1/3$^{rd}$ of the typical seeing conditions. This proposal balances further resolution loss from the pixel dimensions with the other constraints.

In practice, however, tracking, focus and optical issues (as well as Murphy's Law) degrade the image resolution and larger pixels will usually suffice. A more pragmatic approach is to choose a sensor which, in combination with the telescope, has a pixel scale of 1–2 arc seconds (a little less than reasonable seeing conditions). A table of image scales for various focal lengths and sensors is shown in fig.13.

There are other considerations too. If you choose a sensor whose pitch is 1–2 arc seconds per pixel, for a range of focal lengths of 600–300 mm and upgrade later on to scope with a focal length of 900 mm, the system is said to be over-sampled. It is often modest enough to be irrelevant but there are some interesting techniques that deliberately micro-shift the sensor between exposures to improve the effective resolution (dithering).

### Color or Monochrome?

Many sensors are available in color and monochrome versions (without a Bayer matrix). If convenience is the overriding concern, choose a color camera. If you are choosing a monochrome camera, there are further considerations, including the camera flange to sensor distance, that affect whether it is possible to fit in a color wheel (and potentially an off-axis guider) between the camera and the field flattener. If the flange to sensor distance is large, it may also force you to use larger, more expensive filters to avoid vignetting. If you are in an area of high light pollution or plan to take narrowband images, a monochrome sensor is more efficient and versatile.

### Cooling?

If you are able, choose a model with a cooling ability. In the long term, it allows for extended image exposures with less image noise. The cooling ability is expressed in degrees below ambient and since many imagers operate their sensors between -10 and -20°C (20 and -4°F), their ambient conditions set the performance requirement. In my case, it is rare to have temperatures above 25°C at night and my camera is capable of a 40°C reduction. I standardize on -20 or -15°C at different times in the season (which requires me to have two sets of dark calibration frames).

### Other Considerations

As mentioned earlier, the sensor package affects the overall cost and feasibility of the optical system. I know of some very attractive cameras, but it is impossible to achieve the correct flattener to sensor distance including an off-axis guider and filter wheel. The "back-focus" is usually specified and it requires a careful desktop study to add up all the dimensions before purchase. In the case of the popular APS-C size, the sensor to flange distance also determines whether you can get by using 1.25-inch filters or move up to a more expensive 2-inch system.

With these attributes, set by your imaging need, the next filter considers performance characteristics. Some are easier to assess than others. For instance, extended imaging requires reliable hardware and software and for PC users, robust application interface and ASCOM support, otherwise it is incumbent on the imaging application providers to individually write and maintain driver code for each device.

The overriding image quality requirements are governed by two needs: low image noise and capturing a wide range of subject brightness. The two are not entirely exclusive and require a little explanation. We know that image noise comes from two main sources, the sky and the sensor. When we consider the sensor noise, it is quoted on a pixel basis; in electrons for read noise and electrons per second for dark current. The ability for a sensor to capture very dim and bright subjects (a dim nebula with stars sprinkled through it) at the same time is often loosely referred to as "dynamic range" and is indicated by the Full Well Depth (FWD) specification. This is quoted as the capacity of each photosite (aka pixel) on the sensor, in electrons, before it maxes out. This is not the full story, however; the effective dynamic range of each pixel is reduced by the presence of noise, as it messes up the ability to distinguish between very dim subjects. In simple terms, if a pixel can measure up to 10,000 electrons but there is an uncertainty of 10 electrons due to sensor noise, the dynamic range is 10,000/10 or 1,000 levels, equivalent to about 10 stops, for any single exposure. All is not lost; if you recall the discussion in previous chapters regarding more exposure, one way to improve the image's effective dynamic range is to integrate multiple exposures, which reduces all random noise generated from the sky and electronic sources.

The second complication arises due to the pixel size. If you consider two sensors with identical pixel noise and full well depth parameters but with different dimensions, the final image will not be identical. In this example and for any given area, the one with smaller pixels can capture more electrons and the noise will be higher. If one sensor has pixels that have half the pitch of the other and we average 2x2 pixels in the image processing software, the one with the smaller pixels has 4x the full well depth for any given area and twice the noise level, doubling the effective dynamic range. (In the case of a CCD sensor, if we instruct the sensor to combine or bin 2x2 pixels in the chip, before readout, it is possible to maintain the same pixel read noise and still achieve the 4x improvement in full well depth.)

In practice, the pixel dimensions for similarly sized sensors are usually less diverse but do make a practical difference to the image dynamics. Comparing their characteristics for a unit area removes this potential complication. In addition to the dynamic range and noise performance, the quantum efficiency and download time of the sensor are further considerations.

Quantum Efficiency (QE) measures the ability of the sensor to generate an electron from an incident photon and ultimately the exposure time required to capture an image. It is affected by the silicon itself as well as the sensor's optics (for example, the losses from the cover glass, reflections and micro-lenses). Efficiencies have steadily improved over the last 20 years and the latest back-illuminated designs, in which the photodiode elements are not obscured by the support circuitry are peaking at 80%. That may not make it into a tangible improvement in image quality as, while modern sensors in consumer cameras become more sophisticated, it is often in the pursuit of high-speed video. High-speed electronics generate more heat and are accompanied by a dark current penalty. The difference in peak QE between models is modest but their spectral responses may be quite different. Some sensors are optimized for

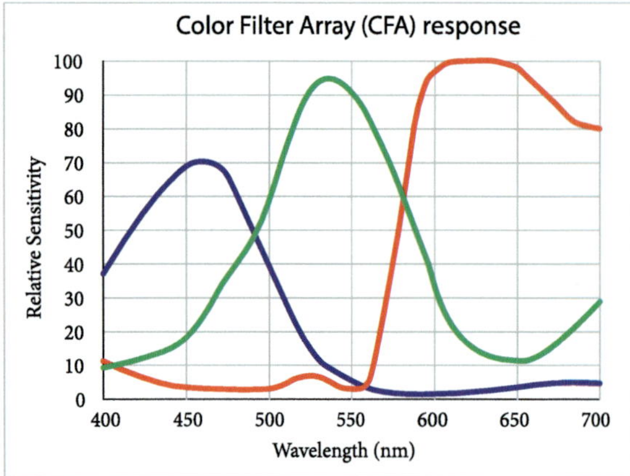


*fig.14 This shows the normalized sensor response through the RGB filters in its color filter array. The peaks of the blue and green wavelengths are lower than the red, as this response also takes into account the sensor sensitivity. The notable feature are the generous overlaps between the colors.*

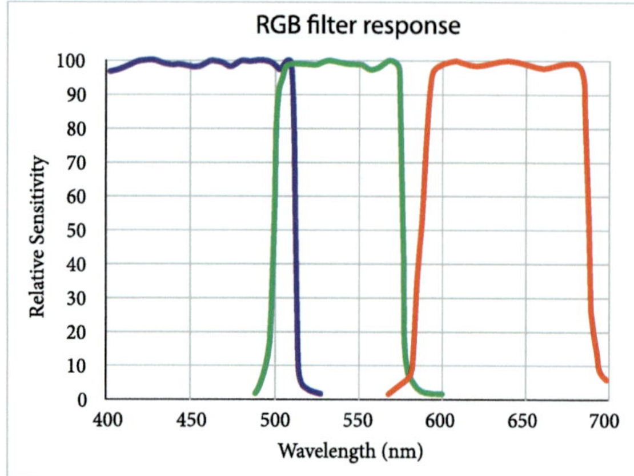


*fig.15 By comparison, this shows the normalized transmission of a dichroic RGB filter set. The pass-bands are better defined and overlap to a lesser extent. Notably, the yellow wavelengths associated with light pollution are blocked (between green and red), as is infrared.*

security camera use and have extended red sensitivity into the infrared. As a result, these sensors exhibit much-improved QE for the critical deep-red colors associated with the ionized hydrogen and sulfur nebula emissions.

For some applications, such as lucky imaging, which takes multiple short exposures, download time is an important consideration. My 8 MP CCD camera takes about 15 seconds to download a single frame. The CMOS equivalent can download over 10 exposures per second!

### Filters and Filter Wheels

A secondary performance benefit of a dedicated camera is the possibility of using a monochrome sensor with separate filters. While a color camera, whose sensor is fitted with a color filter array (often abbreviated CFA) is convenient, a monochrome camera with a filter wheel is more flexible and has performance benefits too, as we will see. In essence, a color image is formed by assigning two or three filtered luminance signals to the color channels of an image file, typically the primary colors for transmissive devices or the three secondary colors for printing purposes. The signals can be of any wavelength though, including those outside of the visible electromagnetic spectrum, including infrared, ultraviolet and x-rays. Consumer color devices use the familiar RGB primary colors for a "realistic" image rendition. Using separate filters on a monochrome sensor, however, facilitates more creative imaging interpretations.

Filter technology comes in two principal forms: colored glass/gel and thin-film (also referred to as interference or dichroic filters). Colored glass absorbs the rejected light, whereas thin-film ones reflect it. You can always identify dichroic RGB filters as, viewed from an angle, they resemble the complementary color. Most color filter arrays on commercial sensors use colored glass; apparent from the generous and sloping pass-bands of the three filter transmission responses (fig.14).

Although RGB filters can be found in both technologies, thin-film filters dominate astrophotography and are a must for narrowband and light pollution rejection. Although thin-film filters are considerably more expensive, they do offer significant performance advantages. They can be made with very precise and complicated specifications, with multiple passbands and with almost perfect transmission at selected wavelengths (fig.15). Light-pollution filters require the benefits of thin-film technology to achieve their complex response curves. There are some other subtle benefits from using separate filters:

- improved resolution – since the full sensor resolution is used with each filter and there is no requirement to de-mosaic the CFA image file (which blends neighboring pixels into a single color pixel).
- improved focus and reduced color fringing – since the focus position can be optimized for each filter wavelength, rather than rely on premium optics to focus all wavelengths at precisely the same position.
- optimized exposure – within any imaging duration, the user has complete control over the exposure for each filter as well as the number of frames. For instance, a CFA sensor has twice as

many green-filtered elements as it does red or blue. Green, however, is not a common color in deep-sky objects and the exposure plan can record more photons for the other colors.

- lower background shot noise – on account that dedicated astrophotography RGB thin-film filters are carefully tuned to exclude the principle yellow light-pollution wavelengths and the incredible selectivity of narrowband filters effectively block light pollution and its associated shot noise.

- improved quantum efficiency, as dichroic filters are more efficient than those in color arrays

These benefits come at a price; dichroic filters are expensive and then one additionally requires a process to insert and remove filters from the optical path. At the simplest level, some T-thread adaptors and refractor focuser tubes have a 48-mm internal thread that accepts common 2-inch filters. This, however, requires a partial disassembly of the imaging system to change them. This is less arduous if you are imaging for several hours with each filter, or on separate nights. More convenient methods load all the filters into some form of carrier, using a variety of rotating wheels or sliding mechanisms to change them over. The most basic is a manual slider similar to an old-fashioned slide projector, but it is rarely seen these days.

Filter wheels are more convenient and come in manual and motorized versions, commonly housing 5–8 filters in 1.25-inch and 2-inch sizes. A filter-wheel housing adds more mass to the focuser tube but is convenient to handle and keeps the filters away from dust and fingerprints (figs.16, 17). As a filter wheel is inserted between the field flattener and the camera system, its depth has to be taken into consideration to ensure the sensor distance is at its optimum spacing. It is always possible to add spacers but a camera with a large flange to sensor distance may be more challenging. Usefully, many camera companies ensure that their camera/filter wheel/off-axis guider systems fit within the T-thread standard spacing of 55 mm. The Starlight Xpress model in fig.17 has an off-axis guider port. My QHY one does not and additionally requires both USB and power connections.

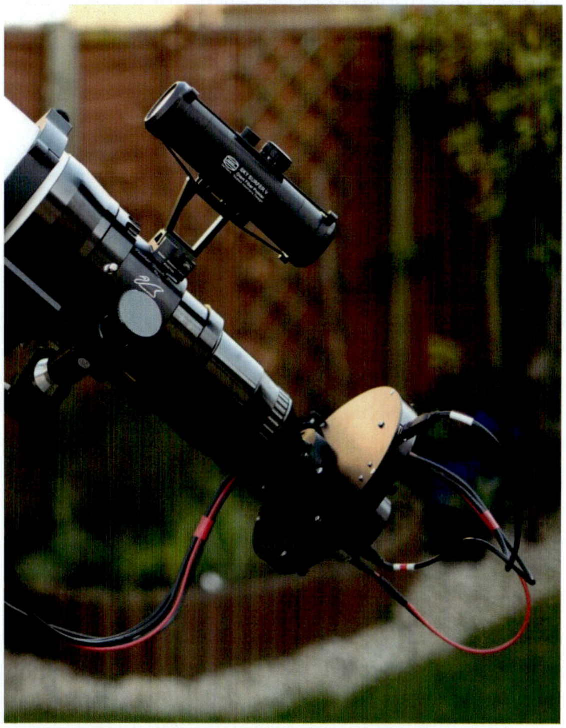

*fig.16 In this setup, the field flattener is screwed to the focus tube and the filter wheel is screwed between that and the camera. The power for the internal motor is supplied via the USB cable. This one can handle 7 x 1.25-inch filters or 5 x 2-inch filters.*

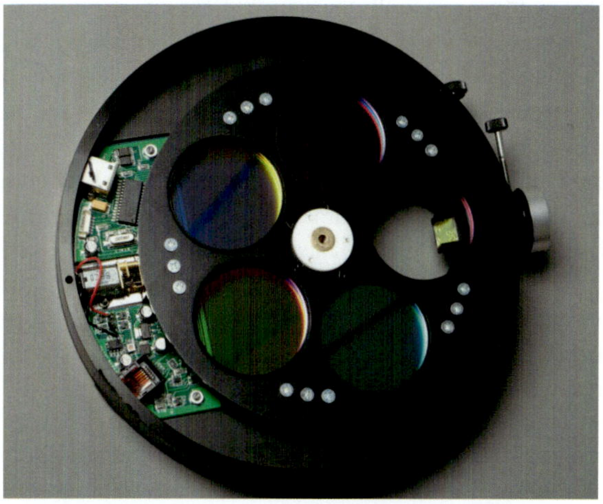

*fig.17 Inside the filter wheel in fig.16. You can see the small mirror in the aperture that directs light to an off-axis guider and the 5 filters, corresponding to LRGB and Hα. The wheel is turned by a small motor and controlled (and powered) through a USB 2.0 interface.*

# Setting Up a Simple Mount System

*Important fundamentals that underpin all that follows.*

Regardless of whether you are setting up a full-blown imaging system or a portable rig, there are some basic things to do at the outset, associated with stability, alignment, balance and targeting. Setting up includes one-time calibrations and adjustments, as well as repeated start-up activities that become second nature after some practice. There is also an element of chicken and egg: a degree of assembly and installation is required for instance before one can successfully do the one-time alignments and it may need a few iterations to get everything just so. We start with the basics and in later chapters, add more exotic equipment.

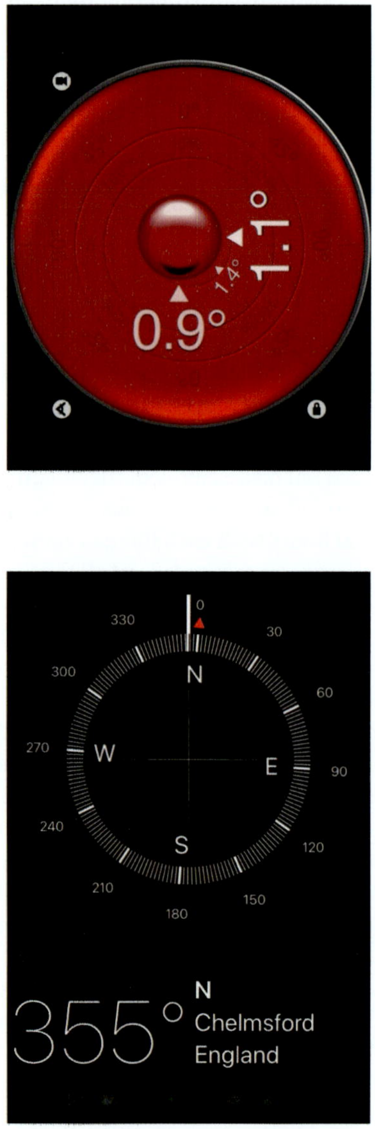


*fig.1* *Smartphones are a wonderful thing. Most have in-built GPS, compass and gyro, enabling a phone to find North, level the tripod and set the approximate altitude of the mount to the latitude. In addition, they have planetarium applications that find an object and indicates its position in the sky.*

### Where Precisely?

The siting of any imaging rig benefits from a little planning. For instance, a clear view of the celestial pole is handy for quick mount alignment and, although there are alternative methods, these take longer and are better suited for a permanent setup. Any installation should be situated on stable ground and if it is in a public place, consider potential trip hazards too. Tripod spikes or hard rubber feet are fine on hard ground but when the ground is soft, they slowly sink in, ruining any alignment and ultimately, images. On the soft earth in my backyard, I found an effective temporary solution by placing a paving slab under each tripod leg and a permanent solution with my ground spikes, described in detail in *The Astrophotography Manual* and on the website. Placing a tripod on decking may appear to be one way to avoid muddy conditions but, without an independent and isolated support, will move and vibrate as you walk about.

Many dedicated tripods are made from steel, a few are made from ash or aluminum. The actual material is less important than the construction; the lightweight versions supplied with less expensive systems are usually too flimsy for imaging work. If you are using a photographic tripod, carbon fiber models are often considered the best compromise of weight and rigidity and have some vibration absorbing properties too, at a cost. The most stable models have just two or three leg-sections. For imaging work, the leg extensions are primarily used for leveling or, in the case of visual astronomy, to raise the telescope eyepiece to a comfortable viewing height. The better tripods have leg braces and clamps at the leg pivot to improve the overall rigidity.

Height is not a priority for imaging and it is best to use a short extension to level the mount and ensure stability. Manufacturers are becoming increasingly aware of the role of the mount support and a number now offer simple portable piers, in which a central tubular pier rests on three extended feet, often held in position by triangulated support struts or cables. The legs usually fold flat to the pier or are detachable and stow inside for transportation. Some have alternative screw-in feet, including spikes, rubber and broad discs, for soft ground. If I am traveling light, I use three 6-inch square plywood squares on soft ground.

### Location

An open space presents a wonderful vista but it is not a necessity for deep-sky imaging. To avoid the worst of the light pollution and the degrading effect of the atmosphere, imaging is usually restricted to altitudes above 30° from the horizon. At an altitude of 30°, the optical path passes through twice as much atmosphere as it does straight up. This not only affects transparency and seeing but the angle introduces some refraction too. (My first eager attempts to image Jupiter at low altitudes and at high magnifications produced terrible color fringing on the planet. I thought I was at the limit of the telescope optics. When I tried again some weeks later, with Jupiter high in the sky, the problem had almost disappeared.) If you are using a single-axis mount, the increasing refractive effect of the atmosphere introduces apparent star drift at low altitude. At 30° altitude, one would see a 5 arc-second drift over a 5-minute exposure. If this drift is along the RA axis, autoguiding may remove it, if it is in DEC, no such luck.

Bright lamps in the surrounding area are another cause of grief: even though your scope may have a long dew shield, stray light from a bright light source will likely flare inside and affect the final image. This is particularly true with camera lenses, whose minimalist lens hoods do little to shield off-axis light sources. This is not a case of OCD, as the smallest flare becomes more obvious as a result of the severe image stretching required to reveal faint details during image processing.

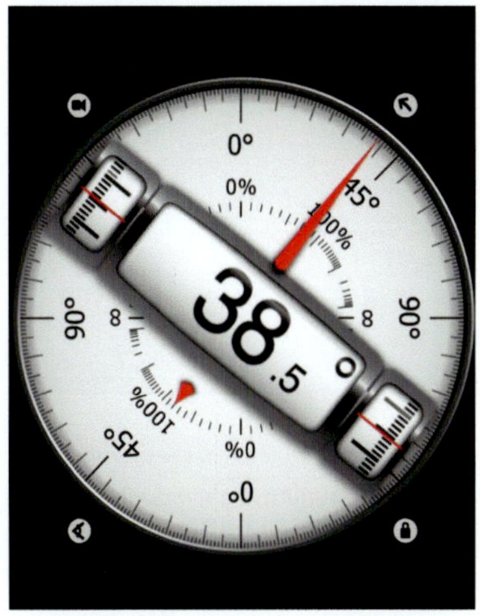


*fig.2* *An inclinometer application on an iPhone, in practice, the RA axis should be inclined by the same angle as the latitude.*

### Tripod and Mount Alignment

It helps the initial alignment procedure if the telescope is leveled and aligned with true north. Even in the case of an equatorial mount or a wedge-mounted fork, although their RA axes simply have to align with the celestial pole, there is some benefit from accurate leveling as it improves the accuracy of the polar scope setting and any initial alignment slews. Some may claim otherwise but, at the end of the day, why not? It only takes a few minutes and is easier if you think of it in terms of east to west and north to south. With single-axis mounts, the center of gravity is biased towards the celestial pole. To improve stability, place one of the legs facing north and away from you and slightly extend all three legs. Place a spirit level across the east and west legs. Adjust one of these legs to level the mount. Turn the spirit level 90° and adjust the north leg to level north–south. You should only ever need to adjust two legs to level a tripod.

I like to set up the mount in my backyard before it gets too dark. After blundering into a steel patio chair in the dark, I use this opportunity to remove all the discarded toys, garden equipment and hose pipes from the surrounding area, as well as the pathway back to the control room! If I am trailing cables back to the garage, I choose light colors, for visibility.

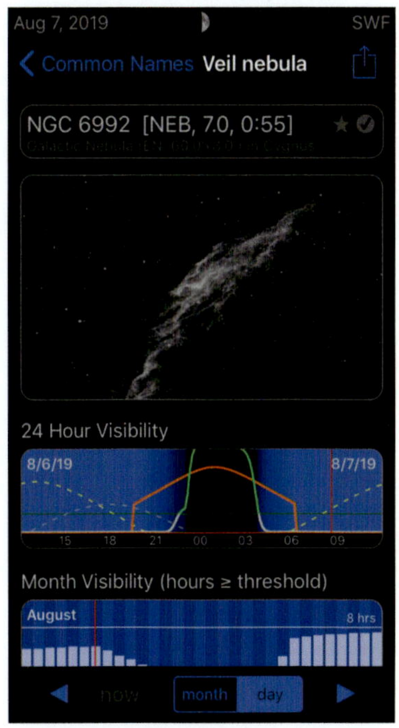


*fig.3* *Smartphone applications are now capable of mapping the sky, searching for any object, and suggesting likely objects of interest and the best time to view for your location and season. A used, inexpensive digital camera, however, is a much better choice for image quality and usability.*

### Assembly

If you have the opportunity, assemble the optics and imaging components together in a well-lit space and at the same time check for any dust on the optical surfaces. With a basic camera and lens, this is trivial, with perhaps only the delicate matter of inserting a light-pollution filter into the camera body, checking the camera menu and exposure mode settings, exchanging the battery for a fresh one or for a battery adaptor and fixing a dew heater around the lens. As things get larger, the assembly and mounting become

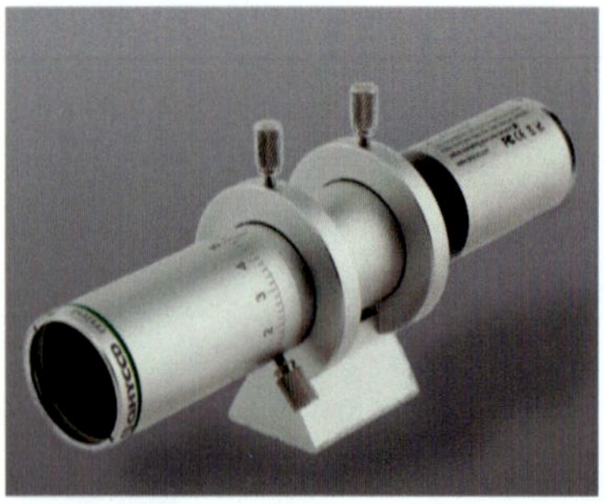

*fig.4 With the increase in the popularity of ultra-lightweight mounts, for use with wide-field imaging, so too has there been an increase in ultra-small guide scope systems. This one has a 130-mm focal length and terminates in a C/CS CCTV lens mount, suitable for the equally growing number of lightweight guide cameras in CMOS and CCD technologies. The assembly, complete with a QHY5L-II camera is just 170 g. With an aperture of just 30 mm, it works best with a bright guide star.*

increasingly delicate. Most telescopes are kept assembled to a dovetail plate for convenience, usually via a mounting plate or tube rings that clamp around the telescope body. Handling a large telescope is trickier, however, especially when it is covered in dew. The combination of weight, length and delicacy requires care. If one is using a single axis mount that has a tracking error, especially if it is with a medium length lens, the assembly will likely need some form of autoguider. This is most likely to be a small lens, with a focal length of 50 mm or longer (for the focal lengths one is most likely to use on a single axis mount) and a tiny camera. The unexpected thing about guider systems is that they can calculate a tracking error to about 1/10th of a pixel and for that reason, it is not necessary for the guider optics to have the same focal length as the imaging system.

As an aside, there is another guider configuration that takes advantage of the compact packaging of dedicated astro cameras. A shallow sensor/flange depth allows one to introduce an off-axis guider (OAG) in front of the sensor and/or filter wheel. In an off-axis guider, a small mirror deflects some main image up to a tiny camera, which is mounted at right angles to the main imaging camera. OAG systems often give the best performance as any flexure in the mechanics or optics equally affects the guider and imaging camera and is automatically corrected out.

There are some very cute guider systems, like the QHY system (fig.4) or an adapted finder scope, with one of the expanding range of small, fast CCD/CMOS cameras, often with 1.25-inch or C/CS mounts (as used on security CCTV systems). If this mini system is going to be dedicated for the purpose, keep it assembled in its in-focus position with the available focus-lock feature. It is useful to know that precise focusing is not required for guide cameras. Several guider applications work best with a slightly out of focus image; their star centroid calculation works better with a slightly diffuse blob and a non-saturated star core.

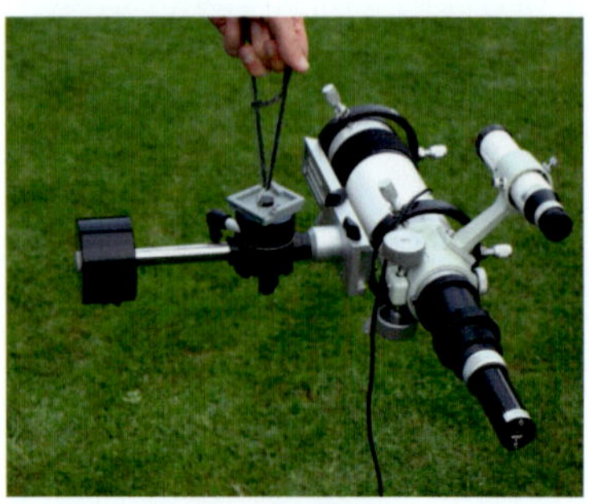

*fig.5 It is difficult to assess the equipment balance on a delicate mount that has some friction on its RA and DEC movements. Sometimes the solution is to turn the problem on its head; literally. Here an old camera mounting plate with a string tied to the back is screwed into the 3/8-inch mounting hole of the counterweight and the telescope assembly and the balance are assessed upside down.*

### Balancing Single-Axis Mounts

These light-weight mechanisms require a balanced load otherwise, they are likely to slip or stall. Depending on the imaging system, each of the mechanical configurations require a slightly different approach. At one extreme, a simple ball and socket head may be sufficient, evolving to longer mounting bars, a counterweight system and/or a side-by-side optical system. Glass is heavy, and as the lens size increases, it becomes more difficult to achieve balance on a simple ball and socket head and it requires various methods of counterbalancing its mass. Traditional balancing techniques assume the user loosens the RA axis (panoramic lock) and assesses any rotational motion, adjusting positions and weight distribution to make the system "inert." That becomes more difficult to assess with lightweight systems and when the system's axes have higher levels of friction and do not move freely. In these cases, a different solution is required.

At some ridiculous time in the morning, it dawned on me (literally). What is a balanced system? It is when the center of gravity is in line with the axes. I realized that there was a simpler way to conceptualize balance, without having to evaluate it with the equipment assembled to the mount. A clue is already in the chapter *Simple Tracking Mounts*, but the concept works with any assembly fitted to a single axis mount. Simply put, for a single-axis mount, the RA rotator screws into a UNC bolt, poking out of an angled mounting plate and, for a balanced system, the center of gravity should be in line with the bolt. When it is, both RA and DEC axes are balanced. To make things interesting, the center of gravity of a ball and socket head system changes as the camera plate is tilted to realize a larger DEC setting. Practically, if the imaging system is light, the imbalances are equally small and are not a problem. It becomes more significant with large professional lenses, heavy full-format DSLRs and telescope systems and these require balancing for best performance.

It is possible to assess and adjust the balance without fitting it to the swing-arm. Several alternatives come to mind; you rest the base on your palm, or suspend the assembly from a screwed-in 3/8-inch bolt borrowed from tripod head and secured to a string (fig.5). In each case, the freedom of movement allows one to assess the balance after altering the position of bars, weights and cameras. In the case of side-by-side systems, such as the one in fig.6, both the imaging and guiding scopes can slide along their dovetails as well as the crossbar side-to-side. To ensure balance for all DEC orientations, check both the vertical (camera pointing up) and horizontal (camera pointing to the horizon) orientations. If those two positions are in balance, everything in-between will be too. When done, it is useful to mark the positions, especially if this is a common setup, as it speeds up the assembly for next time. A pointer of white electrical tape works well. With the balance confirmed, one can either assemble it in its entirety to the mount or if that is too ungainly, in parts, using the marked positions.

## Polar Alignment

More words have probably been written (and wasted) on polar alignment than anything else. It is fundamental to achieving good tracking with both single and dual-axis mounts. Even with dual-axis guiding and expensive dual-axis tracking mounts, they can only go so far, as, even if the guide star or image center is perfectly tracked, polar misalignment can show up as mild field rotation (the image looks like a short exposure star trail). It is also worth mentioning that there is no such thing as perfect polar alignment. There are a set of preferred alignments, within an arc minute of each other which are optimized for drift on a certain axis, field rotation and overall drift. It is important to put polar alignment accuracy into context: if during a 10-minute unguided exposure, at a declination of 50°, the drift is 3 arc seconds (about the same as the seeing conditions) from a polar alignment error of about 1.8

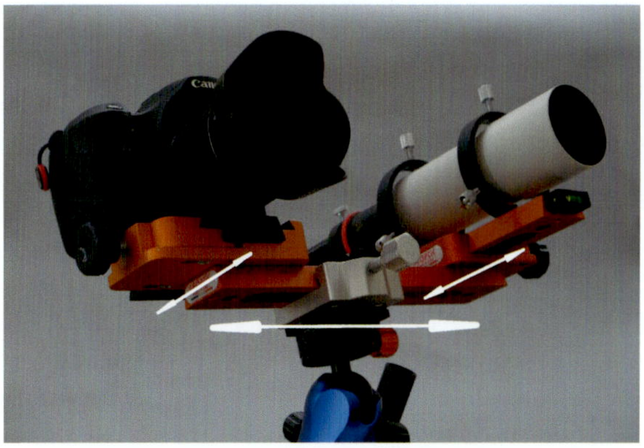

*fig.6 This side by side assembly has a camera alongside a 50-mm aperture guide scope. Each is mounted on a Vixen bar, which is clamped and secured to a long Vixen bar, secured in turn to a Vixen clamp on a tripod head. In this case, it is important to assess balance about the pivot in all directions and it will be necessary to slide the bars in the directions shown to ensure both side to side and fore-aft balance is good.*

arc minutes. In this example, if we assume the tripod feet are a meter apart and one foot sinks by 0.5 mm, this doubles the alignment error (and drift rate)!

Polar alignment techniques place varying demands on the mount and some can only really be done with robotic dual-axis mounts. For completeness, the entire subject will be tackled here, even though some require a dual-axis mount. In general, whichever method you use, the most accurate results are obtained with a fully-laden mount. It may be easier to align the mount before assembling the other components but adding the additional weight and different balance will cause the tripod position to settle, as well as flexure, compromising any alignment. There are several basic methods for polar alignment:

- optical polar scope (all mounts)
- drift alignment (all mounts)
- software-assisted camera polar alignment (all mounts)
- star alignment model (dual-axis mounts)

### Optical Polar Scope

The process uses a small calibrated telescope to align on Polaris (in the case of the Northern Hemisphere). In this device, Polaris is imaged onto a small circular reticle. Since Polaris is about 40 arc minutes away from the North Celestial Pole, it prescribes a small circle every 24 hours (23h 56m to be more precise). If you know the time, it is possible to determine where it

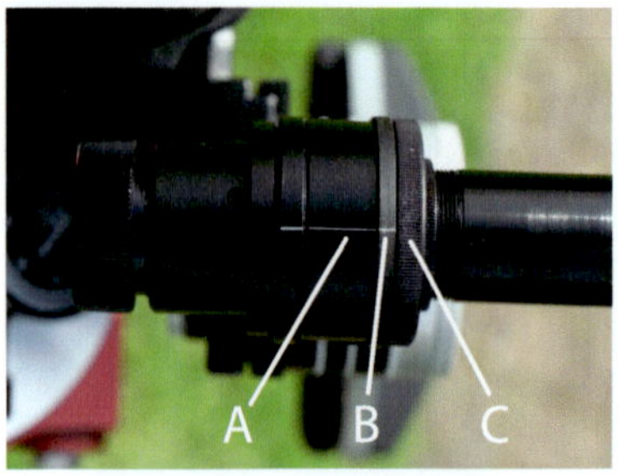


*fig.7 For a polar scope that has a clock-face style reticle, it helps to find its upright orientation and permanently mark it for next time. A thin scratch lines-up the scope (A) with the mounting arm (B). The scope is held in position with a lock ring (C) and with these marks, it takes a few seconds to secure it in the same position.*

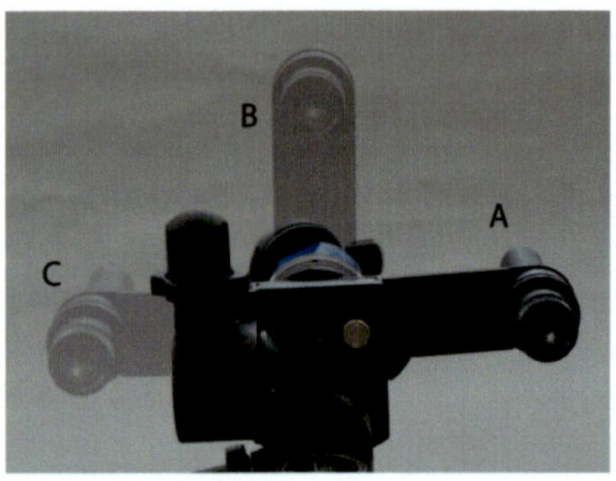


*fig.8 In the case of a polar scope swing arm, it is better to center the reticle by swinging the arm (A–C) rather than rotate the polar scope itself.*

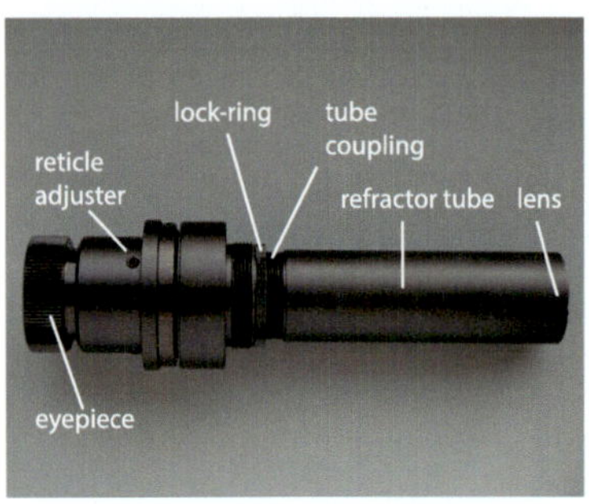


*fig.9 Anatomy of a polar scope showing the reticle adjusters and the lock-ring for the main tube. Check the reticle and star are in focus at the same time, if not loosen this ring to screw the refractor tube until it is (before centering the reticle).*

should lie around the circle. For accurate alignment, one needs to know the rotation angle of the reticle and, that the center of the reticle is aimed in exactly the same direction as the rotation axis of the telescope mount. As the years go by, due to the Earth's precession, the apparent distance between Polaris and the pole changes, slowly making polar-scope reticles less accurate. The better models indicate its position over a number of years and may need an update every 10 years or so. (At some time in the future, Polaris will appear to drift away from the NCP.)

### Polar Scope Centering

In the beginning, we need to do a one-time adjustment and center the reticle so that a centered star does not move off the cross-hairs as the scope is rotated. Centering a polar scope is conveniently accomplished in daylight, using a convenient target, such as a distant object (fig.10). To do this, one lines up the polar scope's center on the target and rotates it, either by rotating the polar scope itself or with the mount, about its RA axis. If the center mark moves off the mark, it needs adjustment. The reticle is suspended between three small grub screws. A tiny screwdriver or Allen wrench is required to adjust these set screws to move it about. The trick is to make micro-movements; loosening one adjuster and tightening another by half the perceived alignment error. This may take half an hour, but once done, it should not require further adjustment. In the case of swing-arm single-axis mount, the polar scope is attached to an articulated arm, which itself may have tolerances that affect centering. Here, the trick is to rotate and fix the polar scope in the arm, so when the arm is at the 12-o'clock position, the reticle is upright too. Mark the alignment on the arm and body (fig.7) for future reference and center the reticle by moving the arm between three positions, rather than rotate the polar scope on its axis (fig.8). In this way, the reticle calibration takes into account any distortion in the arm. This is far more accurate and repeatable than online methods that first center the polar scope reticle and then try to shim the polar scope mounting to attempt to correct for any arm-misalignment. (It is easy to forget that an error of one arc minute is just 1/60$^{th}$ of a degree in comparison of the likely flexures and tolerances of an articulated arm.)

### Polar Scope Mount Alignment

Having centered the reticle, the next step is to align the mount and place Polaris somewhere on the circle, according to the user's longitude, date and time. Again there are several methods:

- three-star centering
- RA angle setting
- hour angle setting.

Each of these typically uses a different reticle. In the case of the three-star setting versions, the reticles look like the one in fig.11. Here Polaris is placed into position and it requires a dark night to locate the two other nearby stars and rotate and shift them into their respective positions. The accessory polar scope on a Paramount uses this system and it is very effective. It does, however, require true darkness (not twilight) to see the two fainter stars. The second system uses an RA angle. This was common on the early SkyWatcher mounts, in which the polar scope is fixed to the RA axis and is accurately aligned by slewing the mount about the RA axis. At first, the mount is rotated until the reticle is aligned vertically and the floating RA scale zeroed on this position. The mount is then slewed to a relative RA position to reflect Polaris' current hour angle. The particular angle is usually given by its PC telescope driver (like EQMOD), a planetarium program or a polar alignment app (fig.12). The drawback of this system is that it potentially requires the entire imaging system to be rotated 360 degrees which may cause collisions between the telescope and tripod. (To avoid this, some invert the polar scope every 6 months, so that they only have to move the RA axis ± 90 degrees.)

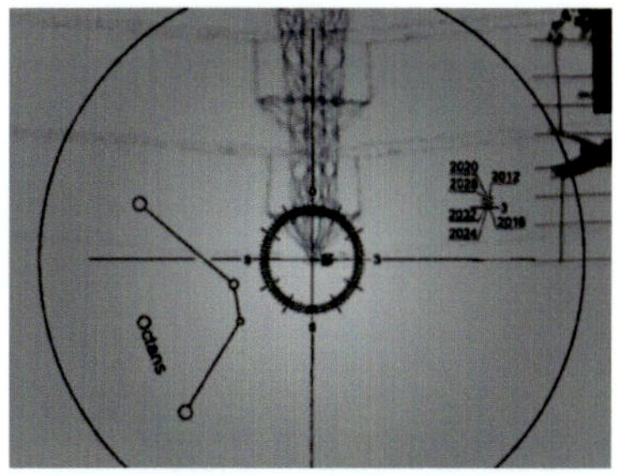


*fig.10 To check centering, aim the reticle at a distant landmark. When the reticle is centered, its aim should not wander as the polar scope is rotated about the RA axis.*

### Hour Angle Setting

This last method is becoming increasingly popular and works on the principle of a static upright reticle similar to the one in fig.10 and fig.13. The mount's handset, or in the case of the non-computerized mounts, a simple smartphone app, indicates where Polaris' current position should be on the clock face. It is moved into position with the mount's altitude and azimuth adjustments. Fornax, AstroTrack, iOptron and the more recent SkyWatcher mounts use a similar system.

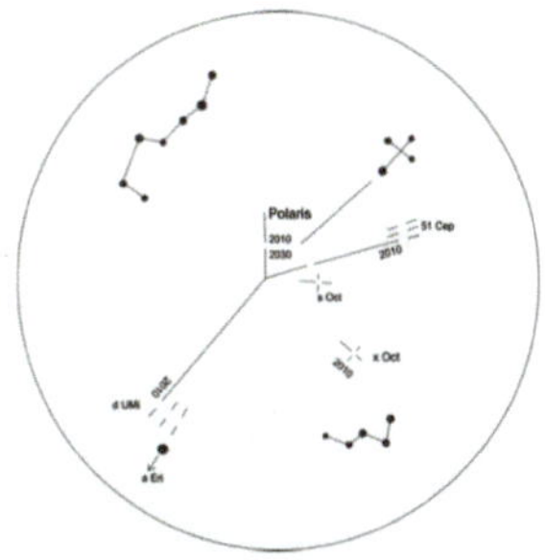


*fig.11 A reticle designed for aligning on three stars, removing the need to know Polaris' hour angle.*

### Drift Alignment

Drift alignment requires no polar scope, just time, patience and an eyepiece or camera system. In essence, it detects the very star tracking issues (drift) one is trying to avoid. It is a case of test, adjust, test, adjust, test. The drift is most readily detected near the celestial horizon and aiming due South or East/West, to set each axis. In essence, for the Northern Hemisphere:

1. If a star on the meridian drifts N/S the mount requires an E/W adjustment.
2. If a star on the Eastern Horizon drifts N/S, the mount requires a lower/higher altitude.
3. Conversely, on the Western Horizon, if a star drifts N/S, the mount requires a higher/lower altitude.
4. In the Southern Hemisphere, the adjustments for 2 and 3 are flipped.

The technique is very accurate and does not require any additional equipment. It is certainly the ultimate test of polar alignment and is often used to fine-tune a permanent installation. In a portable setup, with the imaging camera in place, I find the process a little cumbersome since it is difficult to detect drift from the camera's LCD screen. A more sensitive and speedy drift detection method is to use software to analyze successive images. This requires a PC with a USB-controlled camera. The autoguider software PHD2 does this and has a special polar alignment wizard based on the principle that detects tiny amounts of drift.

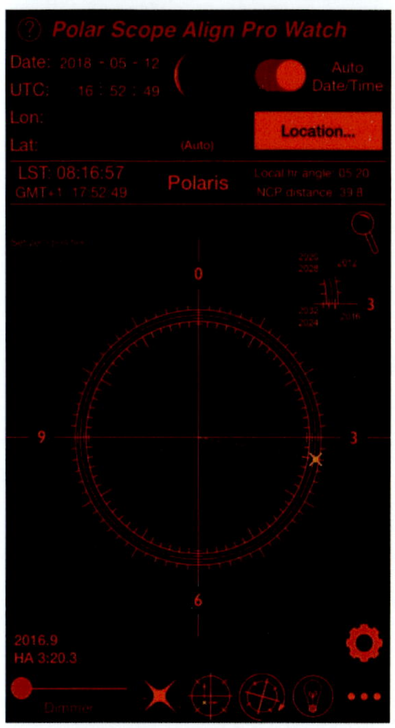


*fig.12 A polar align smartphone app for the reticles in fig.10, and fig.13.*

### Software Assisted Camera Polar Alignment

We are admittedly jumping ahead of ourselves but one may also use a small CCD or CMOS camera and an app to polar align, bypassing the ergonomic issues of a polar scope. In general, a PC or mobile app uses star images to calculate the celestial pole by several methods; QHY PoleMaster and

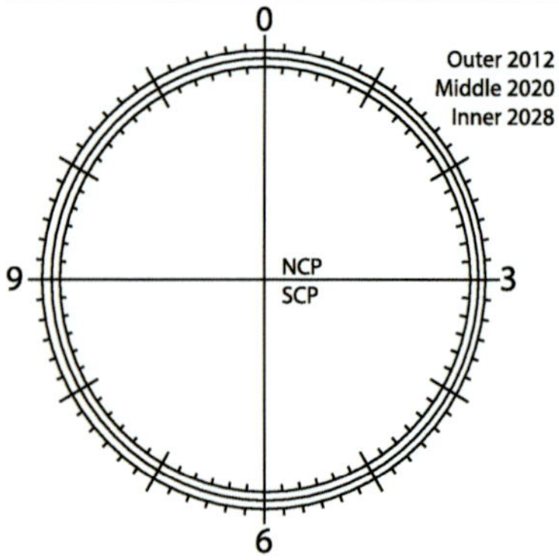


*fig.13 A reticle designed to be vertically aligned and Polaris placed on the circle, according to its hour angle. It is operated in this orientation, with 0 uppermost and a separate application (or the mount handset) uses the current time to determine the position on the clock face. This reticle will work over a 16-year span.*

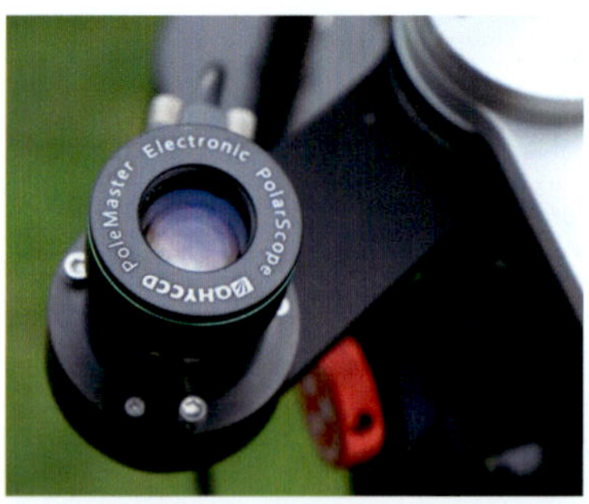


*fig.14 The QHY PoleMaster, fitted to the arm of the Fornax mount. A useful method, if you have a nearby computer, to swiftly get within 30 arc seconds of the Celestial Pole. The application's wizard steps through the alignment process.*

SharpCap Pro work by noting the center of rotation and star positions of rotated camera images, others work out where the camera is pointing and dynamically measure drift. All these systems then give a visual indication of the misalignment as you adjust the mount's bolts, and significantly, do not require the camera/lens system to be perfectly aligned to the mount's RA axis. This makes them ideal for single-axis mounts and portable systems during polar alignment. The internal calculations take any scope-angle error into account and eliminates it from the misalignment measurement.

These systems work with any mount, in both hemispheres and consistently polar align to under 30 arc seconds within ~5 minutes. SharpCap and PoleMaster require the celestial pole to be within the camera frame. This is easy with the PoleMaster unit as its tiny CMOS camera is fixed to a 21-mm focal length CCTV lens with a ~10° field of view. SharpCap (fig.15) is optimized to work with a short guide scope, of about 200-mm focal length and a still- or video-based guide camera. With a typical guide camera, the field of view is less than 2° and the guide scope requires a better alignment, prior to running the software. Keep in mind that while the image resolution is higher in SharpCap, the benefit of sub-30 arc second alignment is entirely lost with the tiniest movement of the support system. (Some of the latest iOptron mounts feature an optional integrated electronic polar scope that functions in a not too dissimilar manner.)

In practice, the accuracy of these systems is as good as the reliability of the mount's Alt/Az adjustment mechanics. If you already have, or plan to use a guider system, the on-cost is a small annual charge for SharpCap Pro. I already own a PoleMaster and although it may seem a small extravagance, I consider it money well spent since I think accurate polar alignment is fundamental to minimize drift (remembering there is no DEC correction on a single-axis mount) and especially if the camera can be used for a dual role. Not surprisingly, the PoleMaster PC software only works with their own hardware but the more curious of us have realized that the unspecified camera model uses the same sensor as their QHY-5L II camera. Interestingly, this camera can be alternatively used as a guide camera (albeit more usefully with a longer lens) and it just requires loading the QHY-5L II device driver and its three ASCOM drivers. SharpCap Pro can access it directly and so

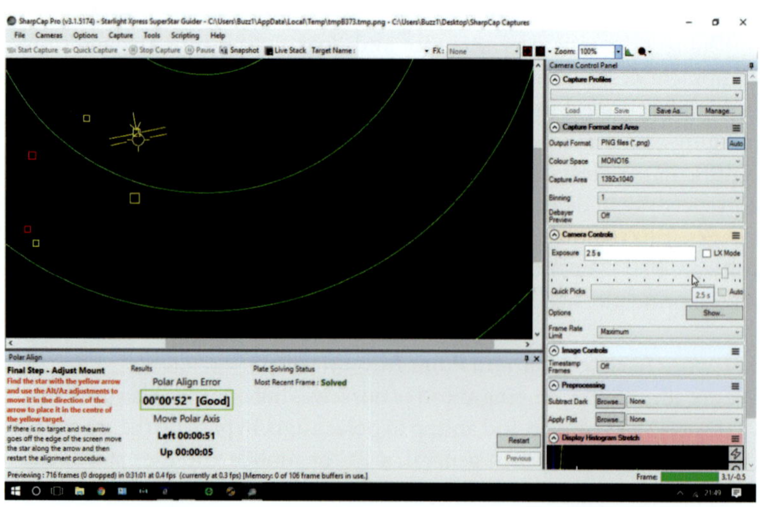


*fig.15 An alternative to PoleMaster is SharpCap Pro, a PC application that uses a guide scope and camera to calculate the alignment error. The yellow cross-hairs are updated to show the direction and magnitude of the Alt/Az alignment.*

can PHD2 guider software, using the WDM-webcam driver. In this way, if one was working on a strict budget, after polar alignment it would be possible to unscrew the PoleMaster camera from its CCTV lens and redeploy it as a guide camera in systems that accept pulse guiding. (The PoleMaster camera has no ST4 Guide port output.) If you must have ST4-based guiding it is possible to use purchase a USB–ST4 adaptor module that accepts ASCOM guider software commands.

Alternatively, with a larger budget, the recent StarAid Revolution (which uses the drift method for polar alignment) additionally integrates mount alignment and autoguider functions into a sensitive guide camera. It uses a mobile app, rather than a PC to set up and control its operation. Significantly, it does not require the celestial pole to be in the field of view. This highly portable system uses artificial intelligence to make it extremely easy to use and set up, at just under $1000.

### Star Alignment Model

To whet the appetite and included here for completeness, rather than a necessity in a simple system, polar alignment can also be achieved by more complex star alignment techniques on goto mounts. This method slews a dual-axis mount to two or three alignment stars in the sky which are then centered in an eyepiece or camera using a computer or the mount's handset. After assessing the star alignment error caused by the polar misalignment, the mount is centered on another star, but this time by using the mount's adjustment bolts. Even more advanced methods take many images at different points in the sky and compare the star positions with those from a stellar catalog, using an automated method called a pointing model. Again, these methods are found on more advanced systems and are not applicable for single-axis mounts. The alignment errors from three or more positions around the sky enable the handset or application to work out the effective polar misalignment. After computing the alignment error, they commonly instruct the user to make a particular adjustment to the altitude and azimuth adjustment bolts, either by a certain knob rotation or by automatically slewing to a reference star and asking the user to adjust the altitude and azimuth bolts to center it again.

## Tuning the Imaging System

At some point, the astrophotographer will make the jump from camera to telescope optics. Simple refractor telescopes typically have two or three glass elements at their front end and a hollow tube that extends to the camera. These are designed for visual use, in which an eyepiece is attached at the back end. As mentioned earlier, however, the focused image forms on a curved plane and an uncorrected image resembles the "jump into hyperspace" look. It requires a field-flattener or reducer inserted between the camera and focus tube to ensure the stars are in focus across the entire sensor. The optical design of a field-flattener contains two or three glass elements and assumes an optimum spacing to the sensor. In many cases, these modules either have a T2-thread coupling or 48-mm thread and adopt the T2 flange spacing standard of 55 mm. There are a few exceptions and you will need to check the datasheet.

Either side of the optimum distance, the focus plane has more curvature and stars become progressively radially elongated at the image corners. Consumer cameras and their associated T-thread adaptors will reliably put their sensors within 0.5 mm of the optimum distance. Dedicated CCDs or CMOS cameras do not comply so readily. They will have an arbitrary sensor to flange distance, depending on whether they have an in-built filter wheel, off-axis guider or an adjustable faceplate. Using the available dimensions you should be able to predict the coupling to sensor distance within a millimeter. Intervening filters will alter the effective optical path length and the optimum spacing will require a little experimentation. For this, you need a method of adjusting the sensor spacing, usually with the aid of a variety of T-thread extension tubes, from 5–40 mm and thinner Delrin washers, from 2.0–0.25 mm. Extension tubes are also becoming available for 48-mm systems.

To find the correct spacing requires a series of test exposures, at different spacer settings and selecting the best one. In practice, for each spacing setup, carefully focus the image (in the middle) and take several short exposures (about 10 seconds). Choose the best image from each set (the one with the smallest stars) and compare these "best shots" for star elongation in the corners on a computer screen. Sometimes the result is obvious or at least can be halfway between two obvious extremes. This is visually challenging and it helps to zoom the image to 200% in order to see the star shapes on the screen.

Of course, a computer can calculate star roundness very easily and not surprisingly there are specialist software utilities such as CCDInspector from CCDWare that are able to analyze the star shapes in an image. From this it calculates field curvature and tilt, as well as contrast, vignetting and focus. At $179, it is not a "budget" item and something to aspire to at a later stage.

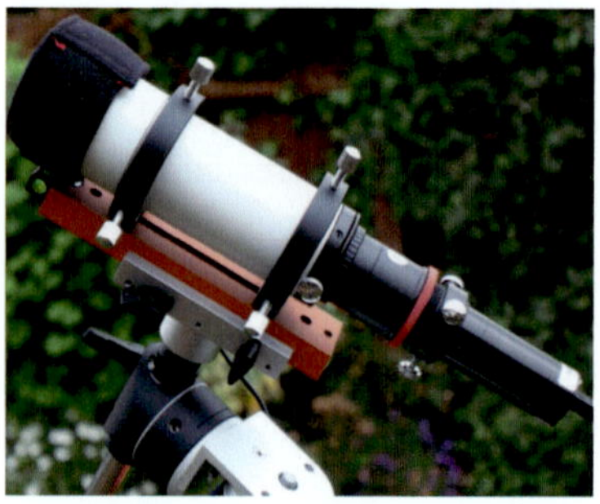

*fig.16 A 50-mm f/4 finder scope, converted into a guide scope with a variable extension tube. The red lock ring secures the helicoid focus mechanism and keeps everything rigid, as do the three fixings on the 1.25-inch fixing at the rear. It is less convenient to set up than the non-rotating helicoid, but once set up, this focus mechanism has much less flex, which is paramount for guiding.*

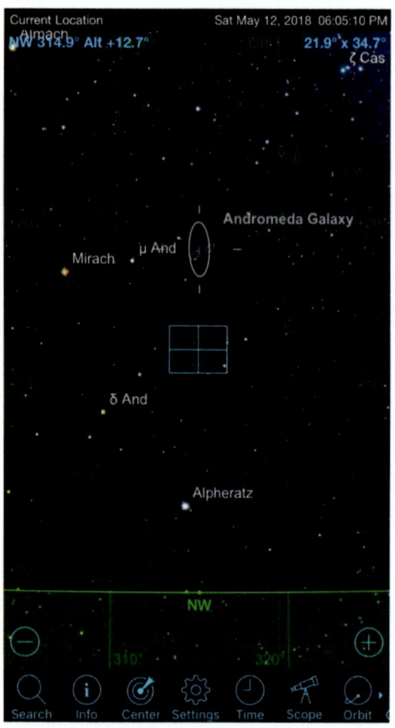


*fig.17 SkySafari on an iPhone, one of several applications that transform casual astrophotography with their ability to align, locate and control a mount.*

On the subject of tilt, it does not make much sense to have a perfectly flat focus plane, if it is not parallel to the sensor. Tilt may arise from the assembly tolerances, focuser sag and in the case of a dedicated CCD sensor, the alignment of the sensor chip to its mounting flange. Unfortunately, the normal 1.25- or 2-inch couplings, used for eyepiece mounting, are not designed for the demands of an imaging system. I'm repeating myself but the reality is that these are not optimized for repeated secure and orthogonal coupling. Where possible, use a screw-thread coupling or, failing that, seek out a coupling system that uses three lock screws and a brass ring, rather than one or two. If you offer up the camera to the flange (so that it is square) and progressively tighten the three screws one at a time, the assembly should be good. I prefer a metal to metal contact to those that use a compressed rubber gland to center up an eyepiece.

When mounting SLR bodies to optics, especially EOS models, I have discovered that not all T-adaptors are created equal: some models are deliberately slimmer, so their spacing can be fine-tuned with spacers in the optical path. Some have an oversized slot for the bayonet pin and do not lock securely, with the result that the camera body moves between or during exposures. There are some premium models out there and sometimes a small additional investment is required to make the most of the overall outlay. T-adaptors are more common that 48-mm adaptors and it is tempting to use a slim step-down ring. Be careful, it is easy to overtighten the thread so that it will not unscrew, even with lubrication.

## Guide Scope Alignment

Although a guide camera system does not require the same angular resolution as the imaging camera for effective guiding, what is important is the rigidity of the system and minimizing any differential movement between the two imaging systems. When I'm using a digital SLR I convert my 50-mm aperture finder scope for guiding. For focal lengths up to 600 mm, a guide scope with a focal length of around 200 mm is sufficient. This little system is quite light and conveniently fits onto a Vixen dovetail bar or a normal finder scope shoe. In this setup, the small diagonal included for visual use is swapped for an optional variable focus tube extension (fig.16). Although the guide camera has a 1.25-inch diameter and can fit directly into the tube, I prefer to screw it into a 1.25-inch to C-thread adaptor. This is then secured, with the shoulder of the adaptor butted up against the focus tube. This assembly is repeatable and does not require further focus checks. The adaptor also has a threaded front end, to accept filters and since the optics in guide scopes are simple doublets, I screwed an old infrared blocking filter to the front of the C-thread adaptor to improve sharpness. It takes a little while to find the right focus position, but crucially, the focus barrel has a lock ring and once set, the assembly is rigid and good for instant focus in the future.

I also tried a larger 60-mm guide scope featuring a helical focuser. This generic unit is re-badged by a number of different OEMs and the engraved focus barrel looks the part, designed to conveniently re-position the focus setting. On my unit, the helicoid thread tolerances were too large and the camera end flopped about in thick grease. The focus lock screw stopped the helicoid moving but did not stabilize the assembly, making it unsuitable for auto-guiding purposes, where rigidity is everything.

## Other Assembly Items

With the mount and optics fully assembled it takes a few power, and signal cables to complete. To avoid snags and reduce drag, I route the cables close to the main pivot with a Velcro™ tie. I also make all the electrical connections before applying any power. This is important, a great many power supplies have floating outputs (i.e. not ground-referenced) and connecting them hot may give a static electricity discharge. For instance, if I power up my monitor and laptop independently and then attach them with the display cable, in the dim surroundings, I can see a spark as they touch! This is potentially damaging to sensitive electronics. In regard to USB cables, the interface has to acknowledge a signal within a certain time. The USB cable length is one parameter that affects the signal delay and is the reason behind the 5-meter single-cable limit. Hubs and extenders allow one to transmit further but not necessarily reliably.

While it is overkill at this stage, with increasing system complexity, it helps to bundle all of the electrical modules together for ease of assembly and robustness. This is one of the practical projects in *The Astrophotography Manual*. The idea is catching on and there is an increasing number of compact modules that offer tightly integrated power, dew heater, USB, autofocus and mount interfaces (Primaluce, Lunatico and others). Some of these are fully-fledged PCs, operated remotely via WiFi from another computer or tablet, others interface with a PC as an intelligent do-it-all hub for power, control and signals.

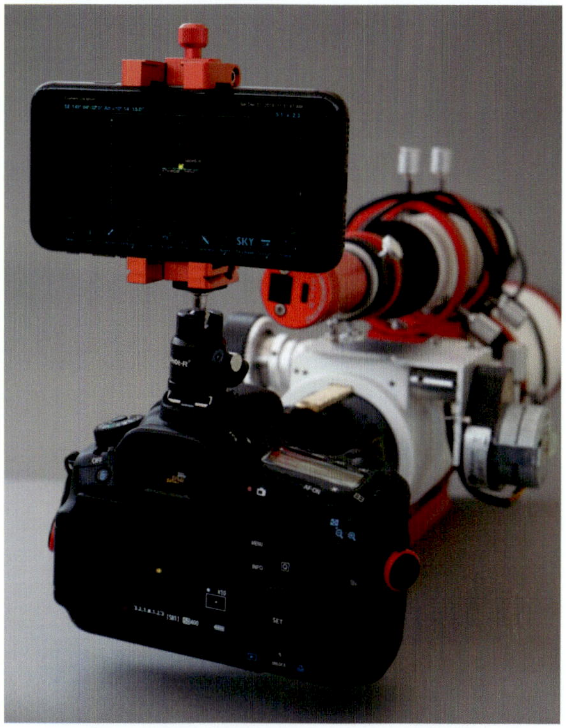


*fig.18 Another smartphone assist for astrophotography. Here, an iPhone is mounted on the hot shoe of a DSLR via a small ball and socket joint. Using an astronomy application with a dynamic display, the phone's compass and gyro in the phone are accurate enough to find an object to within 15 arc minutes, sufficient for modest focal lengths up to 400 mm.*

## Finding the Target!

When one does not have the luxury of an intelligent robotic mount and PC control, the first major practical issue you encounter is the problem of aiming. In many cases, the stars are too indistinct in the viewfinder or are too small to register on a live-view rear LCD. Apart from a few bright subjects, for example, the Pleiades, Andromeda Galaxy and Orion Nebula, precise aiming is virtually impossible without repeated trial and error. Even with a PC downloading images, plate-solving and providing coordinates of the image, it takes 10 minutes or more to nudge the system into position. Often though, we do not have the luxury of a PC and one is forced to judge the alignment from visible stars in the vicinity and check the initial alignment with a 30-second exposure at a high ISO on the LCD display. If the subject is a nebula, it is unlikely to register on a single 5-minute exposure. Necessity is the mother of invention, however, and after a particularly cold frustrating evening, I had an epiphany.

I was holding an iPhone up to the sky to determine the position of a promising target using the SkySafari app in compass mode (fig.17) and noticed that the pointing accuracy was suspect, due to my inability to hold the phone exactly square to my eye. The phone was very sensitive to any angular movement, however, and I wondered how accurate the dynamic sky view really was. To evaluate the accuracy, I mounted a telephoto zoom lens to a DSLR on a tripod and fitted the iPhone to the hot shoe with a "selfie" adaptor. I accurately centered both on Vega, making small adjustments to the iPhone angle so that Vega was on the cross-hair. It only needed a small adjustment; a promising start. Swinging the assembly around 180 degrees to another bright star, Capella, they were still in agreement to within about 0.25 degrees. A few Internet purchases later, I assembled my virtual finder-scope in fig.18. (This studio image is for illustration purposes and the DSLR image is simulated.) The smartphone mount is made of metal, with a screw clamp and is very rigid,

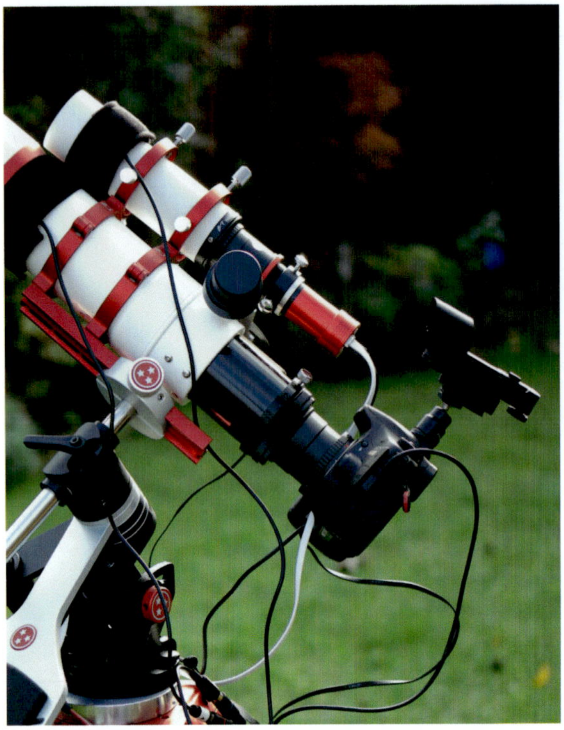

*fig.19 Swing-arm designs have a limited operating time before they need to be reset. When that occurs, the telescope loses its alignment and, without taking more trial exposures, it is tricky to frame up exactly as before. It then occurred to use a lightweight Red Dot Finder (RDF) and fit it to the camera hot shoe with the help of a tiny ball and socket joint. When the camera is initially lined up on the subject, the RDF is aimed at a prominent bright star. Its tiny ball and socket joint is locked and the 90-minute long camera exposure sequence kicked off. At the end of travel, the camera is rotated on its head's panoramic base (or the mount's RA axis) until the star aligns with the RDF again.*

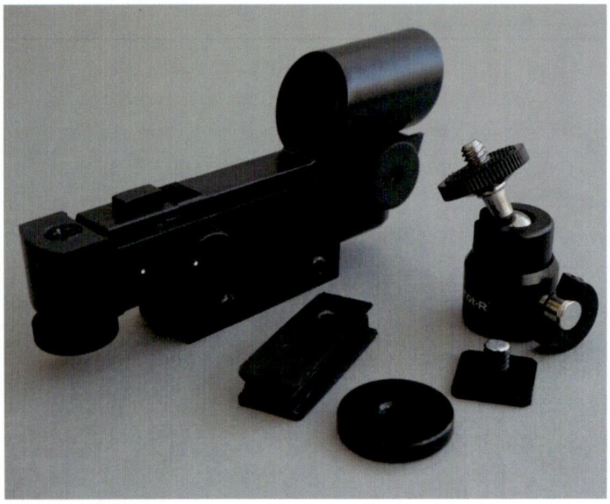

*fig.20 The RDF used in fig.19 is a plastic Celestron unit. I cut down one of its plastic mounting plates and drilled and tapped a couple of 1/4-inch UNC holes to accept the tiny B&S head. This head is designed for mounting video lights on a camera hot shoe and it has a 1/4-inch tapped base and a tiny hot shoe adaptor with a locknut. The whole assembly is very light and inexpensive.*

with both a normal 1/4-inch and a dovetail base plate. It is screwed to a small ball and socket joint that is clamped to the DSLR hot shoe.

To improve the pointing accuracy further, in practice, it is better to calibrate the iPhone orientation to a nearby bright star and then swing the scope assembly to put the target in the middle of the iPhone display. This is best done by first centering the star on the camera's LCD using live preview at 10x magnification and then making a small adjustment to the phone's orientation, with the screen set to be a marginally larger field of view to the camera (in this case four degrees across). If the planetarium app has a field of view feature, set this up for your lens/camera combination and rotate it to frame up the shot. If possible, use the telescope's rotation feature, or other means, to align the camera body with the frame and lock securely.

After this initial alignment, a confirmatory short exposure and a final tweak to the position, the mount is set tracking at the sidereal rate and the camera exposure sequence kicked off (using an intervalometer or computer control). The mount requires resetting, however, before it reaches the full extent of its swing. This act, however, causes a corresponding RA pointing error of the same and the telescope/camera require rotating, to point back to the target. For perfect registration over an extended imaging sequence, this adjustment has to be accurate.

To help this re-alignment and make longer exposure sequences easier, I invented another simple solution that uses a lightweight, articulated pointer system that is fitted to the camera hot shoe (figs.19, 20). Here, after the initial alignment is complete, the smartphone holder is replaced by a basic red-dot finder (RDF). I chose a lightweight and inexpensive plastic one and set about modifying its baseplate so that I could mount it to the camera's hot shoe and point it *independently* towards a bright star. RDFs are designed to mount on guns and they have a type of bayonet fitting. I cut down one of the plastic mounting baseplates and tapped a 1/4-inch threaded hole into it. To articulate it, I used the same tiny hot shoe-mounted ball and socket joint, which I screwed into the RDF baseplate.

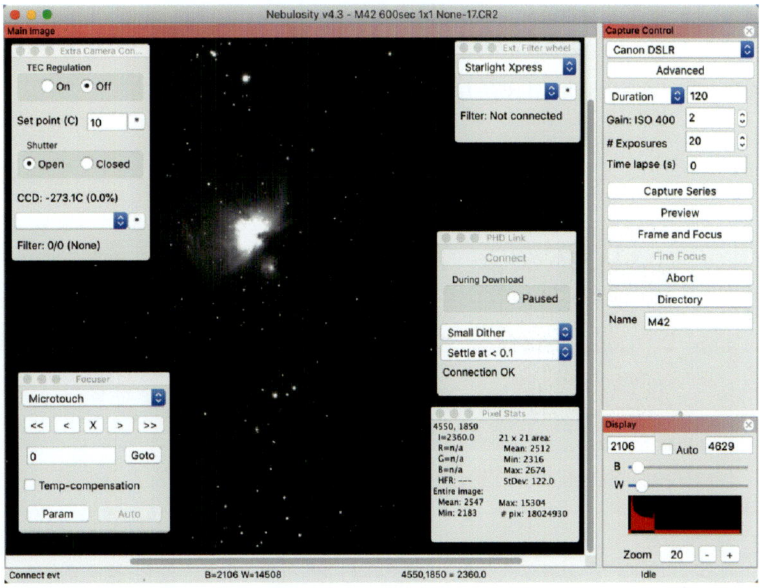


*fig.21 Nebulosity 4 is an often overlooked application. There are Mac and PC versions and it keeps things as simple as possible. The capture screen has optional floating windows and is ideal for photographic or dedicated cameras in simple setups. Sequences and filter changes are possible too and assisted focusing tools are provided. In the middle of the night, simple is good! It also is capable of basic image pre-processing and processing, all the way to the completed image. It does not offer too much in the way of automation but, on reflection, I think that is not a bad thing when starting out.*

In practice, having settled on an initial camera alignment, aim the pointer at a prominent star. When the mount's swing-arm is reset for another round, rotate the camera on its ball and socket head's panoramic base or the mount's RA adjustment, until the same alignment star lines up with the red dot. Over a number of hours, you may notice a small drift error in DEC occurring, due to a small polar misalignment. If this is large enough to be a problem, make a small adjustment in DEC to bring the star back into alignment.

In the case of the Fornax unit, it is possible to mount the telescope/camera in three ways; using a ball and socket head, a counterweight/dovetail accessory aligned to the RA axis (fig.19). I have also tried an optional right-angle bracket that mounts the counterweights to the tracker, to relocate the counterweight and telescope system movement away from the mechanism. This works well for a single sweep but the counterweight clamp's axes do not now correspond to RA and DEC and both clamps require releasing and repositioning after a tracker arm reset, with the potential for misalignment.

## Final Setups

We are almost ready to go but, as mentioned earlier, it helps to double-check your camera setup:

- ensure all your batteries and spares are charged
- or... consider a camera power adaptor
- set aperture mode to manual (on lens/body)
- set manual focus (on body or lens)
- set shutter speed to T, M or B, depending on requirement (and application software)
- set manual ISO setting (on body) begin with 400
- set white balance to daylight (on body)
- disable long exposure noise reduction (on body)
- disable high ISO noise reduction (on body)
- disable image stabilization (on body or lens)
- set remote/microphone socket to remote
- disable mirror-lockup settings when using an external intervalometer (on body).

With the checks complete, it is time to take a few trial exposures. As discussed in the earlier chapters, the type of image (vista/star trail/deep sky) confine the exposure duration and in most cases, one is normally operating at, or near, full-aperture. That leaves ISO as the means to fine-tune the exposure to ensure the images retain some star color.

## Conclusions

Effective setups are a mixture of common-sense, experience and little ideas that make things easier and quicker to accomplish. What is easy in daylight always takes on a different perspective in the dark. These setup considerations are aimed at the first time user with the simplest of equipment. In later chapters, as we introduce additional equipment and automated control over mounts, focusers and cameras, it makes the setup more complex, requiring the installation of various modules, wiring, computers and acquisition software, which brings its own challenges.

# Orion Constellation

*A familiar constellation with hidden depths.*

***Equipment:***
*Fuji X-T20 - 35 mm f/2*
*Tripod with ball and socket head*
*Fornax LighTrack II mount*
*Intervalometer (built-in)*
*Mosquito repellent*
*Small torch and head torch*

***Software:***
*Nebulosity 4 astro software*
*Affinity Photo*

***Exposure:***
*132 x 30 seconds at ISO 200*

The familiar Orion constellation holds a number of secrets that reveal themselves with ever-increasing exposure. Near the sword of the hunter, is the famous M42 Orion nebula, clearly visible from a single short exposure and the running-man nebula slightly above it. At a slightly higher declination are the three stars of the belt that hide the popular horsehead nebula, flame nebula and a number of smaller reflection nebulae. Hidden in the broader sky background are two further expansive red nebulae which usually require narrowband filters to pick them out. One is the possible supernova remnant of Barnard's Loop and the other surrounds the top-most star Meissa. To see the Orion constellation in all its glory, there is no better place to go to than the images on the deepskycolors.com website. The exposures used to create those images are extreme and in our modest setup, the aim is to produce a starry life-like representation of the constellation showing the brighter nebulosity within its boundary. The cropped picture above is not going to win any awards but it does have the right attributes; a subtle dark noise-free background, small sharp stars with good color and shows some of the hidden treasures, even in this modestly stretched state. Up close, there are thousands of stars, though those in the corners have obvious coma at f/2, (fig.2). With a more extreme stretch, the background noise becomes intrusive but it does start to reveal the other famous nebulae, that are the staple of many a portfolio (shown in fig.4). Overall, this assignment is more of a voyage of discovery than a single practical example.

## Setting Up

This is the first assignment that uses a tracking mount and is a good chance to become familiar with the basics of alignment, focus, exposure and image processing. The image is taken with a Fuji X-T20 mirrorless camera, with a short prime lens and the Fornax tracker. The tripod was roughly leveled, North-aligned in the middle of my lawn and pushed down firmly to settle the feet position in the soft earth. I attached the Fornax tracking unit and its wedge to the top and fitted an inexpensive SkyWatcher EQ5 polar scope to the arm. I had followed my own instruction and centered the reticle and arm assembly and confirmed in practice the drift was minimal and good enough to image with modest telephoto lenses over several minutes.

After rotating the polar scope arm so that its clock-face was vertical, I set Polaris to the same position as shown on the PolarAlign iOS app. The camera was

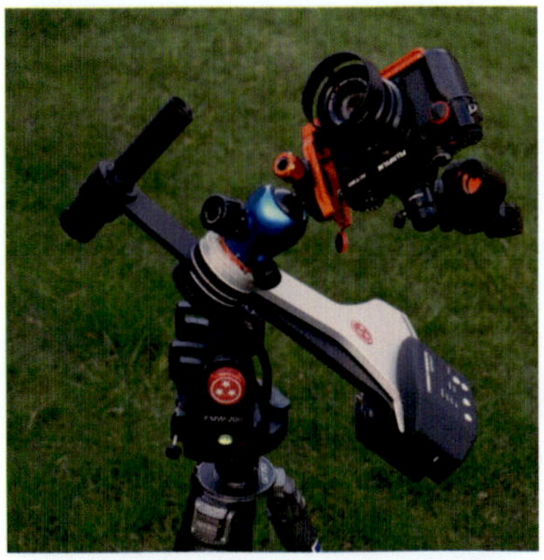

*fig.1 This tracking imaging rig is about as simple and light as it gets. I like the Fuji X-series cameras for general photography but their unique RAW files impose some restrictions on which applications one can use for image processing.*

*fig.2 Stars are extremely unforgiving and show up any lens aberrations. In the extreme corners, at f/2, even this quality prime lens is struggling with obvious coma. Next time around, I will stop down to f/4, extend the exposure time and use a plug-in intervalometer.*

mounted to an L-bracket to provide a better balance for the portrait-orientated image and fitted to a high-quality ball head, that featured a panoramic base (used for resetting the camera alignment after resetting the mount (fig.1)).

The Orion nebula is one of the most obvious constellations in the Northern Hemisphere and needs no electronic aid for location. It is a good choice for this assignment as it is easy to frame up in the viewfinder and its bright stars show up using the x10 focus magnifier feature on the rear LCD. It has a further benefit, the brightest star Sirius is close by, and, if you have noted my red dot finder trick in *Setting up a Simple Mount System*, is an obvious target to realign to after a resetting the tracker. The one drawback with imaging Orion at my latitude is that it appears low in the sky and only momentarily rises above my imaging horizon. The exposure period is limited to the few hours as it passes between two neighboring houses. Imaging at a low altitude has other drawbacks, one of which is a strong sky gradient towards the horizon. Normally, one would remove this in processing. In this case, my intention was to give the image a more natural look and capture a less processed version.

## Exposure

With a tracking mount, the necessity to deliberately shorten exposures to prevent star trailing is removed and we are free to push for higher quality, by collecting more photons. In this case, I selected the lowest ISO setting (ISO 200) and opened up the 35-mm lens to f/2. After a few trial exposures, I conveniently settled on 30 seconds, the maximum exposure time in T-mode, which produced a deep blue background that retained much of the star color. To keep things simpler still, I used the camera's built-in interval timer shooting mode. With the swing arm of the tracker reset, I framed the constellation, set the camera to manual focus and, after confirming all the fancy noise reduction modes were disabled on the camera, focused on Rigel using a magnified LCD view. This lens is not long enough to work effectively with a focus mask and, at the end of the day, it was a judgment call that the stars were as small as possible. Being careful to not touch the focus ring, I confirmed the framing, set the mount tracking and aligned the red dot finder to Sirius. It was not a humid night and I judged that I did not require any dew-control measures for a few hours and started the camera exposure sequence. After 90 minutes, I returned to the camera, reset the mount arm, loosened the panoramic base of the ball head and rotated the head and camera so that the red dot finder was aligned once more to Sirius and set it tracking again. After the imaging session, I took a set of calibration frames (lights, darks and biases) for use at a later date.

## Processing

The X-Trans RAW files from this Fuji model are more challenging than most. It uses an unusual color filter mosaic pattern and not all RAW converters support the most recent bodies. If at some time your favorite image processing application does not support your shiny new camera, for example the present Canon EOS R series, which record CR3 files, one possible workaround is to first convert its RAW files into Adobe's digital negative format (DNG) before processing. The DNG converter is a free utility on Adobe's website and is kept up to date with the latest hardware.

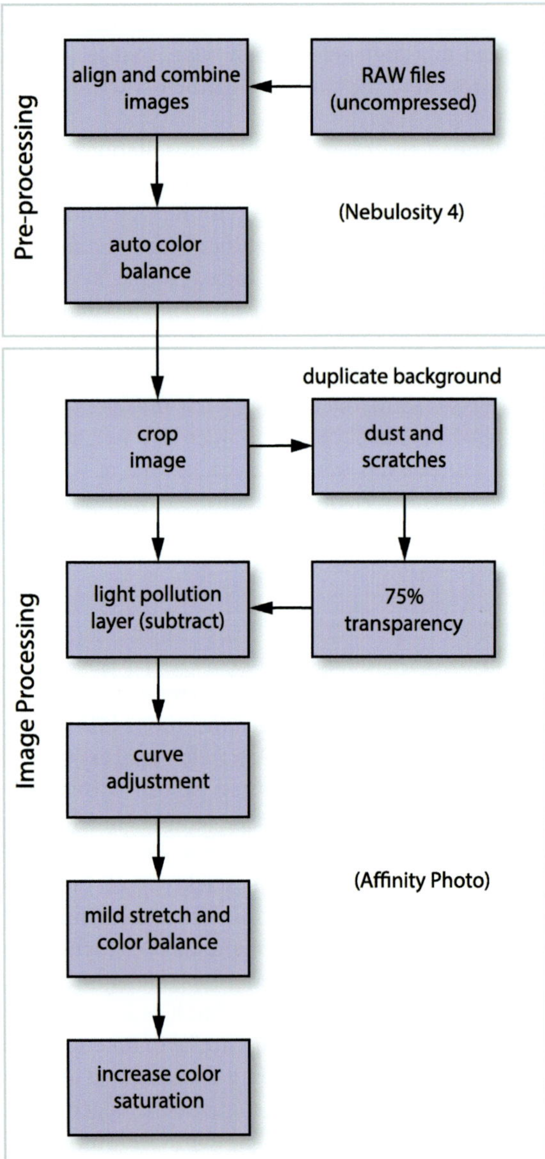


*fig.3 The processing workflow for this image is relatively simple. The secret is to respect the image and not overdo the adjustments. It is often the case that several small adjustments are better than one large one.*

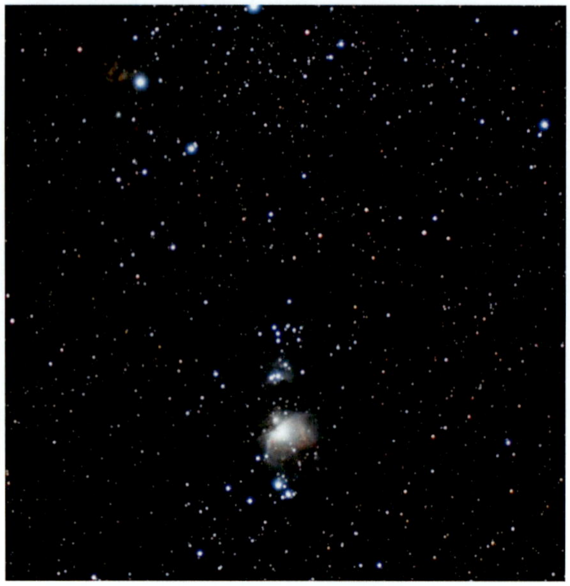

*fig.4 With an extreme stretch this cropped image shows the M42 Orion nebula in more detail, the small nebula above it and at the top left of the image the orange glow of the Flame Nebula and the red nebulosity surrounding the Horsehead Nebula. Although there are 132 exposures, they total just over an hour through a 17.5-mm aperture and it requires considerably more photons to probe the darkest recesses of space.*

In the case of the Fuji X-T20 although it does not have as much sensitivity to the deep-red hydrogen alpha emissions as the Canon EOS 60Da, I did discover its infrared blocking filter is less severe and the sensitivity extends to Sulfur emissions and beyond. (It produces a weak daylight image, when fitted with a 720-nm IR filter.)

While writing this, DeepSkyStacker was in the process of changing its RAW converter from DCRAW to LibRaw (which supports my Fuji camera). LibRaw is periodically updated and supports more than 1,000 formats and it is used by several applications, including the emerging SiriL and ASTAP image processing application. For this assignment, however, I decided to go back to an old favorite, Nebulosity 4. While it may not have automatic pre-processing steps, on the Mac, it works with the Fuji RAW files directly and keeps things as simple as possible. With a single command, it aligned and stacked my 132 image files. Its alignment routine requires one to click on two stars (ideally in opposite corners) and it automatically does the rest. It took a little while, but there again, it was converting, registering and averaging several gigabytes of data.

The dark-blue stacked image had discernible but dim stars and required adjustment to the background and levels to bring it to life. After saving it as a 16-bit TIFF file, the remainder of the processing was carried out in Affinity Photo. After a mild curve adjustment, the image layer was duplicated and the dust and scratches noise filter applied to it (with a radius of 50 pixels) to create a background image. Selecting the subtract blending mode for this blurred layer, creates a black background. To give the sky some tone, I reduced the layer's transparency to 50%. After merging the layers, it

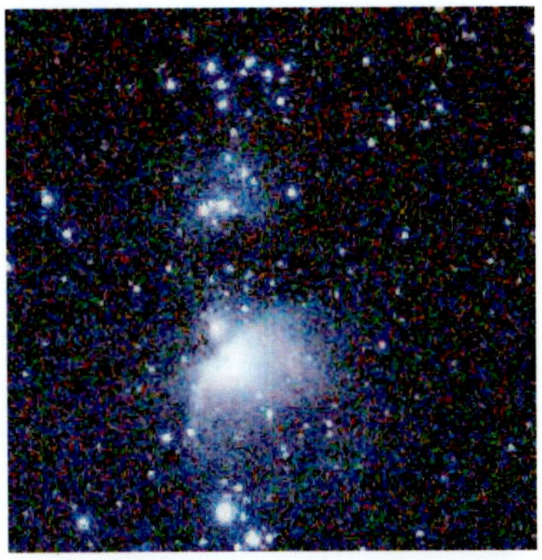

*fig.5 A single 30-second exposure shows obvious noise at the pixel level and the nebulosity is difficult to pick out.*

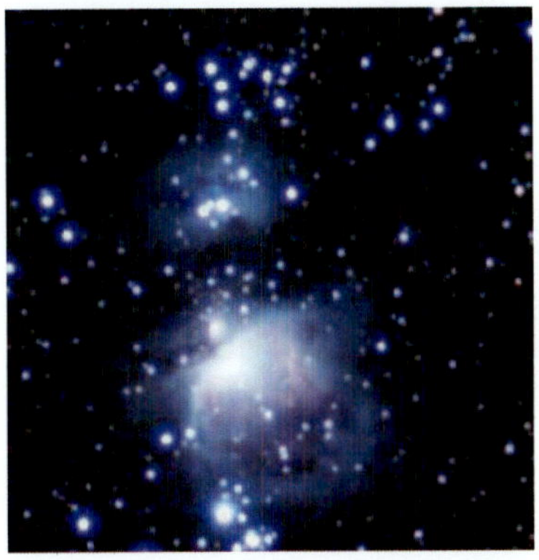

*fig.6 An hour later (120x 30-second exposures) with identical processing and no noise reduction, the benefit is obvious. Not only is the background smoother but the dim nebulosity extends further and has some color too.*

was stretched further, to reveal the less luminous stars. It is often the case that an iterative approach is required. This was the case here, as the stretch exaggerated the background unevenness. Going back, the transparency of the blurred background was increased to 75% and the stretch re-applied. Finally, the star color was boosted, using an Hue/Saturation/Luminance adjustment layer, increasing red and yellow saturation by 30%.

Out of curiosity, I performed the same adjustments to a single 30-second exposure and compared the results up close. The images in fig.5 and fig.6 show the background noise improvement from averaging 132 images. Although 132 images may seem extreme, one has to consider that the overall exposure time was little over an hour and through an aperture of only 17.5 mm. A telescope will typically have an aperture of 70 mm or larger, passing through 16x more photons in any given time. Even so, if one aggressively processes the stacked images, it is possible to detect more of the Orion Nebula and detect the equally famous Horsehead and Flame Nebula (fig.4).

To be honest, after coming from my 30-hour plus colorful narrowband images to this, I was not greatly excited by the result. I have seen better on the web, taken with unbelievably brief exposure periods, albeit at lower latitudes and from dark-field sites. It is important to share this example, however, as it is typical of what many of us might initially achieve with simple means. It significantly uses and confirms our ability to align and track an object over several hours as well as process multiple files.

I waited a year and had another go at the Horsehead and Flame nebulae, this time using an astro-modified Fuji X-M1, an old 200-mm f/4 Carl Zeiss Tessar and a 2-inch light-pollution filter. The camera modification (by a third party) extends the red sensitivity and the light-pollution filter not only removes the yellow-brown haze, it importantly removes the random photons that accompany it. Stopping down to f/5.6 and extending the exposure to a few minutes each, the processed result of a few hours is shown in fig.7. The two bright stars on the left are part of Orion's belt and the slight haze around them is caused by the light dispersion through some high-level cloud. To reveal fainter nebulosity requires yet more exposure, darker skies and ideally, imaged from a lower latitude, to raise the object's altitude.

*fig.7 The result of two hours exposure, taken with a modified Fuji X-M1, fitted with a 200 mm f/4 telephoto lens and a 2-inch light pollution filter taped over the end. This camera had its IR blocking filter removed, improving deep-red sensitivity. In this image, the Flame, Horsehead, Running Man and Orion nebulae are more obvious, as is the extended deep-red background nebulosity. For the image sequence, I controlled the camera remotely, using the ASCOM trigger described in the practical chapter* Cloudy Projects.

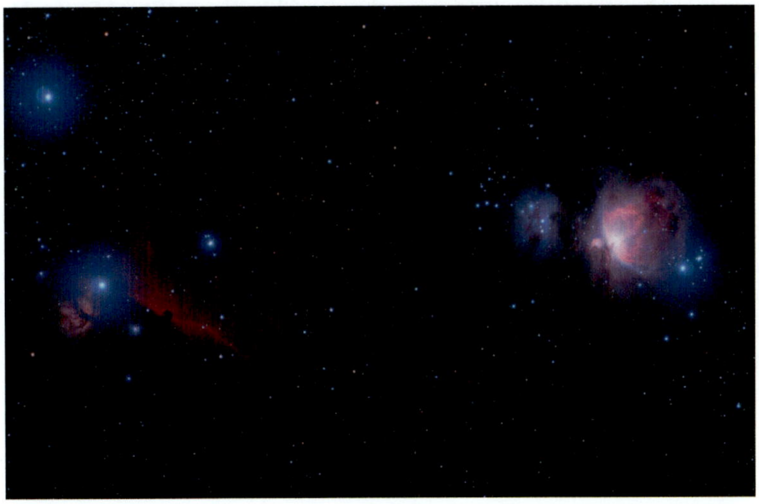

# Improving Focus (part 1)

*Longer exposures and higher magnifications demand better focus.*

In the beginning, focusing a camera required good eyesight, a ground glass screen with a fine finish and patience. In time, ground glass had various optical aids added, including split field and microprisms. It still required patience. With the invention of autofocus, SLR and DSLR viewfinder screens were only required to show the composition and started shrinking, camera lens designs changed their architecture to permit increasingly rapid response times and the whole concept of hyperfocal distance and depth of field was largely forgotten by the masses.

In astrophotography, precise focus is essential; all puns aside, stellar images ruthlessly expose poor technique, as we instinctively know what stars should look like. Poor focus increases star sizes in general and makes them more diffuse. Towards the edges of the frame, other issues, like coma and chromatic aberration become more noticeable too. There are a number of tools and techniques, however, of increasing sophistication, which improve focus accuracy and make it easier to achieve.

Fig.1 shows the cone of light from a star, that shrinks to a minimum at the point of focus and progressively increases either side. One never gets a perfect singularity as diffraction and seeing conditions diffuse the star shape. There is a zone, however, either side of the focus position which will give an acceptable result and for which there is a general agreement. This zone is very small and is called the Critical Focus Zone or CFZ. The following equation is used by astronomers to calculate the zone for any given telescope. In the following equation for an aperture ratio f and wavelength $\lambda$:

$$CFZ = 5.\lambda.f^2$$

The interesting thing is that the focal length is not a parameter, only the traditional aperture ratio (f/stop), which is, of course, the focal length divided by the aperture diameter. This is of no surprise to photographers, who know the depth of field reduces with wide aperture ratios.

It is interesting to see what this means in practice. For a William Optics ZenithStar 73 telescope, the zone is about ± 0.05 mm. At 0.1 mm from the optimum

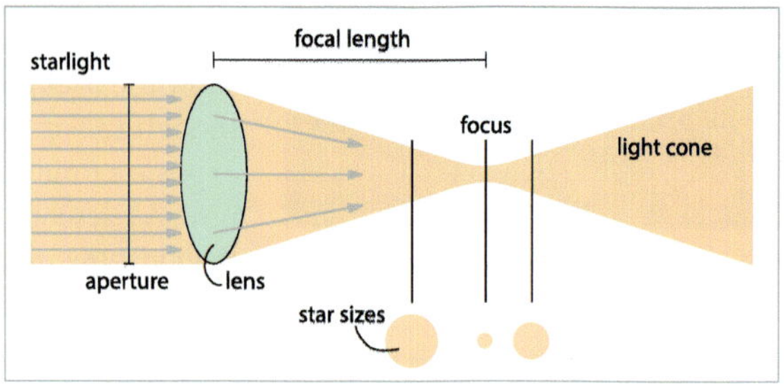


*fig.1 The image of a star changes its apparent diameter with focus position, at a rate governed by the optic's f/ratio.*

position, a star diameter increases from 4 µm (the diffraction limit) to about 17 µm. In the first instance, most of the starlight hits a single pixel, in the second case, they cover 16 pixels, with reduced pixel intensity. The slightest focus tube movement will ruin the end result. As a telescope cools down, its focus position shifts too. On my refractors, one would expect the aluminum tube to shrink and the focus position to move outwards. Not necessarily, as not only does the aluminum contract, so does the glass, decreasing the focal length. Night-time cooling is a problem and with longer lenses, I check the focus every few hours. When I am imaging all night with a telescope, I re-focus for every 1 °C ambient change and always allow the optics to acclimatize before attempting any imaging.

### Visual Focus Methods

The viewfinder of a DSLR is a compromise. Finding the optimum focus is made difficult by the low magnification and the small screen. A mirrorless camera fares slightly better, as they have electronic viewfinders and in common with some DSLRs have a live view feature on their rear LCD screen, typically with a 5x and 10x magnification option. In practice using this magnification aid is better than an inexpensive right angle finder. With care, you can reliably achieve a passable focus. In recent years, many digital cameras have improved their video modes and in doing so, often have a live video feed capability. Connecting this to a small, 5–7 inch battery-powered monitor, is better for assessing focus. In the case of wide-angle lenses (100 mm or shorter), this is often the only viable method and it is sometimes better to manually focus on a very distant object in daylight and use this for the imaging session. It is not without risk, as it is easy to forget to disable autofocus or accidentally touch the focusing ring during the system assembly or adjustment.

There are a few tricks that improve its reliability. The first is to use full aperture and while it is tempting to center a bright star and focus on that, sometimes it is better to find a star closer to the corners of the image and move the magnifier position. Here, the shape of the star changes with focus (due to optical defects) and these are sometimes easier to detect than tiny changes in its apparent size. On my EOS DSLR, some tiny stars in the magnified view disappear at the point of focus too. I assume that when they are out of focus, they cover several pixels but when focused, the image of some misses any sensor pixel. Do not do what I did; I found a really bright star due South and tried to focus on it. The best I could get was an elongated orange blob. I thought the worse of my lens and cursed my bad fortune. Four days later, I realized I had been trying to focus on Saturn!

In essence, while split images, microprism and contrast detection focus methods do not work in astrophotography, in their place we use various techniques to minimize star size either directly or indirectly, using optical trickery. In both cases optimum focus may be assessed visually or via computer methods. In this chapter, we concentrate on visual methods, the most common of which use focus masks.

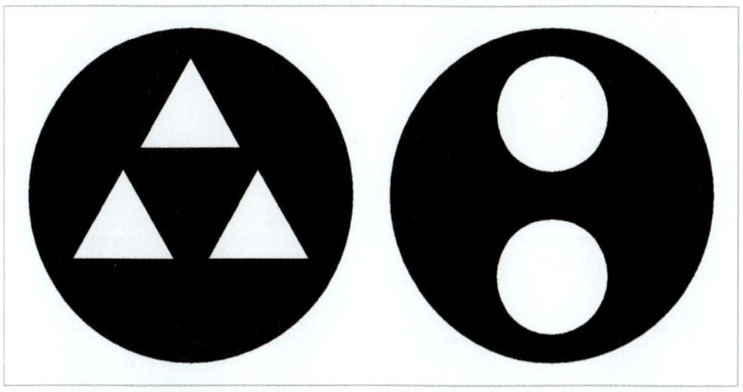

*fig.2* *Two simple home-made focus masks, after Hartmann, which are placed over the lens. The image of an out of focus star mimics their appearance. As the star approaches focus the shapes converge and become a round point at focus.*

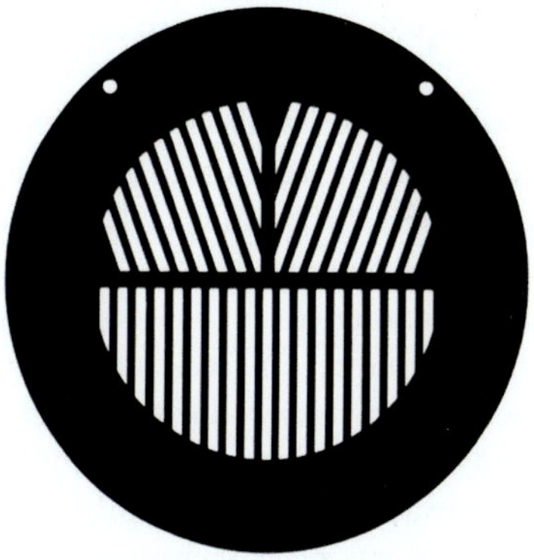

*fig.3* *A Bahtinov mask, available in a range of diameters to suit different telescope apertures. The angles of the grid create diffraction spikes (fig.5) which intersect when the system is in focus.*

## Focus Masks

There are some wonderfully inventive methods to use optical effects to enhance the eye's ability to find the focus point. They all work on the basis that if your aperture is fragmented, for want of a better word, the light from a bright star through these sub-apertures align when the image is in focus. Some of the simplest are Hartmann masks like the two in fig.2, with two circles or three triangles. The diffraction from the straight lines of the triangle mask additionally adds diffraction spikes to the images. The great thing about these masks is that they are easy to make.

### Diffraction Masks

Any mask or aperture creates diffraction. Straight lines cause a star's image to have radiating lines or spikes perpendicular to each edge. This and the superposition of different apertures conspire to create a unique situation that we are able to exploit. These optical diffraction patterns are popularly created with a Bahtinov Mask. This mask has three sets of grids, orientated at different angles (fig.3). While it is possible to make these (an Internet search will locate multiple websites with design wizards) it is not easy to cut the delicate pattern. More conveniently, the commercial ones are available in a number of sizes and with simple mounting pins that locate over the telescope dew shield. Some more advanced ones use 5 sets of gratings, like those from GoldFocus (fig.4), in combination with special software.

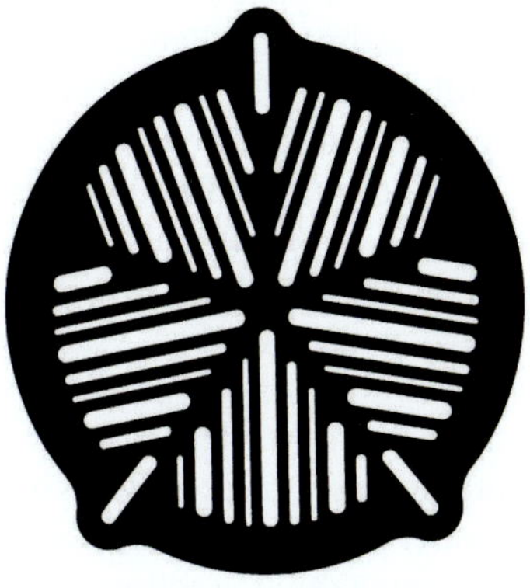

*fig.4* *Other, more complex masks exist too. This one from GoldFocus works best when the star diffraction pattern is analyzed by a computer.*

The multiple slits from these masks create strong diffraction spikes off a bright star and, when the star is in focus, these spikes all intersect at the same point. The eye is adept at detecting things like this (which

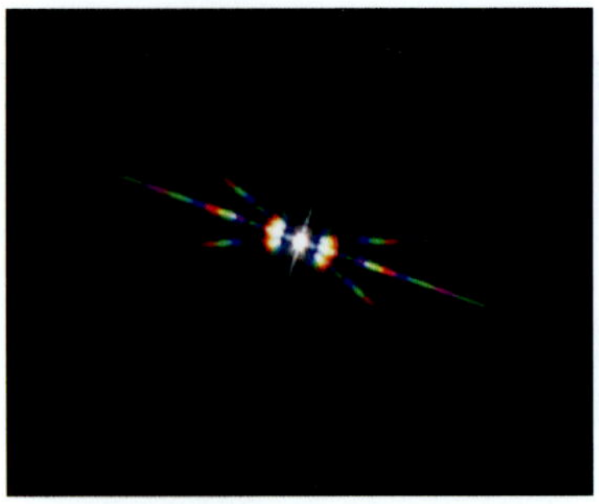

*fig.5 The magnified image of a bright star shot through a Bahtinov mask. If you look closely, in this in-focus shot, the three long lines are spaced evenly and intersect at one point. Short focal length lenses are less distinct.*

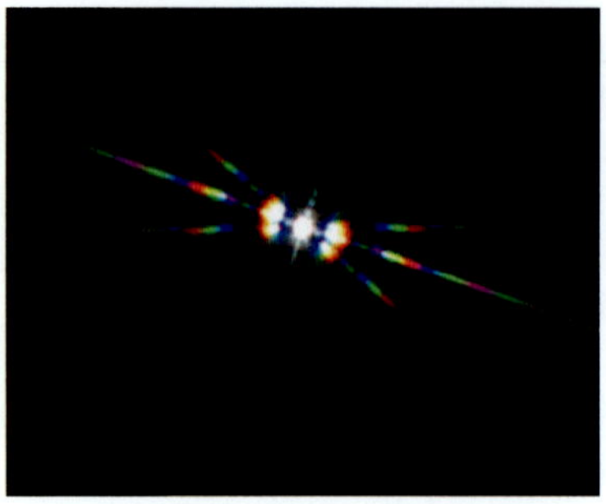

*fig.6 To give an idea of the sensitivity of this method, this image shows the image at the critical focus point, which many would accept as the limit of acceptable focus. On a 10x camera preview, it is just visible.*

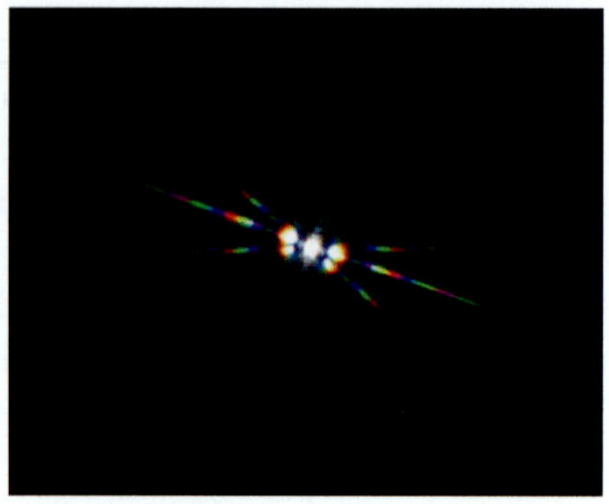

*fig.7 Moving further out, beyond an acceptable defocus limit, the stars quickly become bloated. The middle spike is now clearly off center in this magnified view but still challenging to see in a viewfinder.*

is why Vernier scales are still effective to this day) and it is possible to detect and correct small focus errors. The images in figures 5, 6 and 7 show an in-focus image with two at the limits of the critical focus zone. The good news is that these spikes are readily seen on the rear LCD of a digital camera or in the image acquisition software. The bad news is that this technique is increasingly difficult to use with short focal lengths. The technique works well with my refractors, but the diffraction spikes from medium telephoto lenses are too small to observe. Wide-angle lenses do not produce discernible spikes.

The focus mask technique is inexpensive, effective and portable. Using one on a medium telephoto lens or longer will improve your results over plain sight. Some of the latest refractors even have Bahtinov masks incorporated into their lens cap. My William Optics RedCat refractor has a helical focuser, which does not lend itself to autofocus, but the integrated Bahtinov mask, creates a bright image which is easy and accurate.

Focusing with a mask is ultimately limited by our eye, brain and seeing conditions. Not surprisingly, it is possible to overcome the human limitation by allowing a computer to analyze the image and provide a numerical value. A utility, the "Bahtinov grabber," was developed by the late Neils Noordhoek and is still available for download. The APT image acquisition program has a Bahtinov grabber in its focus toolbox (fig.8).

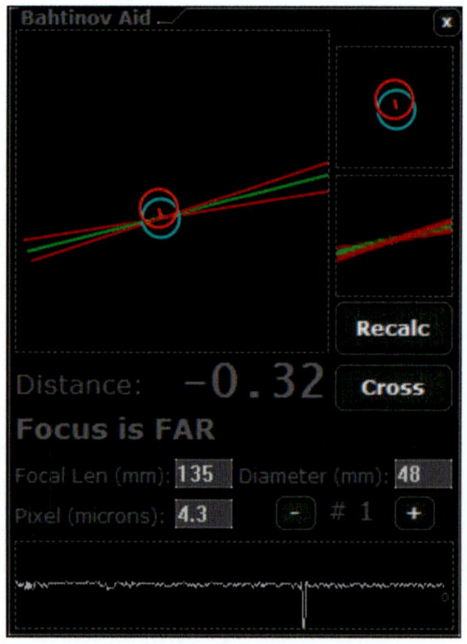


*fig.8 When a computer is used to acquire the focusing images it can additionally be used to analyze the pattern. A computer is more adept at noticing a small visual anomaly. Here, the Bahtinov grabber in the* Astro Photography Tool *application provides a numerical readout of the defocus.*

The angles of the Bahtinov mask grid are not entirely arbitrary. They are set at a value that makes the system sensitive to focus change and yet detectable to the human eye. Other angles increase the sensitivity but require computer analysis at a sub-pixel level to discriminate focus positions. One such system is the GoldFocus™ mask, which claims to be at least 10x more accurate. While it is possible to use these masks visually, as with a Bahtinov mask, full accuracy is only achieved with its proprietary analysis software. A single measurement is susceptible to seeing conditions and to overcome this, the software automatically averages multiple frames to establish a more stable reading. It also can do this at several focus positions, to effect a form of calibration and move an motorized focuser to effect a basic autofocus.

## Mechanical Considerations

While it is possible to use a conventional manual-focus lens coupled to almost any digital camera with an adaptor, infinity focus may be a problem, however, either due to the adaptor depth or, less obviously, if the camera sensor has been modified. In these cases, the infinity stop on the lens may not quite reach infinity focus. Most autofocus lenses do not have this limitation and focus beyond infinity, which allows the autofocus process to home in on the best focus position. The nature of instantaneous autofocus with minimal power consumption requires a mechanism that moves freely; completely the opposite of tight-tolerance mechanisms found in astronomical telescopes. Many low-cost zoom lens barrels wobble. This is of little concern during the short exposures of conventional photography but potentially causes star-shift during longer exposures, often confused with poor focusing or tracking. As focal lengths increase, the rigidity of the mechanism is ever more critical. Not only may the focus position slip or shift with the angle of the lens but in addition, a tilting focus mechanism will cause the plane of focus to tilt too, causing uneven focus across the frame. Some mechanisms have adjusters and most telescope focus mechanisms have a focus lock. With manual focus, it helps to lock the position to improve rigidity during imaging. There is no such luxury on a helicoid mechanism.

The fragility of your tripod and mount also become painfully apparent when you attempt to manually focus. The merest touch of the focusing knob or ring causes a star to dance around for several seconds. It requires a light touch, small movements and patience to analyze the focus and zone in on the best position. Unfortunately, for all but the briefest imaging sessions, focusing is not a fix and forget process. As focal lengths increase, a small ambient temperature change causes a change in the focus position. Many of my first results with a medium refractor telescope suffered from poor focus. Looking at the images in sequence, it was very apparent that the initially good focus deteriorated over time. For best results, most imaging sessions require frequent interruption to confirm or fine-tune the focus position.

## Computer Methods

As we become more advanced and throw a computer into the mix, focusing becomes easier and more accurate. Computer analysis has an immediate performance advantage over human assessment and includes an operational benefit too from the remote operation of a

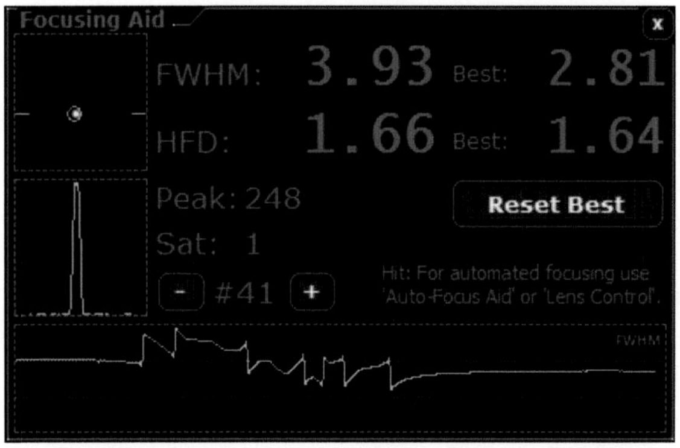


*fig.9 Astro Photography Tool has a number of focus aids. Here it is directly measuring the diameter of the star. The best numerical value helps achieve the optimum focus position. Nebulosity 4 has a similar system, acquiring repeated short exposures and displaying the diameter and a graph of the results. The jagged appearance is normal, due to the vibration imparted onto the telescope with manual focusing.*

focus mechanism. A motorized focuser imparts little vibration into the imaging system and unlike manual focus adjustments, is ready to make an immediate assessment without waiting for the image to stabilize.

At the end of the day, it is star size that counts. Indirect focus methods are great but ultimately, if the stars are as small as they can be, everything else is good. As we mentioned before, our eyes are not the best judge, especially if the star is twinkling. Computers, however, are adept at measuring star diameters at different focus positions and automatically finding the precise position at which the stars are at their smallest diameter, i.e. in focus. Unfortunately, stars are fuzzy blobs and it requires a practical definition for the effective star diameter. Two common definitions are used; HFD and FWHM. HFD, or half flux diameter, is the star diameter within which half of the star's photons fall. The FWHM, or full width half max value, is the star diameter at which the intensity is half its peak value. For practical purposes, they are equally effective and various autofocus applications may use either one or both. Both Nebulosity and APT have readout systems that allow you to monitor the star size electronically while you manually focus a telescope (fig.9). The more advanced GoldFocus system achieves better results with its 5-way spike pattern but in practice, may take longer to realize its full potential in the presence of poor seeing. More advanced computer methods, including autofocus and temperature compensation, as part of a more automated system are discussed in the *Improving Focus (part 2)*.

# Image Processing (part 1)

*If you thought image capture was time-consuming, image processing is just as demanding.*

There is no denying, image processing is a deeply technical and time-consuming business and it is just as well that image acquisition takes as long as it does, otherwise there would be a backlog of data to process. Just as with everything else in astrophotography, there is a progressive path of sophistication and it is better to keep things simple at first and then, after mastering the basics and establishes a "feel" for things, attempt more elaborate manipulations. The image processing chapters in this book fall into three sections. In this first one, we consider aesthetics and the general image processing principles. Rather than a dry, step-by-step instruction, the assignments pick up on selected practical details and the later image processing chapters explore some of the more advanced techniques in support of more challenging deep-sky objects.

I commented in *The Astrophotography Manual* that astrophotography reminds me of traditional monochrome imaging, where the magic begins when you enter the darkroom. Similarly, there is no single interpretation of a negative that is "right" and the same is true of deep-sky images. These are distorted in color and tonality purely to satisfy an artistic requirement. In both hobbies, the steps taken to enhance an image require technical knowledge applied with artistic sensitivity. There is seldom a fix for a poor negative and it is easy to spend a whole day in the darkroom perfecting a single print. Image processing in astrophotography is just as demanding and it takes many hours of patient experimentation to become proficient. Looking back, it is not uncommon that with growing experience, it is worth revisiting archived image capture files and having another go.

In this book, the image processing steps are laid out in graphical form for easy reference. Astrophotography is consistently more demanding in regard to image processing than conventional imaging and it requires extensive manipulation to create the final image. This takes time and cannot be rushed. In the case of the more advanced deep-sky images, the processing often splits into multiple paths (including blind alleys) and may take an entire day to complete a single image.

## What Makes a Good Astrophotograph?

Art is certainly in the eye of the beholder and although astrophotography is essentially record-taking, there is still plenty of room for interpretation to turn multiple sub-exposures into photographic art. These include both technical and artistic attributes. Most can agree on some general guidelines but it is important to note some of the more original interpretations, that break rules, may also work pictorially. A good part of photography is knowing what you want to achieve before you press the button. It certainly is the discipline that was adopted by film photographers in the last century. Without Photoshop to correct their errors, the photographic artist had to carefully consider exposure, composition, lighting, focus, filtration and perspective. (You can still buy a T-Shirt, which declares "Photoshop, helping the ugly since 1988.") Digital cameras have made astrophotography what it is today but I think, at the same time, their immediacy and the ease to correct mistakes with extreme image manipulation undermines discipline and encourages a culture of impatience and carelessness.

### Technical Considerations

The technical aspects of a good astrophotograph have less room for interpretation; stars should be tightly focused and circular, all the way into the corners of the image. A well-exposed and processed image should retain some star color; from red through to blue. Bright stars appear larger in an image, as the exposures required to reveal faint nebulosity often turn bright stars into diffuse white blobs. Image stretching during processing causes stars to bloat further and progressively washes out the color in the dimmer stars. We instinctively know what a star should look like and so its reproduction ruthlessly reveals poor focusing, tracking and optical problems. The subject matter of a conventional photograph is more tolerant of the same aberrations. A star's image also highlights any image registration issues between the multiple exposures. While, some of the more advanced processing techniques reduce star bloat and re-shape stars, these do not entirely cure the problem and it is better to use good practice to avoid the issue in the first place.

The sky background is another area in which there is a general consensus. It should be neutral in areas where there is no nebulosity and very dark grey (but not black). For printing purposes, I adjust the darkest regions to measure ~94%K. Putting aside nebulosity for the moment, the background should be evenly illuminated throughout and show little noise. Image processing increases the visual appearance of noise in the darker areas and there are some very clever tools that reduce it. If these are taken too far, however, the image takes on an unreal, plastic look. There is a degree of subjectivity here and just like film grain, a little noise adds a touch of reality to images. The "right amount" is something that can only be determined by the display medium, repro scale and your own judgment. In addition, pure green is not a color that appears naturally in a deep-sky background and is normally neutralized (with the exception of false-color images and comet heads).

Sharpness, resolution and contrast are interrelated in a complex triangle of visual trickery. A high-contrast image can give the appearance of sharpness and conversely a high-resolution image may not look sharp. For a long while it was a long-running debate between the small film format and digital photographers. Fine-grain monochrome film has approximately twice the spatial resolution of a 24-Megapixel DSLR, yet the digital images look "sharper.".The original sharpening tools in conventional imaging programs were not optimized for astrophotography; a well-processed image needs careful sharpening at different scales to tighten stars, without creating black halos, as well as to emphasize structures within galaxies and nebulosity without creating other unwanted artifacts.

Image resolution is often limited by seeing conditions and imaging technique rather than optics. Good image processing makes the most of what you have. The trick to successful imaging is to trade off between sharpness, faint details, noise and resolution to arrive at an outcome that looks "natural." Too little processing is often better than too much. There are a lot of judgment calls and it is easier said than done.

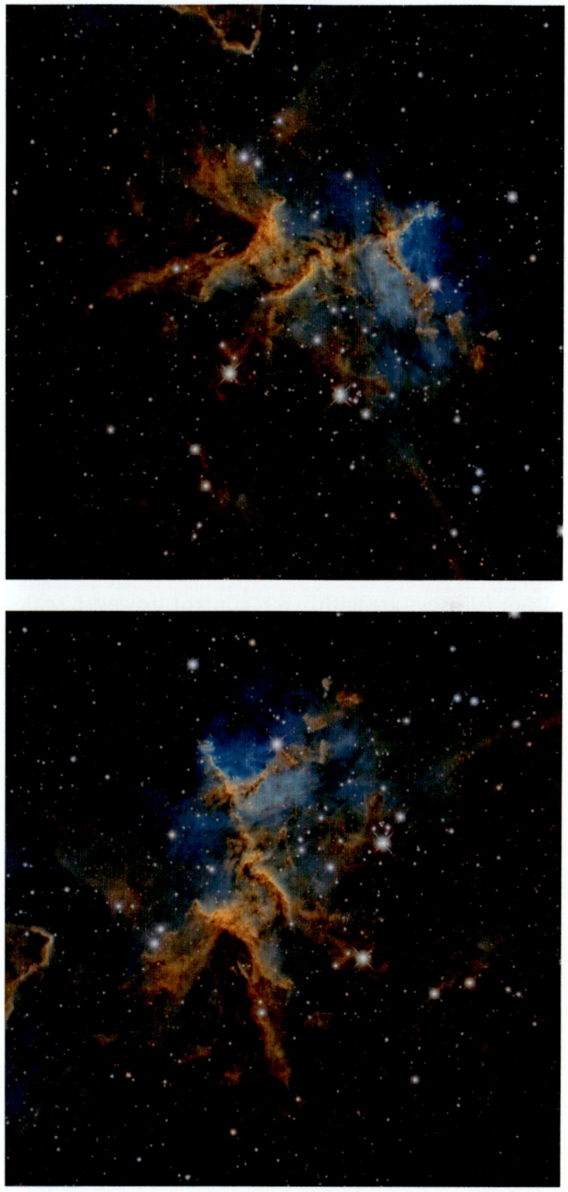

*fig.1 Something as simple as the image orientation changes the dynamic and reading of an image. I prefer the one at the bottom, it supports a notion of a rising column of hot dust.*

## Aesthetics

Photographers and artists often have an innate ability to compose images. The choices they make consider orientation, scale, framing, the position of the center of interest, balance and directing or confining the view. In this regard, astrophotography is no different from any other form of art and the guidelines are similar. The main difference between them is you have no control over an object's relative positioning and the images are two-dimensional in so much that everything is in focus with no foreground or background. Many excellent images follow well-established artistic guidelines but they do not have to. When I used to judge photographic competitions I would sometimes discover a compelling image deliberately broke the "rules." A memorable image will often generate a strong emotional response and formulaic adherence is less important.

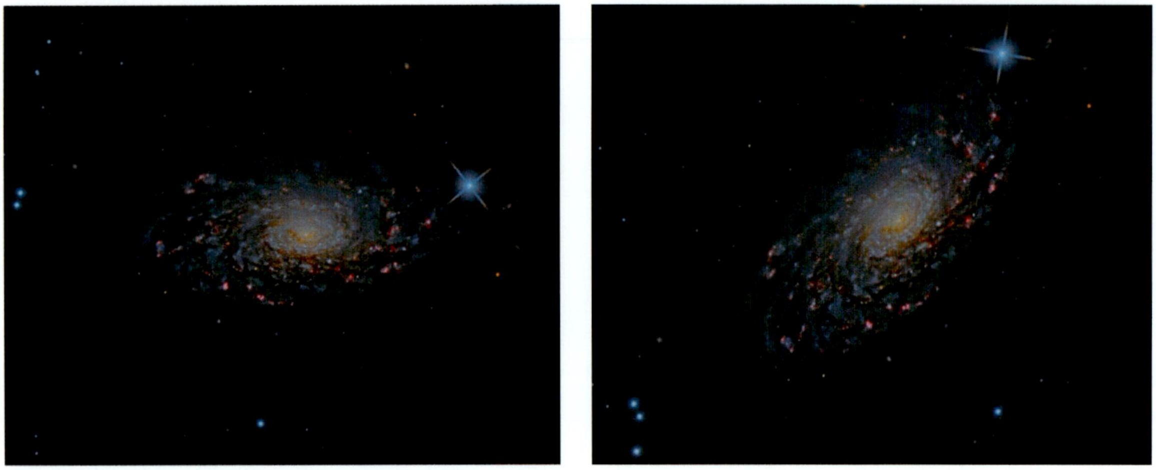

*fig.2 By tilting the galaxy at an angle, it appears more dynamic. Convention suggests a diagonal from bottom left to top right but it does not have to be that way. It helps to experiment first by trying out alternative framing on a downloaded image.*

Some things are within our control, however, scale, framing and orientation are something that should be considered before exposure. This is particularly difficult for the astrophotographer since a short exposure reveals little detail other than bright galaxy cores and stars. Most systems allow the camera to be rotated but, at the same time, one should consider whether the final image is better in a landscape or portrait format and heretically, more compelling reflected about a vertical or horizontal axis too. During the planning phase, I look up others' images on the Internet and use these processed images to consider my own framing, or I may use one of the better planetarium programs that show the boundaries of catalog objects. There are many guidelines in general photography, some of which are the rule of thirds, avoiding distractions on the image periphery and image dynamics.

For reasons that are unclear, placing an object of interest on the intersection of thirds has a pleasing effect, especially if there is some balance in the image to offset the main attraction. This does not always work for an image with a single object (a cluster or galaxy) and sometimes a square image, with a central object, is more powerful. The brain automatically seeks out bright areas and areas of high contrast. A distracting object near the edge of an image draws the eye away from the center of attention and can be particularly troublesome in deep-sky images. I sometimes crop the final image to exclude a particularly bright star on the periphery and to ensure the border does not bisect a small galaxy or bright object.

Image dynamics are another interesting area; some orientations and placements feel better than others. For instance, a landscape orientation is considered passive compared to a portrait orientation. There are fewer portrait-orientated astrophotographs. The few that exist convey power, of which the famous Hubble Space Telescope's vertical image of the "pillars of creation" is a memorable example. An object's orientation also generates different emotions and mirrors some portraiture tricks of the trade: if you consider two similar portraits; in one the eyes are level and in the other, the eyes are tilted, they provoke a different reaction. The angle and direction of the tilt also have a surprising effect. If an object has an axis, in so much that it is not an amorphous blob or perfectly symmetrical, tilting that feature enhances the dynamism. For example, try reversing an angle; it has a surprising effect. In the West, it is said, our brains "read" an image from left to right, like a book; and in addition, a swooping diagonal from the bottom left corner to the top right feels more natural than the opposite.

All these suggestions are purely subjective. The intent here is to make you aware of them and to improve an image through conscious decisions and experimentation. An experiment with a single familiar image in different crops and orientations is very informative. The difficult part is to realize this with the faint image during image capture; on more than one occasion I have spent too many hours acquiring an image, only to decide I wished I had framed it differently.

## What is Image Processing?

Image processing is everything that happens to your captured camera files after acquisition. This variously includes terms used in other texts such as

"Pre-Processing," "Calibration" and "Manipulation." This very broad definition divides into three main activities:

1. sorting and calibration
2. registration and stacking/averaging/integration
3. image manipulation

Calibration, sorting, registration and stacking are mechanistic in nature and are in effect an automated precursor to manipulation. Sorting removes the substandard exposures before image calibration or stacking and can be semi-automated based on a set of criteria or from a simple visual evaluation and rejection of poor exposures. In some references, these actions are referred to as pre-processing or even processing. The outcome is a color image file, or a set of monochrome ones, ready for image manipulation. Image manipulation (sometimes also called processing or post-processing) is itself a substantial activity to accomplish various aesthetic needs. These manipulation activities enhance the calibrated and stacked image. In no particular order they:

- remove unwanted background color and gradients
- enhance star shape and color
- repair cosmetic defects and improve composition
- reduce image noise
- sharpen and enhance structures
- enhance faint details and increase their contrast
- manage color hue and saturation

### Image Processing Software

Astrophotographers use an array of tools for image processing. Thankfully, there is a good deal of file compatibility and an image may use multiple applications before completion. Calibration and stacking are best achieved with specialist programs. Image manipulation software includes dedicated programs (Maxim DL, Nebulosity, AstroArt and PixInsight are prime examples) as well as general purpose editing programs such as Photoshop, Affinity Photo and GIMP. In addition, there are specialist utilities and plug-ins that stack images, remove stars, counter gradients and create star trails. The use of general purpose imaging programs divides the community; some purists believe that all processing should be done by mathematics and regard manual manipulations in Photoshop as "painting." Others do not care, as it is the final result that matters.

Astrophotography has very distinct requirements that are optimally met with specific imaging controls and although general imaging programs may deliver instant gratification, they require elaborate techniques to produce a high-quality image. Even so, there may still be a need for specialist processes and this is where dedicated software will help. These applications are daunting at first; the terminology is unfamiliar and there are numerous controls. It is not always obvious what to do, when and by how much. With care, however, these achieve high-quality images; reducing noise, improving color and refining detail.

Having said that, many astrophotographers use Photoshop almost exclusively for their manipulation and it is amazing to see how some have adapted its layer and blending modes to their needs. It can be a trap though; while many Photoshop users are confident with the basic controls, it is easy to underestimate the sheer complexity of the manipulations for an astrophotograph. In comparison, using a dedicated tool may appear foreign at first, but it very quickly becomes more efficient. Bit-depth is an issue; some conventional editors have restricted 32-bit file support while dedicated applications manipulate 32-bit or 64-bit images (for high-dynamic range images).

I currently use PixInsight for my image processing but, in the context of this book, it is a little intimidating and expensive. Working to a strict budget, the cost of renting Adobe's Creative Cloud products soon mount up too and I suggest trying Nebulosity and Affinity Photo for editing (or GIMP, which is entirely free) and DeepSkyStacker, ASTAP, SiriL or Nebulosity for image calibration and stacking. They do a similar job but check it supports your digital camera's RAW format.

### Image Processing Order (Workflow)

Throughout this book, the image processing steps (workflow) are laid out, in graphical form, for easy reference, like that in fig.3. Unlike conventional digital photography, astrophotography places high demands on image processing and each image often takes extensive manipulation to create the final image. This takes some time and, for more advanced deep-sky images, the processing follows multiple paths and it is not uncommon for it to take an entire day to complete. There is a general consensus on the initial processing steps, after which, the content of each image usually requires a mixture of generic and unique manipulations. After a while, one develops a sense of what might work, tries out an informed guess and then lets the results prove whether it is an improvement or not. Image processing requires patience and it takes many separate actions to improve an image, regardless of the application.

There are many alternative workflows and crucially, tools and processes may be applied globally or selectively. A selective application is perhaps one of the most significant differences between imaging applications and is covered in the advanced processing chapters. The reason it is important is that many manipulations, when applied to the entire image, fix some issues but at the same time create new ones (the extent of which is very dependent on the subject). In general, experimentation is the key to success and it is essential to save work at various stages of completion. Not all programs have a multiple undo feature and it may be necessary to back-track and try a different approach. This is where the more recent non-destructive photo editors are wonderful, as are adjustment/filter layers. PixInsight goes one step further, where actions are entirely separate and can be applied to other images.

There is one other thing to watch out for; over an extended period, it is easy to lose a sense of visual judgment and over-do things. It is not uncommon to return to an image processing project on the following day and realize it has become coarse and undo the last few actions. I save new file versions at a few key stages during image processing:

- stacked files before manipulation,
- after background, gradient and color balancing
- after the first major stretch
- luminance and RGB color files before combination
- 16-bit TIFF image version in Adobe 1998 color profile for printing
- 8-bit JPEG image in sRGB for Internet use

In addition, there are several guiding principles that will help with image processing, at all levels of sophistication:

- use a calibrated monitor
- it is much easier (and better) to process an image with a good signal to noise ratio
- poor acquisition technique is rarely recoverable
- each image requires a different approach, based on content, contrast and the exposure quality
- many small adjustments are better than a single big one
- some manipulations solve one problem and create another
- keep notes on the steps and settings
- if it looks right, then it probably is
- store partially processed images along the way
- do not be afraid to experiment

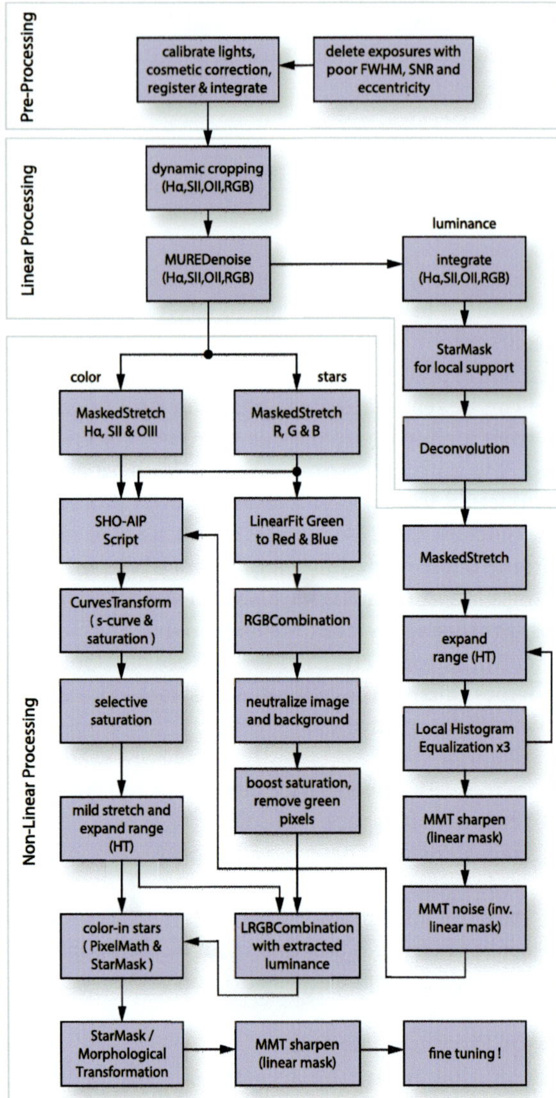


*fig.3 This is a complex example image processing workflow from* The Astrophotography Manual *for a nebula imaged through separate filters and using a dedicated astrophotography image editor. Simpler editors are not as adept at linear processing and the order may change.*

- small errors before image stretching become big errors afterward and are considerably more difficult to fix
- it often pays to be ruthless and throw away poor exposures (focus, tracking etc.)
- keep original exposures as they may have a second life as your processing talent improves
- different parts of the image will have different signal to noise ratio and will almost certainly benefit from selective processing with each step

- remember, there is more than one way to achieve a certain look
- consider if a particular manipulation is better applied globally or selectively, to the entire image, or just luminance or color information
- there are no rules!

As such, this first processing chapter explains the concepts and introduces global processing, in which most manipulations are applied to the entire image. The second and third image processing chapters discuss more complex techniques and trade-offs to improve deep-sky images and finally discuss selective and specialized processing.

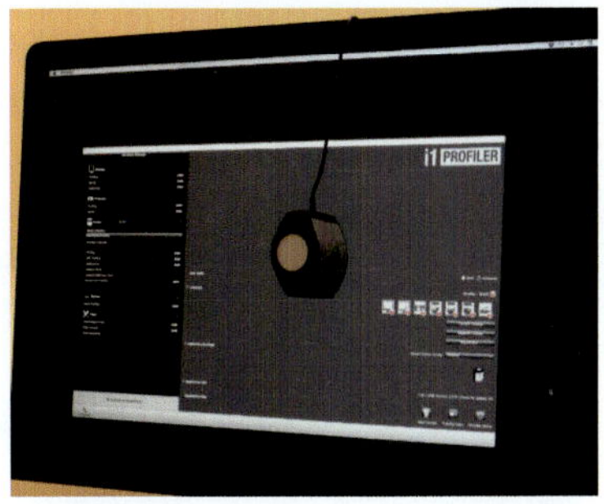


*fig.4 Image processing works best with a calibrated monitor. This not only ensures that the colors are accurate but crucially, the black point is too. Here, a USB-powered calibration device is centered over the screen, ready to read colored tiles to build up a monitor profile. Note too the grey desktop. Our color perception is influenced by our immediate surroundings. A neutral plain desktop is best.*

### There is Nothing There!

The first major challenge we meet is that most images appear to be featureless black before image processing (even after calibration) and require extensive stretching to reveal faint details (fig.5, fig.6). This manipulation not only boosts faint imagery but it also makes the image background (and noise) more noticeable. Getting the background level just right is tricky. The best way is to use a numerical readout to set the level but even knowing that, it is human nature to regard the computer monitor as a trustworthy output device. It rarely is, however, unless it has been accurately calibrated with a monitor calibration tool (like those from X-Rite, Gretag and Datacolor) so that its intensity and color of all tones and especially the darkest tones are accurately rendered (fig.4). These devices plug into a USB port and the accompanying utility program compares the color response from the display surface with a reference table. At the end of the measuring cycle, this software creates a device color profile for the Windows or Macintosh operating system (saved as .ICC or .ICM files). While not perfect (it never can be, as prints work by reflected light from CYMK inks and monitors by transmitted light from RGB LEDs) it dramatically improves the monitor's compliance to one of the various reproduction standards.

*fig.5 An image, as it comes off the camera, is less than exciting. There are often just a few white pixels corresponding to the brightest stars.*

I prefer to do my editing on a high-quality LCD monitor and with its desktop set to a neutral grey, so it does not affect my color perception. Its appearance does not appreciably change with viewing angle and the output is stable. Laptop monitors are a mixed bag. On some, the color and intensity vary with the precise viewing angle and these models are less suitable for editing purposes. One has to choose two key parameters for screen calibration, color temperature (white point) and intensity. I now set the intensity to 100 mcd and the white point to 5800 K, down from 120 and 6500 K, for better print/monitor correlation. When all is said and done, the resulting image file will be printed, displayed or shared on the Internet. Those receiving devices need to know the assumptions (or interpretation, if you like to think of it in that way) of the files' RGB values. These assumptions, which include gamma, white point and RGB gamut are collectively referred to as the color profile and the key information is normally incorporated into the PSD/TIFF/XISF/FITS or JPEG file header.

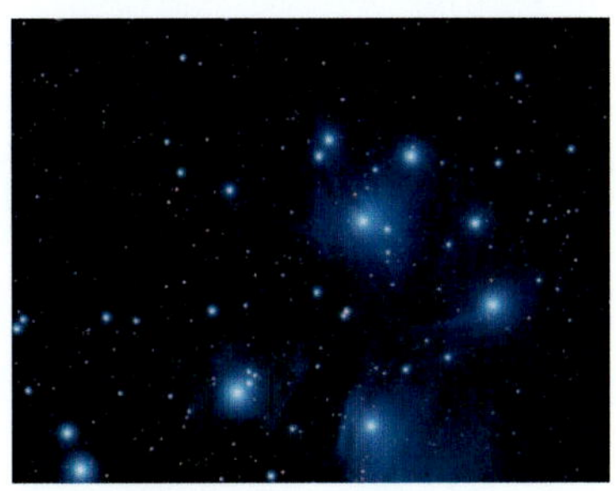

*fig.6 After stretching, however, the image in fig.4 shows more potential, the nebulosity is revealed and the fun begins.*

Color profile choice is an important consideration in the final image file. Files for printing often use a 16-bit AdobeRGB profile and those for Internet use, are converted to the consumer standard 8-bit sRGB profile. Even with a calibrated monitor, our brains play tricks on us and an image can change its apparent appearance depending on ambient lighting and our accommodation. This is particularly true for the extreme tones and (if

I remember) I use numbers (%K or pixel value) rather than judgment, to fine tune the final background dark tone level and peak star intensities. If it is too light, it has the potential to show up image noise.

Extensive manipulations pose several challenges. As mentioned earlier, the image stretching required to enhance the faint structures may, at the same time, ruin the brighter stars and features into blank white areas. To begin with, it is better to start out with global manipulations and apply modest stretching to sympathetic targets.

## Getting Started

While there are no rules and there are several paths to reach the same goal, image manipulations fall into several broad categories that are carried out in order (fig.7). The reason for this is that some manipulations are considerably easier earlier on; for instance, some manipulations work best on raw camera data or after basic color calibration. Image calibration and integration require raw camera data too, as do more advanced techniques such as deconvolution. Some noise reduction techniques and basic color calibration work best with linear data, before stretching. The workflow in fig.7 deviates from classical ones, because conventional image editing applications struggle with near-black linear images, which require a temporary screen stretch tool to show up the issues and tools that give numerical as well as visual feedback (for instance, shadow values, clipping and lurking subtle gradients).

### Image Preparation

To an extent, image preparation or pre-processing is not exciting and much of it can be automated. It should not be underestimated though since at the same time, it underpins the quality of the image. While each image may be processed with standardized settings, it is often the case that alternative settings may improve things slightly, principally around optimizing the signal to noise ratio and rejecting outlier pixels caused by singular events e.g. aircraft and satellite trails.
The principal steps are:

1. sorting and calibration
2. registration
3. stacking (integration)
4. cropping
5. noise reduction
6. deconvolution
7. simple color balance
8. background equalization

### Sorting and Calibration

These steps are already described in detail in the chapter *Improving Quality (part 2)*. Suffice to say that it is tempting to skip the sorting step. Looking forward, with the fewer frames typically acquired by portable setups, the impact of a single bad frame on the averaged stacked image is greater and statistically speaking, it is more difficult to reject the less obvious rogue pixels too with a smaller sample set.

### Registration and Stacking (also called Integration)

In its simplest form, stacking calculates the average (mean) value of the multiple image frames to create a less noisy image. Each of the images, will have slightly different framing, however, and before one can combine them, they require alignment or registration. Simple averaging reduces noise, but at the same time, it only reduces the impact of a singular issue on one of the averaged images.

For high-quality images there is a more sophisticated calibration process that allows one to remove the effect of singular events (for example, satellite trails, aircraft trails, meteors and cosmic rays) without increasing noise. This process evaluates aligned pixels from each image and, using statistics, automatically rejects pixels with very different values to the others. In this way, only the bad pixels are discarded and the rest of the image's pixels are used. This is just as well; I live on the approach to one of the London airports and about half of my exposures have problems.

I have mentioned cosmic ray hits a few times without explaining what they are; the Earth is constantly being bombarded by high-energy radiation from outer space. When this hits the Earth's atmosphere, it triggers a shower of high-energy particles. During the prolonged exposures in astrophotography, it is highly probable that a few will hit the sensor, causing multiple electron generation in a few pixels. These appear as little light squiggles but, since they never occur in the same place on each image, the algorithms built into the advanced calibration tools effectively make these pixels stand out and ultimately ignored in the final stacked image.

### Cropping

It is rare to show an image at full crop. The reason for this is that the separate image frames rarely overlap perfectly. The borders then have fewer pixels to average causing them to have high noise or missing data. In the case of dedicated editors, apart from tidying up the image, cropping is useful for some manipulations that

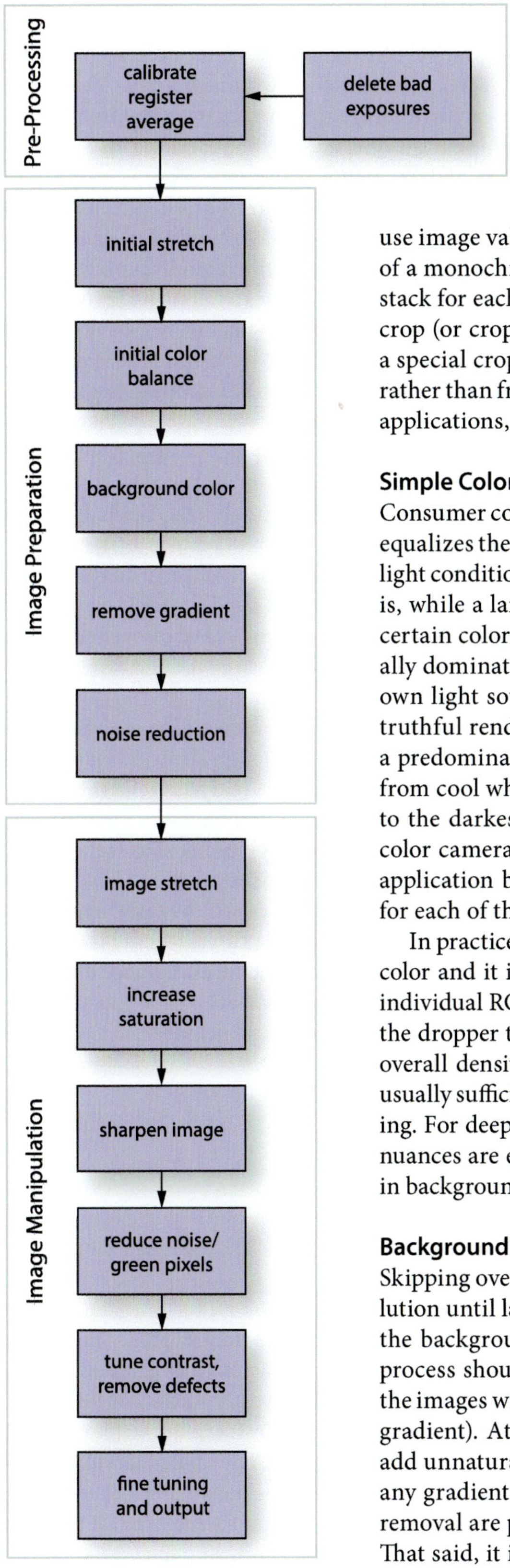


*fig.7 A simplified workflow for a color camera, assuming global processing and a mixture of dedicated utilities and conventional image processing software. Here the color image is processed in its entirety. More advanced processing extracts and manipulates the color and luminance information in separate workstreams and selectively stretches, sharpens and reduces noise.*

use image values to determine their adjustment parameters. In the case of a monochrome camera with separate filters, one will have an image stack for each filter. Here, each image stack requires an identical image crop (or crop the combined RGB image). This is normally achieved by a special cropping tool in the processing software that has a numerical rather than freehand setting. Alternatively, and when using conventional applications, crop the near complete image later on.

### Simple Color Balance

Consumer color cameras have an automatic white-balance feature which equalizes the RGB values for the image highlights. It works well in bright light conditions but for astrophotography, it falls short. The specific issue is, while a landscape or studio image is lit by a single light source of a certain color temperature, in astrophotography the background is usually dominated by orange colored light pollution but the stars are their own light source with a very different bias. The goal is not to achieve truthful rendering, but to eliminate the light pollution effect and have a predominantly dark, neutral background with a range of star colors from cool white to red. This requires two color range adjustments, one to the darkest tones and another to the lightest. With a conventional color camera, this is most easily done in a conventional photo editing application by individually adjusting the highlight and shadow levels for each of the RGB channels.

In practice, a fixed daylight balance will come close to the desired star color and it is the background levels that requires adjustment with the individual RGB shadow sliders. Our eyes may fool us and it is best to use the dropper tool and its readout to set a neutral value (R=G=B) with an overall density resembling dark grey at about 90% K. This approach is usually sufficient for basic vistas that do not require radical shadow boosting. For deep-sky work, accurate color balance is more critical as subtle nuances are exaggerated by the any stretching process, as are variations in background color from subtle sky gradients.

### Background Equalization

Skipping over the more complex topics of noise reduction and deconvolution until later, the last essential step in image preparation is to make the background even as well as neutral. While the image calibration process should remove the effect of optical vignetting and dust spots, the images will still show a variation in background illumination (or sky gradient). At the same time, light pollution and color calibration will add unnatural color to the gradient. When these images are stretched, any gradient is made more obvious. While a color cast and a gradient removal are possible in a stretched image, it is easier to fix beforehand. That said, it is difficult to do this with a very dark image in a standard

*fig.8 In the presence of modest light pollution, an image stretch of any image off a color camera will likely reveal a dirty orange background.*

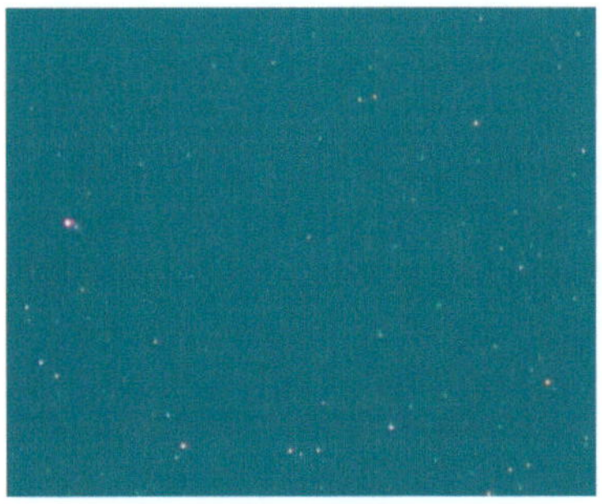

*fig.9 A light-pollution filter will remove most of the orange glow. The subtraction of the yellow light also benefits image (shot) noise. The outcome, however, when stretched, may now emphasize the slight blue cast of the filter!*

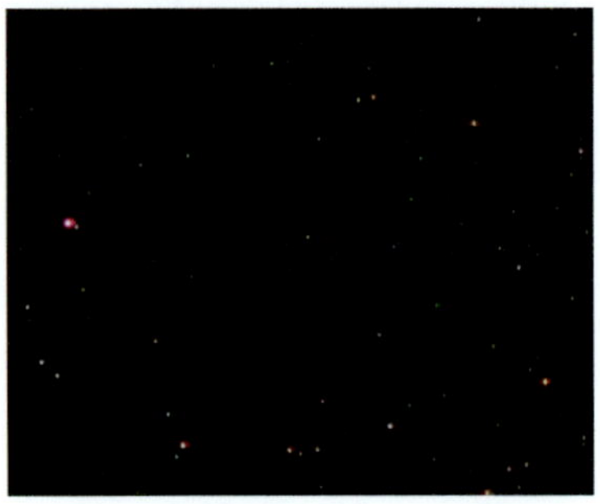

*fig.10 All is not lost; after background subtraction, as described in the text, the background becomes neutral dark grey (not black) and the stars have a range of colors.*

editing application that lacks a screen-only stretch that reveals issues. One workaround is to iteratively fix these issues after each stretch.

The various processes require sampling an image across the frame and creating a smooth gradient image to subtract from the image. It is possible to achieve with conventional editing applications and is made considerably easier by the $50 GradientXTerminator Photoshop plug-in (that also works with Affinity Photo). Specialized gradient removal on high-bit depth images is a standard feature of dedicated astro-editing applications. For instance, a manual approach (which removes background gradients and color cast at the same time) creates a duplicate of the image and removes the main subject with the healing brush before removing all the stars with a generous application of the Dust and Scratches filter. This creates a blurred background image. This blurred image is applied to the original using the "Subtract" blending mode at 100% opacity and an offset around 35, to give the background a non-zero value. An example is shown in fig.9 and fig.10.

This general approach works well with simple star fields but may eliminate the interesting gradations in a Milky Way vista or large scale nebulosity. This is because the manual process uses the entire image to infer a background level whereas the more advanced applications sample background levels selectively, avoiding interesting features and colored elements.

## Image Manipulation

Image preparation is the bedrock of the more entertaining manipulation processes in which the image comes to life. Although fairly mechanistic in nature, without it, image manipulation can be a real hassle. This is where the fun starts and the key to this fun is to experiment and then experiment some more. While it is easy to copy dialog boxes into these pages, with settings that work for the example image, they will be of little value to the reader, another one of my own images or even the same image in a year's time. After processing dozens of images I can safely say that while the principles and workflow are generally consistent, each image demands respect and empathy that encourages a fresh look at the manipulations to achieve the result you want. These new processes include:

- stretching
- noise reduction and sharpening
- color saturation
- enhancing structures
- image repairs
- fine-tuning

### Stretching

This is the magic moment when the image comes to life. This is especially true of the deep-sky images that feature dim nebulae or galaxies which, up to this point, are virtually invisible. We have all used a curves tool of some kind to change the relative brightness of the image tones for pictorial effect. Stretching in astrophotography, however, is at another level. To give some idea of the degree of manipulation, a mid-tone in one bright nebula has, before stretching, a 16-bit value of 640. After stretching, it is about 20,000, about 5 stops brighter. At the highlight end, the values are virtually unchanged, to prevent bright stars from clipping to pure white blobs. The basic ways of stretching an image in a conventional photo editing suite use the levels and curves tool and lately, HDR toning. Dedicated editing applications have more advanced methods based on image scale and local contrast to boost faint details without blowing out the brighter features.

For high-quality images, it is useful to have a 32-bit or higher image prior to stretching. (Keep in mind that Photoshop's 16-bit mode appears to be only 15-bit and even with a 14-bit RAW file, averaging just 8 images effectively creates a 17-bit result.)

### Noise Reduction and Sharpening

These two processes are two sides of the same coin. Sharpening processes in astrophotography differ from those in conventional photography. Astro images have no lines to resolve and there are few sharp edges. The common tools, such as Unsharp Mask, are best avoided and it is better to use tools that gently boost local contrast for a natural look. Sharpening accentuates image noise and, as such, with the simpler global processing that one starts off with, it is best to be less aggressive. If the sharpening is carried out on an adjustment layer, a layer mask covering the brighter areas allows a stronger application or preferably, several lesser applications.

Noise reduction, on the other hand, is best applied to areas that need it most; the darker areas. Its effect is to reduce local contrast and color variation. Again, selective application will prevent noise reduction softening interesting details. In both processes, stars get in the way. Sharpening stars causes them to become brighter and potentially clip. Reducing noise makes them diffuse blobs as if you had imaged through a toy telescope. The more advanced tools and techniques select stars in an image using their bright, small-scale properties and effectively exclude them from any noise reduction.

### Color Saturation

Image saturation is usually required to enhance the subtle color of a dim nebula and surrounding stars. A simple color saturation tool may not be effective and it has to take other image characteristics into account. For instance, as a pixel's brightness increases, the color saturation reduces. At the limit, it clips the R, G & B values and the result is a neutral white. Once a pixel has equal RGB values, it can only be white, grey or black. Color saturation application is most effective on mid-tones where there is headroom to differentiate the RGB values. At the same time, the user must be mindful that image stretching and sharpening increase the brightness of interesting features, with potential to clip shadows and highlights. In my deep-sky work, I set my exposures to clip the cores of only a few of the brightest stars and then give some headroom in processing, before sharpening using the levels (or equivalent) tool to shift the peak values down to a value of about 80%.

There are numerous methods to improve color saturation using layers and blending modes. Most work by blending a slightly blurred boosted color image with itself, using the color blending mode. The blurring is essential, as it reduces the risk of introducing color noise into the image. The principal differences between the methods are how they achieve the boosted color image. There are several possibilities: combine two images with the soft light blending mode, use the Vibrance tool or exploit the LAB color mode and apply a subtle S-curve to the a and b color-difference channels. Keeping in mind that the processes that sharpen and boost details reduce saturation at the same time, it is best to boost saturation a little at various points in the processing workflow.

### Enhancing Structures

Faint structures are typically enhanced by changing local contrast in the region. The HDR Toning tool in Photoshop and Affinity Photo produces some dramatic results and usefully, works on 32-bit images. Experimentation is key, typically using the Local Adaptation method and small adjustments to the edge glow settings to define the image scales that benefit from the contrast enhancement. Another common tool that enhances structures in general editing applications is the high pass filter. When this is applied to a duplicate background layer, it produces a faint grey image. This image picks out structures in the image, defined by the radius setting. The magic occurs when the blending mode is changed to "Overlay" or for a more subtle effect, "Soft Light." Both have the potential to make stars white and the more advanced user applies this with a mask, to obscure the stars (or remove the stars) from the image, before application.

### Cosmetic Repairs

This hodgepodge of adjustments are used as required and include a number of fixes for improving star appearance and pixel defects. We instinctively know that stars should be circular colored dots and in practice, they are not forgiving of optical defects or poor tracking. After taking an image of a star field, with what I considered to be a premium digital camera lens, I never looked upon it again as favorably. Conventional, plug-ins and specialist applications offer tools to reduce color fringing and there are a few tricks one can do to reshape stars too. In the case of color fringing on stars, if the color is unique, it can be selected with the Color Range tool and neutralized by adding the opposite color. In more complex images, the chromatic aberration removal adjustments (often found in the lens correction filter menu) may work better.

Pixel defects that escape the calibration and integration processes may be fixed using the clone tool or a dose of the Dust and Scratches filter, with the scale turned down. Photoshop and Affinity Photo also have content-aware cloning tools. It can be as easy as selecting the general area and choosing the Content-Aware option (Inpainting) tool. Smaller blemishes, especially round ones, are fixed with the spot healing brush with the "proximity match" option selected before clicking on the problem area.

Another common problem is the presence of green pixels. Green does not exist as a natural color in astrophotography (with the exception of comets and a very few bright nebulae). Our eyes are sensitive to green light and it is best to de-saturate these rogue pixels or replace them with an average background level. A combination of a color range selection tool and the curves tool, found in conventional imaging applications, is usually sufficient to select a range of green pixels and neutralize them. Note, we are not reducing all green pixel values in the RGB file, only those pixels that are predominantly green. The dedicated editors have a specific tool to replace green pixels with the average background value.

Accurate tracking is a common issue with lightweight equipment, as it is more likely to be operating with less-than-perfect polar alignment, flexure and stability. If the stars are slightly elongated, by up to a few pixels, it may be possible to use a simple technique to improve their appearance. This simply blends a duplicate image layer with the original, using the Darken blending mode. The correct appearance is obtained by moving the top layer one pixel at a time in each direction. This is quite effective when applied to star fields but may degrade the appearance of galaxies and nebulae. A more sophisticated method rotates the image so that the elongation is parallel to one axis and selects a few bright stars with the color range tool to select them all. The selection is enlarged and feathered by a few pixels using the "Refine Edge" controls and the layer blending mode set to "Darken" as before. The layer is offset a few pixels this time using the "Offset" filter until the desired result is achieved. The layers are then flattened and the image is de-rotated to its original orientation.

### Fine Tuning

As with any image, after the general processing, there are always a number of things to tidy up. This may be cropping off a bright star near the edge of an image, altering the aspect ratio to suit a particular purpose and fine-tuning the color balance, mid-band contrast and shadow levels. Images for printing are best left in a high-bit mode and with a color profile like Adobe RGB. Images for the Internet are best down-converted to an 8-bit RGB JPEG file and converted to the sRGB color profile. This is the default color profile for many consumer devices and gives a better chance of the right public perception. The biggest issue surrounding image distribution is one of viewing on an uncalibrated, overly-bright monitor. This will make what appear to be perfect dark backgrounds appear grey and noisy as our perception of small tonal variations increases with brightness. One cannot prevent a viewer using an overly bright monitor, but you can, temporarily, check for potential lurking issues. For printing purposes, even though my monitor and A2 inkjet printer are calibrated, my prints always appear a little darker and with less impact. Color matching between different device types is a complex subject but suffice to say that I find a better perceptual match is obtained by applying a small, temporary gamma boost (using the mid-tone slider in the "Levels" tool) to the image, prior to printing.

### Processing Software and Utilities.

In essence, there are two main approaches to image processing software. For those who have experience of image processing with Photoshop and equivalent packages, there is an understandable reluctance to abandon familiar territory but, at the same time, a recognition that these applications are not capable of the specialist tasks at the beginning of the processing workflow, especially those that manipulate the raw image before stretching:

1. sorting (essential)
2. calibration (essential)
3. registration (essential)
4. stacking (essential)
5. deconvolution (optional)
6. initial non-linear stretch (essential)

After these activities, the image is in a state that is recognizable and about 80% there, from which 16-bit image processing with traditional tools takes over. Other astrophotographers, especially those pushing the boundaries of feasible manipulation, prefer dedicated applications that take the initial exposures through to the final result. The table in fig.11 shows this distinct division between those utilities that supplement Photoshop and Affinity Photo and those applications that are able to fully process and in some cases, control equipment and acquire an image too. Many of the utilities are provided free of charge by generous individuals. Several of them, in turn, rely on free utilities such as RAW converters. Checking the update history of each, I noted that some of the RAW conversion utilities had not been updated for several years and those with the latest digital cameras will find their RAW files are incompatible (for instance CR3 files from Canon's M-series). An increasing number of utilities are also written in JavaScript that additionally allows them to run on a Linux system. Some utilities do not manipulate 32-bit files. I suspect these are targeted at DSLR users, where there is little point using a 32-bit file space to integrate 12- or 14-bit camera RAW files.

In general, the dedicated applications anticipate the full range of general purpose and dedicated cameras and process 16-bit depth images and more often than not default to 32-bit or higher, for composite images with a high dynamic range. This is essential for the initial image preparation and later manipulation tasks, especially with dim subjects and high exposure counts. They come at a price though and one pragmatic approach is to start off with the free utilities and an existing photo editor. When your budget allows, a dedicated application should be part of the upgrade path, to match the performance of the imaging system.

Of the utilities, I defaulted to the popular DeepSkyStacker (DSS). The present version has just switched from DCRAW to LibRaw to handle the latest digital camera RAW files. While it can read and convert my Canon 60Da files, it is confused with those from a Fuji X-trans sensor. In some cases, the workaround for a new RAW format is to use an interim step to convert them to Digital NeGatives (DNG) with the free Adobe converter utility. The Fuji, however, has a unique mosaic pattern and it does not comply with convention. In comparison, SiriL correctly processes Fuji RAF RAW files. It appears to be under rapid development and, according to the website, this started life as an adaptation of the well-regarded IRIS software for the Linux platform. In doing so, it is equally compatible with Mac and Windows platforms. As DSS is well established and there are numerous available instructions and tutorials on the Internet, I thought it interesting to use SiriL for some of the assignments. It has some shortcomings and hopefully they will be addressed in future updates.

| Title | Price ($) | PC Mac | Calibrate | Stack | Process | Notes |
|---|---|---|---|---|---|---|
| **Full Processing** | multi-purpose packages | | | | | |
| PixInsight | 250 | PC/Mac | yes | yes | yes | full-on |
| AstroArt | 165 | PC | yes | yes | yes | acquisition/ processing |
| Prism Advanced | 500 | PC | yes | yes | some | + acquisition |
| Images Plus | 245 | PC | yes | yes | yes | some acquisition |
| Nebulosity | 95 | PC/Mac | yes | yes | yes | keeps things simple |
| IRIS | free | PC | yes | yes | yes | last update 2008 |
| **Utilities** | stand alone and plugin utilities | | | | | |
| ASTAP | free | PC/Mac | yes | yes | yes | + plate solver |
| SiriL | free | PC/Mac | yes | yes | some | active development |
| StarStax | free | PC/Mac | yes | yes | | star trails |
| StarTrails | free | PC | yes | yes | | star trails |
| StarMax | free | PC | yes | yes | | star trails |
| Regim | free | PC/Mac | yes | yes | | augments Photoshop |
| DeepSkyStacker | free | PC | yes | yes | some | augments Photoshop |
| FITS liberator | free | PC/Mac | | | yes | |
| GradientXterminator | 50 | PC/Mac | | | | gradient removal plugin |
| Background Subtraction Toolkit | free | PC | | | | gradient removal |

*fig.11 In addition to the conventional image processing applications there are a number of specialized utilities, to make up for their shortcomings, as well a range of full-feature astrophotography applications too.*

### Detail Instructions

Image processing is a complex subject, both for conventional imaging and astrophotography. There are many books on Photoshop manipulations including some specific to astrophotography. The various specialist astro-image processing applications have similar tools but they work in subtly different ways. The manipulations in astrophotography are more extreme and each image is different. While it is possible to create a step-by-step process guide, it is counter-productive. The temptation would be to use the same tool settings on your own images, rather than understand the principles and experiment for yourself. There are no shortcuts and even with advanced tools and some experience, I may still progress down a processing path, only to have to undo several hours' work and try a different route. Over time you develop a feel for things; for what might work and what might not and it is a danger to become too prescriptive, as it does not encourage the practitioner to develop their skills. Having said all that, I do highlight different specific processing steps in the practical chapters, to overcome interesting challenges.

# Pleiades Loose Cluster (M45)

*Easy to find and difficult to do well, this cluster is a magnet for astrophotographers.*

***Equipment:***
*Canon EOS 60Da, PoleMaster*
*73mm f/5.9 refractor (430-mm $f_L$)*
*field flattener*
*Fornax Lightrack II, counterweight kit*
*Mosquito repellent*

***Software:***
*Astro Photography Tool (Windows)*
*SiriL (or ASTAP) pre-processing utility (OSX or Windows)*
*Affinity Photo (OSX or Windows)*

***Exposure:***
*10 x 10, 20 x 120 and 20 x 600 seconds at ISO 200*
*IDAS light-pollution filter (mounted in camera throat)*

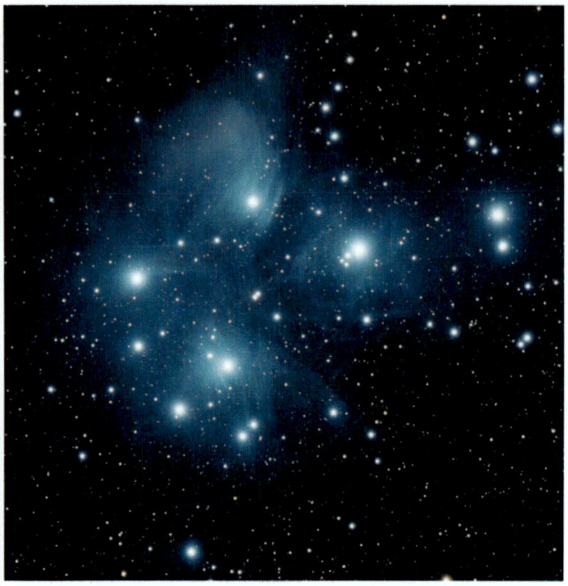

The Pleiades loose cluster is a striking object in the night sky. It is the most distinct cluster of bright stars that are visible to the naked eye. Alternatively named the Seven Sisters and more recently, the Subaru, images of this frequent ancient documents and earlier still, cave paintings from about 17,000 years ago. This then is an ideal early target, as it is easily located without the need for a planetarium or robotic mount and the bright object responds well to shorter exposures.

The hot stars of M45 hold a surprise; they are surrounded by a blue reflection nebula, which becomes more apparent when the exposure depth is increased and careful stretching boosts the background levels during processing.

## Equipment Setup

This is our first image taken with a refractor. The imaging camera was an EOS 60Da fitted with a light-pollution filter in the camera body. The Fornax mount has an imaging duration of about 90 minutes after which it has to be reset. To re-establish the alignment, I used my red dot finder trick, mounted to the hotshoe on a small ball and socket joint. The long focal length demands a little more care and attention in regard to the polar alignment and stability of the mounting system. With shorter focal lengths, one can place the tripod on virtually any surface and the effects of movement or flexure will be mostly unnoticed. With a focal length of 430 mm and an image scale of 2 arc seconds/pixel, this setup is more sensitive to drift or flexure. In this case, the Fornax mount was bolted to a sturdy aluminum tripod whose spiked feet were located into three metal rods buried in my backyard (the details of which are set out in my other books and the website). Prior to this session I had made some tests of the Fornax unit and measured its tracking accuracy. With a measured periodic error of less than 1 arc second, I decided to make this image without the assistance of an autoguider system. This direct drive system outperforms a worm drive in regard to periodic error but as the weight of the imaging equipment increases, it is necessary to balance the telescope and camera with a counterweight system. This reduces the strain on the mechanism and any propensity to slip. The counterweight system is fitted with a Vixen clamp, which works well with the William Optics telescope. Using the techniques from earlier chapters, I balanced the camera and mounted the assembly to the Fornax unit. With the assembly in place, I carefully polar aligned, first with a fully corrected EQ5 polar scope fitted to the articulated arm and then with the

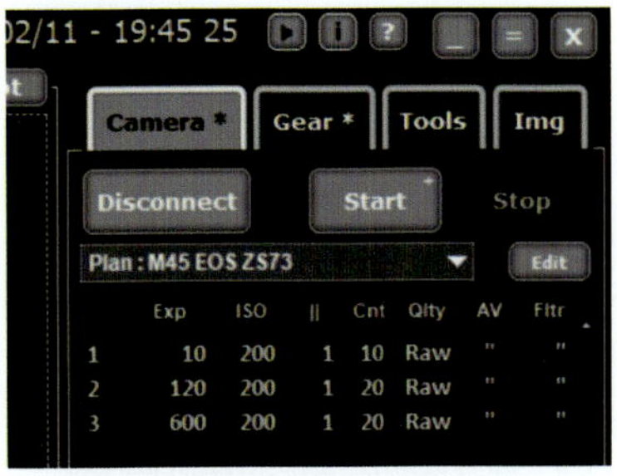


*fig.1 A section of APT showing the exposure plan, comprising of three durations, to try and capture the entire dynamic range of the subject. In the end, I only used the long 600-second exposures.*

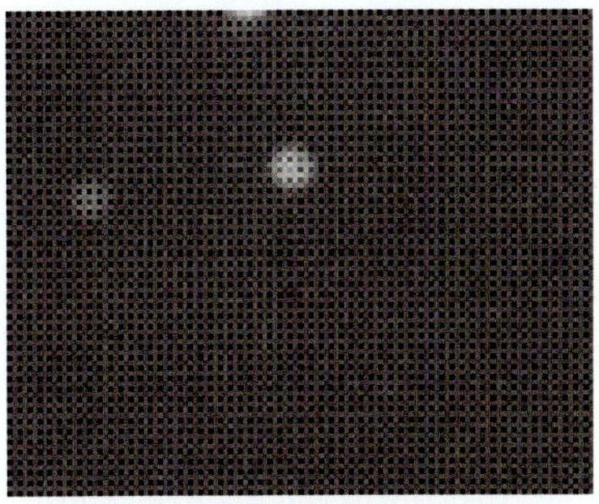

*fig.2 Up close, a FITS file, generated from a color camera, is a mosaic of different intensity pixels, corresponding to the differing filtered intensities.*

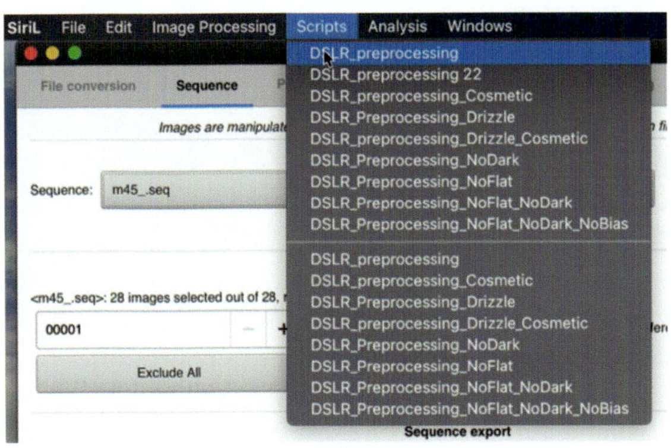


*fig.3 SiriL has a number of prepared scripts for image pre-processing that work on groups of like file types in a defined folder structure. There are downloadable scripts from their website that can process star trails and other basic processing functions.*

QHY PoleMaster accessory. In my setup, the two are very well matched and the slight advantage of the PoleMaster is useful for unguided operation for exposures over 5-minutes duration.

## Acquisition

The difference in intensity between the stars and nebulosity is extreme. I knew from prior experience that to detect the faint nebulosity requires a long exposure but, at the same time, this would cause the brightest stars to clip by some margin. Hedging my bets, I took three groups of exposures at 10 seconds, 120 seconds and 600 seconds, with the thought of blending them together at a later stage. This idea uses the longer exposures for the background and progressively substitutes in the others for the bright star cores. In this case, although I could have easily taken the images without a PC by using a handheld intervalometer, I decided to make the first use of the exposure planning feature of Astro Photography Tool. This application has many more advanced features, such as autofocus and centering that are more essential for more demanding subjects. In this case, I focused manually; with a rough focus, I framed up M45 and rotated the camera (using the telescope's rotation feature) to get the required composition. I then pointed the telescope to nearby Capella and used a Bahtinov mask to set the focus more accurately on this bright star, reducing the chance of focus errors creeping in from manhandling the rotation control. I then swung the mount to its start point, set up the composition as before and aligned the red dot finder to Capella as a reference. I chose a night when M45 was high in the sky and there was no moon during the longer exposures, improving my chances of picking up the faint blue nebulosity. Even so, I knew from experience that processing this image would not be easy.

## Processing

This is the first occasion where we are processing a deep-sky image and trying to reveal faint details from the gloom. There are several compounded factors that make this particularly challenging. First, the exposure is modest compared to the deep-sky images that grace websites. This ultimately limits the degree of stretching that the image can endure before the background noise becomes objectionable, especially with my local light pollution. We are in new territory using new processes to align and stack the images. Processing camera RAW files throw some unique challenges in our path that dedicated camera files neatly avoid. For instance, just as with traditional image processing, any RAW file requires conversion into a different format for editing and, contrary to traditional image processing, the file has to remain linear.

This last point confuses many newcomers. The image preview on a digital camera is deliberated boosted using an image stretch, called a gamma correction, to match our perception of the subject. Similarly, when a RAW file is viewed in a traditional image editor, it does the same thing and once it is converted to an editing format like JPEG, TIFF and PSD, the stretch is permanent and the image is said to be no longer linear. In astrophotography, we stretch files a lot further than in conventional imaging. More importantly, there are some initial image processing steps that require a linear image, including image calibration and stacking. The confusion arises when one views the partly processed image in one of these applications. Whereas the camera LCD showed a clear image, the preview looks to be entirely black, possibly with a few lighter dots, corresponding to bright stars. It is not uncommon to find forum posts from concerned users, claiming their images have disappeared after stacking. Two potential causes are the missing gamma correction mentioned earlier and the way in which the image bit-depth is translated into a bigger bit-depth file in which some applications do not appear to scale up a stack of 16-bit files to 32-bit, causing peak highlight values of 65,000 in a possible 4,300,000. The image is still there, it is just very dark, promise!

There is a further complication that requires a little up-front explanation. The camera RAW files originate from a sensor that just measures photons. Its individual photosites are filtered with a red, green and blue mosaic filter but until it is decoded (de-mosaiced/de-Bayered), the image is a strange patchwork (fig.2). The de-mosaic process combines a group of neighboring pixels to form a single colored pixel. In essence, the color information is drawn from four pixels and the intensities too. At the same time, any sensor defects and anomalies are at the photosite level and eliminating them through calibration has to be done before the image is converted to color. At the same time, aligning or registering images causes minute image shifts before the stacking process. These images have to be colored for that purpose, otherwise, it is likely that photosites with different filtering are combined, with ugly consequences. In summary, the first steps of image processing for a DSLR or mirror-less camera follow a general pattern:

1. conversion of RAW files to FITS/TIFF
2. (calibration, explained in next section)
3. de-mosaic FITS/TIFF
4. image alignment (registering)
5. stacking
6. …

The various image processing applications differ in their ease of use. DeepSkyStacker is highly automated and the pre-processing steps happen behind the scenes, with a few pop-up opportunities to customize the operation. SiriL and Nebulosity have a more hands-on approach, requiring the user to process sequentially, though the latest versions of SiriL now include a script option that automates the pre-processing steps (fig.3). At the time of writing this chapter, DeepSkyStacker relied upon a free RAW converter utility, DCRAW which did not support the RAW files from my more recent digital cameras. (At the time of the final edit, this limitation has gone, as DSS has changed over to using LibRAW, which is kept up to date.) For this assignment, however, I used SiriL, a relative newcomer that additionally offers versions for Linux, Mac OS and Windows platforms. It uses different RAW conversion tools and although this software is new to many, it additionally offers some core image-manipulation tools too.

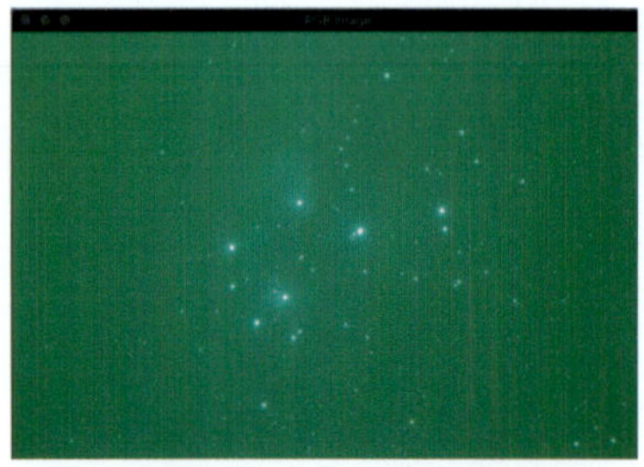

*fig.4 After registering and stacking the images, an auto-stretch of the image produces an unpromising green image in SiriL with very little detail.*

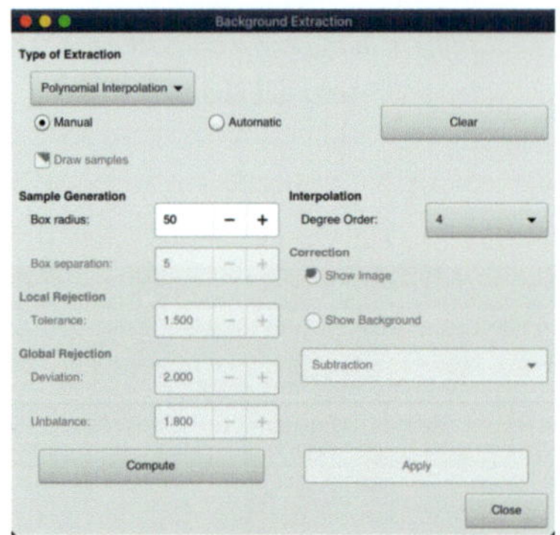


*fig.5 To remove the obvious color issue in fig.4, the color calibration tool is used to remove the majority of the color cast by sampling a blank area of sky. It has a further control to neutralize the highlight colors but, in this case, it was not necessary.*

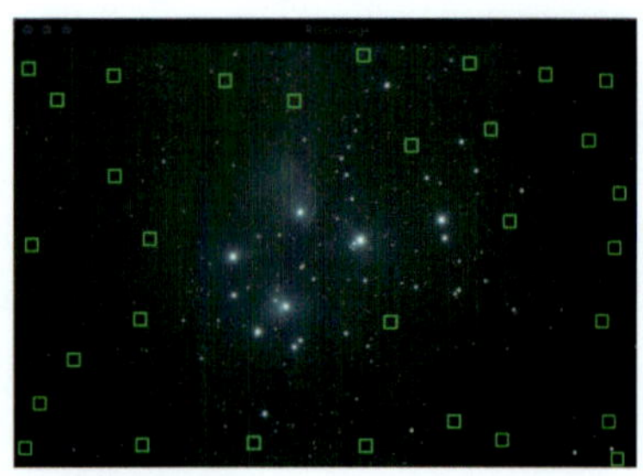

*fig.6 Following color calibration, the image is starting to take shape. The uneven background is sampled by clicking areas of plain background (without nebulosity) and running the Background Extraction tool in SiriLto subtract the background from the image.*

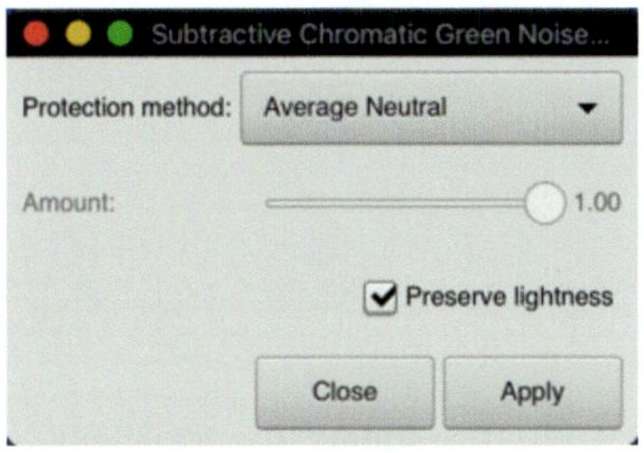


*fig.7 Green pixels rarely occur naturally in astrophotography. SiriL has a tool do replace these with neutral colors, producing the result in fig.8 (after a screen-stretch).*

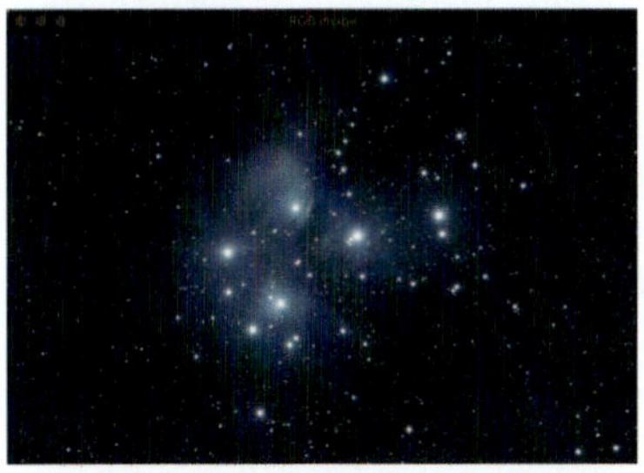

*fig.8 After color calibration, background extraction and neutralized green pixels, the image is now showing obvious nebulosity. The background is still too light and the color saturation is a little weak.*

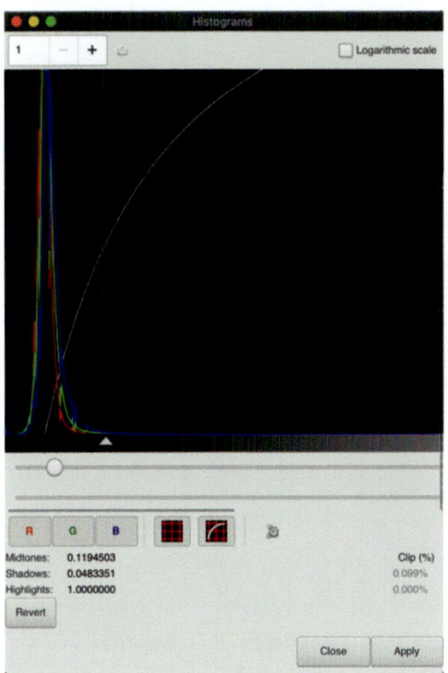


*fig.9 The Histograms tool operates like a Levels tool and is the main device to stretch an image. It takes several incremental stretches to achieve the desired result.*

Before processing, each of the light frames was checked for exposure and focus issues. A few were discarded over a period during which a vapor trail drifted past the target and reflected moonlight, using SiriL and the script labeled "DSLR_Preprocessing_Noflat_NoDark_NoBias" from the drop-down menu. This script simply runs through the list above. After registration, it selectively averages the registered files, weighting them by the average background level and rejecting pixels outside $\pm3\sigma$. Even so, the initial result was disappointing after an automatic stretch, with three horizontal parallel lines running across the frame. I initially thought this may be a sensor issue, but, when I went through the individual calibrated files, I realized it was the faint trace of an aircraft trail on one frame that had not been completely removed during the rejection and stacking process. I performed the stacking manually and the lines persisted with a $\pm2\sigma$ threshold. In my normal deep-sky work, I stack many exposures and the statistical exclusion of rogue pixels is almost perfect. The valuable lesson shared here is that the faint margins of the aircraft trail are more difficult to discriminate from the generally higher noise levels that occur when one only compares a few uncalibrated frames. After removing that frame and stacking the remainder the initial result (after a temporary stretch) is shown in fig.4.

This screen-only stretch emphasizes all the imperfections, including color balance, vignetting and noise. While not pretty, it does indicate what we need to do to clean up the image. The next step removes the color cast and reduces the variability of the background. The green cast is not unusual as there has been no attempt to calibrate the individual R, G and B channels and probably not helped by the light-pollution filter in the camera throat. This has to be done with a little care; the sky is rarely black, not even dark grey and more often than not it is a very dark red color, especially in the vicinity of emission nebulosity. In the case of this reflection nebula, it is a dark blue color; the interstellar dust is simply reflecting the light from the hot blue stars in the cluster.

In SiriL, it requires two further steps to achieve a recognizable image for processing; Color Calibration and Background Extraction. In the first case, a small area of blank sky is selected (without detectable nebulosity) and the Color Calibration tool is used to remove the green color cast (fig.5 and fig.6). Following that, the Background Extraction tool is opened and after carefully clicking on background areas away from nebulosity and stars, the background level is computed and then applied (subtracted) from the image. The image still has some green tinges, which are removed by the Subtractive Chromatic Green Noise tool (fig.7). The auto-stretched screen image in fig.8 is looking more reasonable and after several applications of the Histograms tool, carefully changing the shadow and mid-band sliders, the result is showing a darker background and obvious nebulosity. The saturation of the blue nebulosity and red stars were next selectively increased with SiriL's Color Saturation tool. After saving the file as a 16-bit TIFF file, the final fine-tuning and cropping were accomplished in Affinity Photo, including color profile tagging and mild noise reduction.

In addition to DSS and SiriL there is another free calibration and stacking utility, ASTAP. It is available for PC, Mac and Linux operating systems. It is a relative newcomer and is developing fast.

# Improving Quality (part 2)

*Longer and multiple exposures demand full image calibration and integration to improve image quality.*

By now, one is starting to realize just how challenging deep-sky imaging is and the limitations of any imaging system. It is interesting to put these challenges in perspective, which makes one realize just how remarkable it is that a digital camera produces anything at all. A good deal of emphasis was placed on image noise mechanisms in previous chapters and since this is not usually a big deal in conventional photography it may seem like one is making a lot of fuss about nothing. For that reason it helps to understand this in context and to illustrate this, it is illuminating (no pun intended) to compare the demands of a studio portrait with that of a deep-sky image of a dim nebula.

In the case of the studio portrait, a pixel on a flash-lit face has a 16-bit value of about 40,000. Assuming a typical sensor efficiency, this represents about 20,000 electrons at the pixel photosite from an incident 30,000 photons during a 1/1,000$^{th}$ second exposure. During this exposure, a photon lands on the photosite every 33 nanoseconds (0.000000033 seconds).

By way of comparison, I measured a patch of red nebulosity in a 20-minute deep-sky image exposure before stretching. In this 16-bit image, the nebulosity registered 50 counts over the average background level. Working backwards, this is equivalent to 1 photon every 36 seconds or so. This deep-sky subject is an incredible billion times dimmer (1,000,000,000) than our studio portrait. Clearly, a little noise in the dimmer image is going to be more obvious but the studio portrait is not noise free by any means. If you recall the earlier discussions, when the subject is bright, its image noise is dominated by the random nature of light itself (shot noise). Shot noise is well defined and its value is the square root of the average light level. In the case of the portrait, the signal to noise ratio is about 170 (the square root of 30,000). The sensor noise is present, but at a much lower level. In the deep sky exposure, the SNR is considerably worse, with a value of about 4. The signal level is much lower and over the longer exposure the small amount of image shot noise is dwarfed by a healthy contribution from sensor noise mechanisms. While sensor performance is a dominant force in astrophotography, to make matters interesting, there are also other sources of "noise" that need consideration and remedy. What follows is unavoidably technical but it is important to understand and I will try to make it as painless as possible.

### Noise Mechanisms

Astrophotography is a constant battle against image defects. It affects much of what we do and not surprisingly,

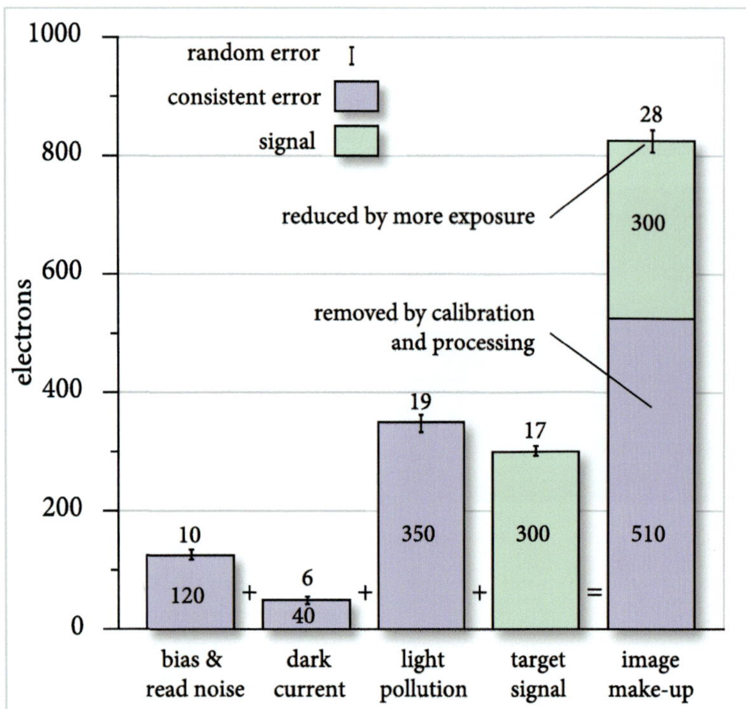


*fig.1 The make-up of an image is much easier to explain with a diagram. The 4 left-hand bars are the individual electron contributions from unwanted and wanted sources that make up an image pixel. The short bars indicate the random element; approximately the square root of the average level.*

many issues have already been mentioned in previous chapters. It is now the time to view the problems and solutions more holistically. On one level, noise is everything in the image that you don't want, including the effects of cosmic ray hits, satellite, aircraft or meteor trails, dust on optical surfaces and optical vignetting, in addition to the randomness of the recorded image on the sensor. For each of these there is an effective strategy for removal or reduction (fig.1). The random events are tackled by increasing the overall exposure and/or averaging multiple images and those that are ever-present, even if not immediately obvious, by image calibration. There is a third category, a lurking malady on some exposures, caused by poor focus and tracking. For images affected by these, the only recourse is to examine each image (initially visually, but increasingly using computer analysis) and discard the offending frames with a sorting process. On some nights, when nothing seems to go right, I have discarded most of the captured frames. With care, however, I usually only discard one or two. In the hope that only a few have issues, a quick glance of each frame in succession makes it easier to detect the exceptions. In the early days, I did not appreciate just how much the focus can drift over time and I was forced to accept somewhat sub-standard frames with diffuse stars. My most frequent culprits now are tracking issues, caused by a passing cloud or windy conditions. Although the averaging (integration/stacking) process follows image calibration, it is worth discussing it briefly to distinguish it from the sorting process.

## Sorting and Rejection Processes

The sorting process discards an entire frame, based on a widespread issue, like poor focus, tracking issues, mist and cloud cover. Deciding what to reject and what to keep is a balancing act, as each averaged exposure improves the overall image signal to noise level. There are many averaging processes, not all of which calculate a mean value but suffice to say, simple averaging processes add all of the corresponding image pixels and divide by the number of frames. (By corresponding, we are referring to pixels that belong together in the final image, not necessarily in the exact same position on the sensor.) In the case of simple averaging, any rogue values, caused by some special event are averaged too and while reduced, still have a ghostly presence, especially after stretching during image processing. Stacking image exposures in Photoshop and Affinity Photo exemplify simple averaging. Dedicated astro-image applications use more advanced averaging processes. These perform selective averages of corresponding pixels in each frame. The key word here is *selective*; during the averaging process, each pixel is compared to its corresponding colleagues and if it is considered an outlier, it is excluded from the average. This only applies to the outlier pixels in any exposure and the averaging process still includes all the "good" pixels elsewhere in the exposure. There are also special circumstances where averaging (mean) is replaced by a median calculation. This is quite good at removing singular events, at the expense of a slightly higher noise level.

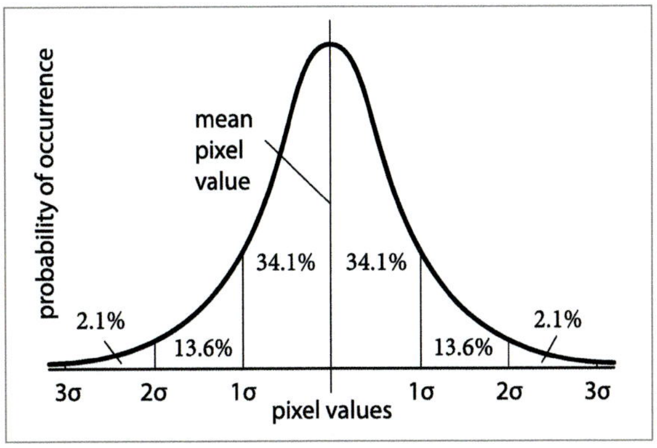


*fig.2 If you examine pixel values in the presence of random noise, they are distributed around a mean value, with decreasing probability that large deviations are present.*

The better pre-processing applications calculate the mean value and variation for each image pixel. It applies a cut-off value and excludes all values that exceed that variation. These use statistics to determine what is considered an outlier, typically evaluated against a user-set value, which turns out to be a multiple of the Standard Deviation (SD) of the pixel data. SD, which is also sometimes written as Sigma or its Greek character δ, is a measure of the distribution of values around an average or mean value. In its definition, and for our purposes considering the pixel values for a particular part of an image, 68% of pixel values fall within ±1 δ, 95% within ±2 δ and 99.7% within ±3 δ (fig.2).

In simple terms, if in 9 exposures, a pixel has a value in the range of 20–30 with respect to the background level but in a tenth, it is 80, it is a safe bet that this pixel is rogue and can be ignored. That does not mean the entire exposure is discarded, just the affected pixel(s) in the averaging process that follows image calibration. The actual math is more complicated, as it additionally has to work with images with different background levels (especially during twilight and dawn) and exposure lengths too. Rogue pixels can be caused by singular events, like cosmic ray hits, aircraft trails and so on but they can also be extreme (and rare) noisy pixels, outside the expectations of the sensor specification. The good news is that, with the right rejection criteria (usually set to around ±3 δ) these anomalies disappear from the final image. Whether by design, or accident, if the separate image frames do not align, this will also reduce the appearance of hot pixels, as they occur in different places of the image and are rejected as singular events. When you hear of astrophotographer's dithering, this is not a sign of indecision but a deliberate instruction to move the image frame by a small random amount between each exposure for just such a purpose.

> Cosmic rays are caused by high energy particles from outer space that form a shower of secondary particles when they hit Earth's atmosphere. They trigger electrical events at the electron level and a single cosmic ray can disable a satellite if it hits the exact right spot on a circuit. When they hit a sensor, they cause a number of adjacent pixels to fill with electrical charge. The effect is a small light "squiggle" on the image, the occurrence of which increases with exposure time. Fortunately, like lightning, they rarely hit the same spot twice and the randomness of their location on each image allows one to statistically distinguish and eliminate them during calibration.

### Constant and Random Defects

Constant and random defects are dealt with in very different ways. In essence, constant defects are measurable and are fixed in each of the individual image files through the calibration process whereas random defects are reduced, through averaging the calibrated exposures after registration.

Constant defects include:

- dust spots
- vignetting and collimation issues
- hot, cold and warm pixels
- light pollution
- sensor pixel gain
- sensor pattern noise
- sensor dark current
- sensor amplifier glow

Random defects include:

- image shot noise, cosmic ray hits
- light pollution shot noise
- sensor read noise
- sensor dark current noise
- sensor amp glow noise

In the case of the duplications there are both constant (mean) and random (noisy) contributions from the same photon/electron source and these are addressed separately, according to their nature. For instance, light pollution may give a sepia pallor to an image, perhaps with a sky gradient too. Even though you can neutralize its effect, this only removes the mean level and image calibration still leaves behind its contribution to the random shot noise that accompanies all incident light.

The constant defects are not all treated equally; vignetting, dust and pixel gain effectively adjust the sensitivity of a pixel, whereas the others contribute to its level. In simple terms, a straightforward subtraction (at a pixel level) of the mean level error removes it from the image. A multiplication with a correction factor fixes those defects that reduce or increase pixel sensitivity. Both these processes occur during the calibration process.

## Calibration

The chapter *Improving Quality (part 1)*, introduced the benefit of averaging multiple images to reduce their random noise level but did not go into the details of image calibration (also known as pre-processing or linear processing). This chapter recaps and then goes into more detail. In our earlier, casual wide-field images and star trails, the short exposures are dominated by shot and sensor read noise and calibration is not really a necessity. It becomes more than a nice-to-have once we move from wide-angle images at wide apertures and image with telescopes with moderate apertures and extended exposure times to capture dim objects. Having said that, although calibration is an essential activity for deep-sky imaging, at the same time, it can be useful in any situation where a sensor is operating at its limits. For instance, we know that CCD- and CMOS-based cameras convert incident photons into an analog electrical charge that is converted into a digital value. This conversion, however, generates minute but consistent differences between individual, rows and columns of pixels; each of the millions of pixels experiences a very slightly different conversion experience. These variations repeat with each exposure and leave a very faint underlying pattern in each frame.

These anomalies are usually obscured by random noise and a plentiful signal; a few prominent hot pixels

*fig.3 A single bias frame from a Canon EOS 60Da. The image is stretched to show the dominant random noise.*

*fig.4 The average of 50 bias frames, with the same stretch as fig.2 shows less random noise and reveals the underlying linear pattern noise.*

*fig.5 The average of 50, 60-second dark frames, has a higher level of random noise but still shows pattern noise, which is removed by calibration.*

may be all that you see. The faint pattern becomes more obvious when you average many exposures with very few image photons to play with. The calibration process resolves these repeating errors by characterizing the optical transmission path and sensor readout for a specific optical configuration (including aperture, filters and camera rotation, sensor temperature, ISO setting and exposure duration). It does this by evaluating three special calibration file-types and then applying a mathematical correction to each image pixel in each exposure. These three calibration file types are known by the names bias, dark and flat frames. These exposures are taken under very specific conditions that allow simple mathematics to determine the image pixel corrections. Since all exposures have random noise, it requires one to average many calibration frames of each type to accurately identify the underlying consistent pixel errors from the random rubbish.

## Bias Frames

These are blank exposures, with no light and zero-length exposure (practically, anything faster than a 1/100$^{th}$ second). If you take a single bias frame, all you can see is random noise (fig.3), often referred to as read noise. Calibration though is not interested in the random noise, but in any underlying pattern. Averaging 50 bias frames significantly reduces the random content and leaves behind the bias pattern, caused by the minute differences in the sensor electronics between pixels, rows and columns, as shown in fig.4. This bias pattern is remarkably consistent and once you have an averaged set of frames, it can be used for a year or so. As it is a one-time event, for deep-sky work, astrophotographers will invest in 50 or more frames and keep an averaged "master" bias file to use later on for multiple calibration sessions with different targets and optics. As the exposure time is short, the temperature of the sensor is mostly irrelevant. The gain (or ISO) setting is important, however, and should be the same as the other calibration and image exposures. With a CCD, there may be only two gain settings. With a digital camera or CMOS-based camera, there are many more and it keeps things simple to standardize on a few ISO or gain settings that cover most eventualities.

## Dark Frames

As the name implies, dark frames are taken in the dark, ideally with a metal lens cap over the optics as some plastic lens and body caps are transparent to infrared and cause strange results. This time around, these exposures are taken at the same temperature and exposure duration(s) as your image exposure(s). Again, they stay remarkably consistent and for my deep-sky work, since one can control the camera temperature, I take 50 frames for a full set of common exposure times in one go. It takes about a day but the result is worth it. It is difficult to control the temperature of a DSLR or mirrorless camera and so, in practice, I take 10 exposures before and after my image exposures, or stick the camera in a plastic bag in the refrigerator to approximate cool nighttime conditions.

The data in these frames is a combination of read and dark noise. Dark noise is thermally generated over time and as the time or temperature increase, so does the overall noise level. This noise is always "positive," it effectively adds random electrons to the sensor pixels. Again, a single frame looks noisy but an average of 50 frames reduces the random elements of the noise,

*fig.6* *The average of 10 flat frames, taken through a translucent diffuser show the slight vignetting of the optics as well as a few dust shadows on the sensor and filter. There is less apparent noise on this image as the light frames have short exposures of a bright evenly-lit target, similar to levels in conventional photography, to form a mid-grey tone.*

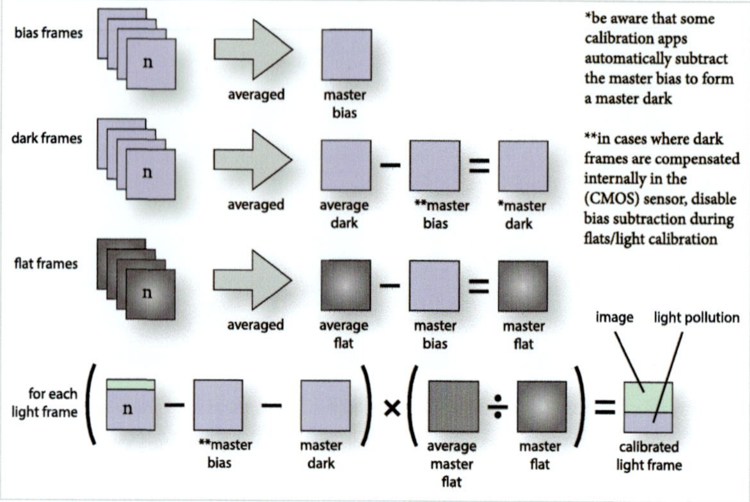


*fig.7* *Most calibration programs do all of the above manipulations automatically. This workflow assumes the flat frames have short exposures and only require calibration with a master bias image.*

leaving behind a combination of pattern noise and hot, cold and warm pixels. When the exposure exceeds about 30 seconds, thermal noise at room temperature is more dominant than the bias noise. The noise level approximately doubles for every 6 °C increase and increases linearly with time. On some CMOS sensors, another time-dependent noise, called amplifier glow, also adds to the exposure. Here, additional thermally generated electrons are triggered by local warming on the sensor. Its appearance resembles that of classical film-fogging in a camera, from a perished seal on the camera back. If the image and dark frames have similar amp glow, it is possible to calibrate it out of the final image. The averaging process creates a master dark file, one for each unique exposure time and sensor temperature combination. Some applications create a master dark by subtracting the master bias from the average of the dark frames. This is not always the best approach, especially with CMOS cameras and is best avoided to avoid unintentional black pixels in the final calibrated image.

### Flat Frames

Some consider flat frame calibration as optional, for the reason of keeping the process simple. It often depends on the issues you see during image processing. If, after stretching, your images show vignetting, uneven illumination or dust shadows, then flat field calibration is the solution. Flat frames differ from the previous two, as they actually allow light to hit the sensor. Here, the user takes an exposure, equivalent to a mid-grey of an evenly-illuminated target. Our eyes are not good at determining what is "even" so this usually requires a diffuser placed over the lens or aiming at an artificial target in the form of a light box. Some infamously drape a plain white T-shirt over the front of telescope and point it at the sky at dusk. For smaller lenses, I clamp a small piece of opal plastic over the lens and point it towards a small tungsten lamp. (White LED lamps are all the rage these days but for astrophotography purposes, their red wavelength emissions are weaker than that for blue and green and, if you are going to calibrate for narrowband filters at a some point, you will need a hotter source.)

An alternative is an electroluminescent panel (and high-voltage power supply). These thin light-emitting sheets are already sandwiched in-between white plastic. You can also make your own at less cost. These are often used for advertising and backlit signs. In my observatory, I hang a home-made A2-sized electroluminescent panel on the wall and point the telescope square-on to the middle. The sheet fits neatly into a picture frame, along with a white plastic diffuser. Its deep-red emissions are just sufficient for flat exposures through narrowband filters. Exposures are typically short, ideally less than 20 seconds and so the temperature (and dark noise) of the camera is less relevant. It is essential, however, that they are taken with the exact optical system used for the image exposures, including any filters, aperture setting, zoom setting, rotation, focus setting and lens hood. In the case of photographic cameras, it also should use the same ISO setting (or gain setting, in the case of a dedicated astro camera).

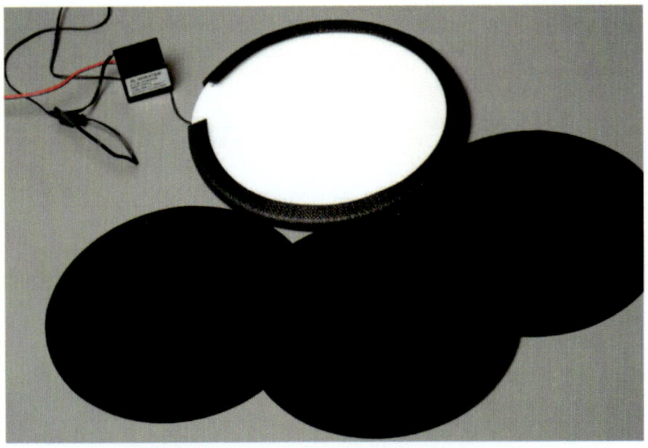

*fig.8 An 8-inch electroluminescent panel with some neutral density lighting gels for creating flat calibration files. This is powered from 12 volts, via a small voltage converter. Some shuttered sensors do not give even illumination at short shutter speeds and the gels are then used to extend the exposure time to a few seconds.*

In combination with dark frames, flat frames give an indication of two forms of error; namely differences in gain between pixels and any shading effects caused by dust or vignetting in the optical system. Although dust in normal photography is rarely an issue, especially at wide apertures, it is not the case in astrophotography. Here, we apply strong tonal stretches to reveal faint details and any slight vignetting or dust shadow become a tunnel and a black hole respectively. In the case of dust shadows, as dust is closer to the sensor the shadows become smaller and more intense. When you think about it, while vignetting is constant, dust may occur at any time and so astrophotographers often take flat frames at the end of an imaging session, especially if the optical system is open to the atmosphere, as in the case of some reflector designs.

In practice, since flat frames have more light exposure, the SNR is already high and it is only necessary to take about 10 frames. If you are using several filters, however, it requires a set of flat frames for each combination, as the vignetting and dust may be different in each case. Fig.6 has a small few spots from dust on the sensor cover glass. They may look indistinct here, but recall that this image has not been stretched to any great extent.

The calibration process uses flat frames to effectively change the gain of each image pixel to counteract any issues. It should not be, however, a reason for not taking care; it is always much better to keep lenses, filters and sensors scrupulously clean. When I use a refractor for a number of imaging sessions, I leave the camera attached and keep the front optic clean. By doing so, I make one set of flat frames with the configuration and use it for several sessions, until I change the telescope or camera angle.

## Calibration Process

Although a conventional image processing application can perform simple image stacking, the calibration process requires more complex pixel maths, in which it subtracts, divides and multiplies pixel values together. It also requires the image to be linear, just as it leaves the sensor. The following calibration process works for CMOS and CCD sensors and follows a simple flow (shown schematically in fig.7):

1. average the bias frames into a "master bias"
2. average the dark frames to form a "master dark" for each set of exposure conditions
3. subtract master bias from the light frames and average these to form a "master flat"
4. calculate the average pixel value in the master flat
5. divide the average pixel value in 4. with the master flat to generate a gain correction image
6. calibrate each light frame by subtracting the master dark and multiplying by the gain correction.

Confusingly, there are some variations; some older applications choose to automatically subtract the master bias from the averaged darks. If this is the case, the light files require both master bias and master dark subtraction, before averaging. These and the older texts assume the user is imaging with a CCD sensor and they reiterate the generic calibration in fig.7. There are some subtle differences in the optimum calibration process between CMOS and CCD sensors. CMOS cameras are increasingly the economical choice, however, either in the form of a conventional camera or astro camera and so, as you become more proficient, it helps to understand image calibration for CMOS sensors in more detail to yield the ultimate quality.

This is now the time to dispel any remaining thoughts that RAW files are unadulterated. It is quickly apparent in astrophotography that CMOS sensor data is often manipulated before transmission. One of these manipulations is an automatic, rudimentary dark-frame subtraction within the chip, using masked, peripheral areas of the sensor. With a CCD exposure, we expect to see the general background level increase with exposure time but in a CMOS device, an extended dark-frame exposure may have a lower background level. As a result, it is not uncommon for a bias frame to have a higher average level than a long-exposure dark frame. This completely messes up the traditional calibration process which subtracts bias frames from dark frames as well as more advanced techniques such as dark-frame scaling. Several applications have not caught up with this yet and the developers do not fully appreciate the CMOS differences. To sidestep this issue requires closely matching the dark frame and light exposure times and temperature. If a calibration routine automatically subtracts the master bias from the master dark (which on a CMOS sensor, may cause black pixels) try to disable this or, add a constant offset (or pedestal) to the dark frame before the bias subtraction, if the application settings allow. The first three steps of the calibration process glibly mention averaging the three calibration file groups. In practice, for high-quality results, it is a little more involved. While much of this is going on behind the scenes, it is interesting to note just why your computer takes so long to finish its computations!

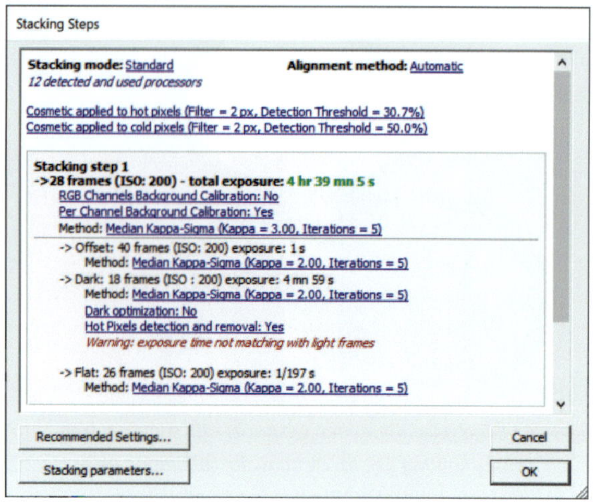


*fig.9 At first, the calibration process appears complicated and intimidating for the novice. Fortunately, applications like DSS assume the most likely settings for any given set of files and has the option to override them if you choose.*

### Calibration Stacking

Stacking is the process by which similar exposures are combined to reduce the randomness of their pixel values. This occurs on all frames (bias, dark, flat and light), but in slightly different ways. Before the images are calibrated, the dark, bias and flat frames are first stacked independently to reduce their pixel variation and extract their underlying patterns, such as sensor uniformity, hot pixels and dust/vignetting effects. At its simplest, this is a simple arithmetic average (mean) but it can also be selective too. With this, the corresponding pixels are compared and the obvious "wrong-uns" are discarded before their colleagues are averaged. This removes the effect of just those pixels affected by an unwanted special event, such as a cosmic ray hits or the improbable event of a rogue pixel value. The rejection threshold is usually set around the 2–3δ, eliminating events of less than 5–0.3% likelihood respectively. Flat frames have a further complication because there may be frame-to-frame exposure variation. This confuses the statistics and skews the averaging process. The calibration algorithms neatly side-step the issue by normalizing the flat frames before analysis, making them statistically compatible and again, in a slightly different way, before averaging them. Thankfully, this process is usually handled automatically behind the scenes (once the application identifies the files are flats). In each case, the output of the stacking process is a single "Master" calibration bias, dark and flat file which are then used to calibrate the individual light files.

### Image Registration

It is necessary to align or register all the images prior to stacking. There is no such thing as perfect tracking and there will always accidental misalignment between frames and sometimes, deliberate shifts too. The registration process takes one reference image and then aligns all the others to it. This alignment can variously shift, rotate, scale and even distort images to line up. The registration process in some of the simpler applications is automatic and not even called out. All the methods calculate star centroids for each image and align them to one another. The way they do this can vary considerably. At its simplest level, Nebulosity asks the user to manually select a common star in each of the calibrated images, preferably in one corner, and then does the same for a second star, usually in the opposite corner.

More sophisticated methods do not require user input and have software algorithms to match star patterns. Some can even match image sets with different scales, allowing one to combine the contribution from two imaging systems of the same subject. Some applications provide some optional registration controls, usually around star-detection parameters, so that the software does not accidentally confuse hot pixels for stars. These are usually in the form of a minimum star diameter and relative brightness over the background.

When using color (CFA) sensors, while calibration can technically work on both monochrome RAW files (before de-mosaic/de-Bayer) and color RGB files, registration (and stacking) works more effectively on color images. The reason is that the registration process makes micro-pixel translations that land stars onto different sensor pixels, affecting any subsequent color decoding during the de-Bayer process. Some applications perform the de-Bayer process up front, to convert the camera RAW files to an editable format. Some do not, however, and if there are separate tabs for calibration, registering and stacking (as in SiriL), you need to check the image is in the right mode before proceeding.

If the registration process is carried out separately, rather than as part of an automatic process, it usually generates another set of images, one for each original exposure. In other cases, it is done behind the scenes and the temporary images are deleted after stacking. There are now potentially three image sets; original, calibrated and calibrated/registered. After the next step (image stacking) is complete, I usually just keep the original files and the final image stacks and delete the rest to conserve drive space. This is then archived to a second drive, just in case.

### Image Stacking (aka Averaging, Integration)

The earlier assignments and chapters introduced us to the concept of stacking images to reduce the random noise level. Again, at its simplest, this is an average of corresponding pixels and this process works well enough with vistas or wide-field shots that do not require extensive manipulation. For more demanding subjects, however, the stacking process is applied to the calibrated/registered image files. These image files still contain our rogue pixels and singular events. It is possible to exclude these pixels by comparing them statistically with their colleagues. There are, of course, a few unique challenges to overcome as, unlike calibration frames, the image files are not constant; background sky levels, Moon illumination and image intensity vary between individual exposures. This shot-to-shot variation confuses any simple statistical analysis and the better applications automatically deploy a more sophisticated process to remove rogue image elements and minimize image noise.

The solution is similar to that used for stacking flat frames; the image files are initially normalized so that they are statistically comparable. Once the statistics have identified the rogue pixels, the image files go on to be averaged. In the more advanced applications, these files are not all treated equally and, based on how noisy they are, their contributions are scaled before adding them together. This is done by a background process that first performs a noise analysis on each calibrated file and ensures the noisy ones contribute less to the final image. On a typical session, lasting several nights, it is usual to see the process log recording a range of contributions between 80–120%. Fortunately, all that is required of the user is to select a few options that instruct the stacking program to reject outliers and normalize the files before stacking. The better applications produce a 32-bit result, though there are still some free utilities that bizarrely scale back to 16-bit.

### Linear Images

This is an ideal time to make a small diversion. Humans perceive light differently to a digital sensor. For a hundred years we have accepted, without questioning, the logarithmic relationship between exposure and image density on film and print materials. In that world, we perceive (both on a print and a monitor) equal steps of exposure, when in effect, the actual light intensity is doubling or halving each time. When the computer age came about, as monitors and sensors worked differently to photographic emulsions, the designers took advantage of our non-linear perception of light to redistribute the pixel values in an 8-bit image file. For instance, when a RAW file is loaded into a computer, an automatic manipulation, boosts shadow and compresses highlight contrast to give a perceived natural look. This is often referred to as applying a gamma curve, similar to applying a curve adjustment to an image using a math formula. This formula uses a power coefficient that we call gamma. The most common value is 2.2, which is used for common consumer devices and photographic output. (A linear RAW file has a gamma of 1, i.e. no correction.) This gamma manipulation makes best use of 8-bit and 16-bit image files, distributing the perceived exposure evenly throughout the available digital values. For instance, an exposure of a Kodak Grey Card, with a reflectance of 18%, will recreate a perception of mid-grey, but with a grey scale value of about 50%, or 127 out of 255 in an 8-bit image. In a linear system, a mid-grey would require a 50% reflectance too.

In deep-sky astrophotography we commonly download 16-bit file sizes (albeit the sensor may internally be only be 12- or 14-bit). As soon as we start stacking multiple images and manipulating, it is better to use a 32-bit image container, with 65,000x more light levels than an 16-bit image. The reason for this is that the individual files are all slightly different, from noise and other factors and the averaging process creates intermediate values, improving tonal definition.

For instance, consider the simplest case of a 1-bit pixel. It can be 1 or 0. If it is twice more likely to be a 1 than a 0, the accumulation of 100 pixels will be about

66. If you have a bigger bit depth, you can keep that 66/100 value as a fraction, otherwise it simply gets rounded up to 1.

While 8-bit images have 256 levels, a 16-bit image has 65,536. That may seem plenty but in many astrophotographs, the interesting nebulosity may occupy less than 1% of that dynamic range. Image stretching may well cause the background to break up into bands (posterization). A 32-bit image has a much finer tonal resolution and does not require gamma adjustment to re-arrange the image values. As a linear image though, it still has many more levels (and hence finer tonal resolution) per stop of exposure in the highlights than it does for one in the shadows. For some high-range images, some applications even use 64-bit files temporarily, before compressing down to a less extravagant file size.

### Calibration Software

Calibration is an arduous task made easy by automation. It is not present in standard photographic imaging products and worse still, by default, these convert incoming RAW images into gamma-corrected color files, which then mess up any calibration process that assumes unadulterated data. While it may be possible to find a work around, that converts to a linear profile and then converts back again, it is better to use an application that automatically works in linear gamma and in high bit-depths. These applications are commonly designed for astrophotography or scientific study. Some are freeware, of which the most popular is DeepSkyStacker (DSS). DSS is a Windows program and relies upon a free program utility called LibRAW to convert camera RAW files. As the name suggests DSS stacks images, but it can calibrate and align images first before stacking, and more besides. Two other more recent and open-source applications are SiriL and ASTAP, that work on multiple platforms. Both SiriL and ASTAP have a number of other free utilities on their webpages. ASTAP has a fast and useful free plate solver, which is being picked up by the image acquisition programs (for centering). Both are still in active development and allow one to have full control over the calibration and stacking process, or use supplied scripts that batch the commands for convenience. In SiriL, the present version scales my 16-bit Fuji files back to 14-bit, the same as the sensor, losing the tonal resolution benefit that accompanies integrating multiple images. That should hopefully be fixed by the time of publishing. These applications output files for editing by other programs.

One can also use Nebulosity (which has the essential elements of acquisition, calibration and image processing). It was the first program I bought when I started out on my Mac and carried it over to the PC with the same licence. Nebulosity is typically less automated than the other applications, though it does have some scripting capabilities to reduce manual intervention and works with its sister autoguiding application PHD(2).

### FITS for Purpose?

File format choice is another thing to watch out for. Each digital camera manufacturer has a unique RAW format, which now include compressed versions, as well as the ubiquitous 8-bit JPEG output. The astro acquisition applications commonly use the FITS specification but this open standard has a number of formats and customizations that give rise to some variation between acquisition applications. It can be the case that the calibration and image files, made with different packages, are not immediately compatible. My recommendation is to shoot all your calibration and image files with the same application, using the same file option. Acquisition applications interfacing to monochrome cameras typically default to FITS but color cameras (also referred to as Color Filter Array or CFA) may have options for saving as camera RAW (without compression), color FITS or a FITS file for each color channel. All these file formats are usually 16 bit. The better applications manipulate these files in 32-bit or higher. One thing is clear, JPEGs are not suitable for data acquisition but are for Internet use, typically assuming the sRGB consumer color profile (the default for many consumer imaging devices).

### Disk Cost

I normally image with an 8 MB CCD and I have been slowly filling up a 1 TB drive. With DSLRs and mirrorless cameras, they have considerably more image pixels. The increased file depth during calibration and manipulation has further storage implications that would fill a laptop hard drive from a few years ago and looks a little crazy. A fast drive is essential for processing with modern cameras and I use a striped RAID SSD array connected via USB-C as a working drive. Once the basic image processing is complete, I remove the interim files and back up my originals and processed image stack to a slower high-capacity drive.

Disc drive speed has a big impact on processing time. My 5,400 RPM HDD achieves 75 MB/s, a SSD RAID array achieved 800 MB/s and the internal ePCI SSD in my Mac mini was closer to 2.4 GB/s. (My first computer only had 0.000004 GB total memory!)

# Deep-Sky Imaging

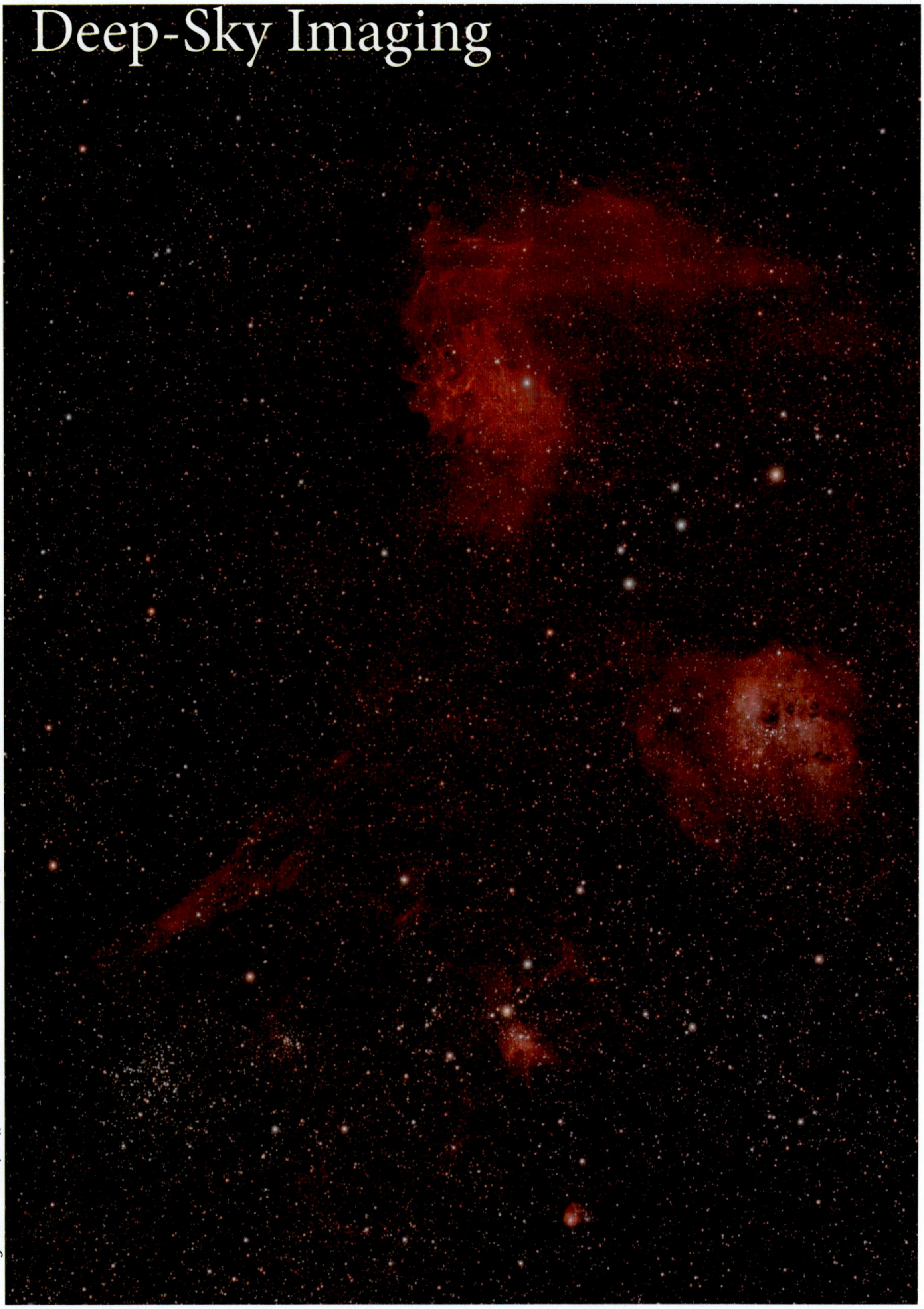

*Flaming Star Nebula (C31), SH2-236 and the Starfish Cluster (M38) taken with color CMOS camera.*

# Raising the Bar

*Imaging dimmer and smaller targets challenges everything that has gone before.*

This chapter marks the boundary between imaging with lightweight systems and more substantial means. While simple trackers and short focal lengths work well with landscape vistas, star trails and the few large, bright deep-sky objects, there will come a point when the imaging bug truly bites and one needs to take the next step to progress to smaller and fainter targets. These are considerably more challenging to image and stress every part of your acquisition and processing ability, as well as your patience and budget. It is the equivalent of a black hole's event horizon, from which there is no return! When a faint nebula is a billion times dimmer than a daylight image, a few hours imaging is insufficient. The reality of imaging dim nebulae and galaxies and at higher magnifications conspires to frustrate and amaze you in equal measure.

In a typical semi-rural site only the brightest stars are visible to the naked eye and only those nebulae or galaxies with the largest surface brightest through a telescope. One must overcome a number of obstacles to image a dim target. The impact of the smaller and dimmer targets is compounded by the speed of the optical system. It is common to use a refractor with a focal length of 350 mm or more. Unlike premium prime lens optics, with aperture ratios of f/2–4, refractor apertures start around f/5.6. A few are faster but these usually have an unhealthy price tag and anyway, anything faster than f/4 will potentially degrade dichroic filter performance at the image periphery.

The solution is to increase the overall exposure, to capture more photons. Longer individual exposures additionally place more stringent demands on polar alignment and tracking accuracy. While it may have been possible to image unguided with lesser targets, now, it is a necessity. At the same time, longer exposure lengths are more likely to pick up unwanted intrusions from meteors, satellite and aircraft trails. Camera (sensor) performance will be more critical too, as longer exposures have more thermally generated image noise, fighting with the few photons from a dimmer target. There is an upper limit for an individual exposure length, set by the highlights and so the obvious solution is to take many more exposures. For a quality result, one clear night (especially in the summer months) may not be sufficient. Once, a sequence exceeds 4 hours or so, it is not much fun waiting around for it to complete. This immediately prompts one to investigate robotic dual-axis mounts, PC control, autoguiding, sequence automation, autofocusing, image centering and potentially, the use of specialist filters and dedicated cameras too.

The various targets now present different challenges. The common deep-sky catalogs list an object's apparent magnitude. This figure is sometimes misleading, as it is indicative of the total light given off by the object and, of two objects with the same apparent magnitude, the smaller one will have the higher surface brightness. Some objects, like the spiral galaxy M33, have a high magnitude but a low intensity over a large field of view. Stellar targets, including loose and globular clusters, require less exposure than a dim galaxy or nebula. That is not to say they are easy; globular clusters are one of my favorite subjects but require very precise focus and tracking to achieve the required delicacy and, if you want to show the full extent, which can be a long way out from the central core, more exposure again.

Dim targets present an additional challenge; they are also more difficult to locate unless there is a stellar signpost near by. Without some assistance, locating a dim target will be hit (and mostly miss) and will not register on a camera's LCD display, even after a 5-minute test exposure. Precise framing and re-framing are very tricky indeed and my previous red dot or iPhone finder ideas are not sufficiently accurate on a single axis mount and we really need other means for centering.

While these challenges are sufficient to keep one busy on their own they omit the fact that the greater number of exposures and the hidden detail in each will certainly require a dedicated astro-acquisition application as well as better image calibration and image processing applications too. It is almost like starting afresh, as all these new systems require assembly, installation, debugging, tuning and learning, so they become second-nature and their operation is reliable and enjoyable.

I love a challenge; that is why the end result is so rewarding. This section explains how to get there.

# Dual-Axis Telescope Mounts

*Choosing and using an equatorial telescope mount.*

The Fornax LighTrack mount I bought for wide-field images was a good choice and consistently performs better than expected. I have not been as fortunate with the half dozen or so dual-axis mounts over the years. This chapter discusses robotic mounts; those with motorized axes that are capable of "goto" operation, in which they are commanded to slew to a particular coordinate in the sky.

The choice of mount is more important than that of the optics. It ultimately limits the quality of your images through their ability to track accurately and support the telescope assembly without wobbling. The mass, maximum load and price often correlate and for portable setups, a conscious compromise is required. If the budget allows, my advice is to buy better than you initially need, so that the next telescope upgrade does not trigger an expensive mount change at the same time. One caveat is to buy something that you can lift and assemble without giving yourself a hernia; my first mount was a used Meade LX200. The previous owner underestimated its sheer bulk and the strength required to carefully carry it from the house to the back yard and assemble it without damaging it. I did not entirely learn from his experience. I more recently bought a premium mount and after a year I found it had really caused a hernia, prompting me to build a permanent installation and use my electronic skills to design an automatic observatory.

Weight is not the only consideration for portable use. Some mounts do not have a provision for a polar scope, to facilitate quick alignment, others have no handset and require a PC to operate (which is less of an issue for an astrophotographer). Whatever the price, the mechanism sets the performance; comparisons of weight, maximum load and mounting options are a good place to start. Maximum load specifications should be treated with caution; the maximum load for imaging purposes will be less than that for visual use and the supplier's specification may include the counterweights as well as the optical system. For similar reasons, some models are more sensitive to accurate telescope balancing than others, which pop up later as a tracking or guiding problem. The lightweight designs are also likely to be more sensitive to cable drag or external disturbances.

The most common mount type used by amateurs for astrophotography are German Equatorial Mounts or GEMs (as shown in fig.2 and 3). These feature a swinging counterweight bar that balances the weight of the telescope and imaging system. At some point, as the mount tracks over the meridian, these designs require a 180-degree turn-around on both axes to continue tracking. This is referred to as a "meridian flip" and is designed to stop the back end of the telescope or camera clashing with the tripod legs or a pier. There are other design architectures that can track past the meridian, most notably some premium models from the company Avalon, though these are outside our self-imposed budget.

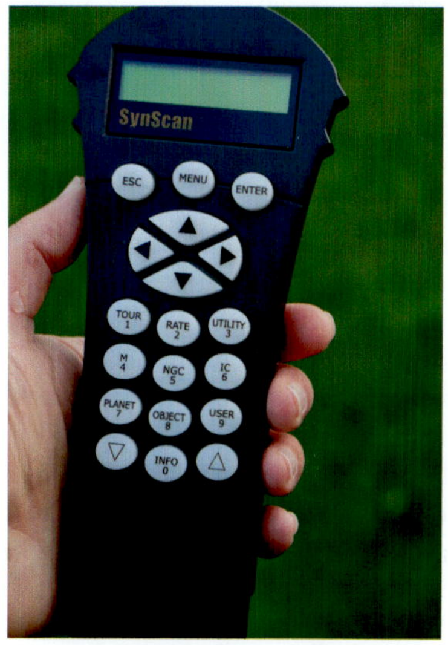


*fig.1* *Many dual-axis goto mounts, sometimes called robotic mounts, have a handset loaded with alignment protocols and catalogs of major deep sky and stellar objects. This one for the SkyWatcher mount series has a second serial connector port that allows for external control.*

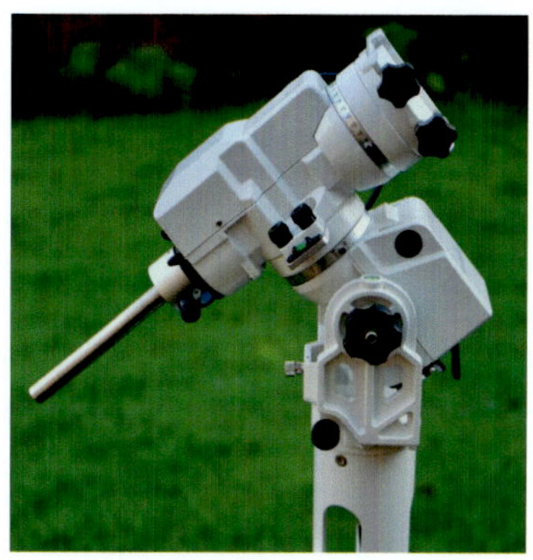

*fig.2* *The later SkyWatcher mounts have more refinement than before, and dispense with the spur gears in favor or belt-driven worms.*

## German Equatorial Mount

This is perhaps the most common all-purpose mount for amateurs. Here, "amateurs" refers to small to medium mounts; the very largest professional observatories use fork mounts and camera rotators for economic and practical reasons. The GEM is compatible with both refractor and reflector telescope designs and has the useful capability of simple polar alignment through a polar scope, aligned with its motor axis. An advantage of this simple polar alignment is that the tripod does not have to be precisely leveled and orientated towards the celestial pole. The telescope is bolted to a dovetail plate which is then clamped to the mount. Dovetails come in two standard sizes; the smaller Vixen (31 mm wide) and the larger Losmandy size (75 mm wide). Since the mount, tripod and telescope are separate, the smaller, lighter components also facilitate transport and storage. Price, size and mass vary enormously; the lower-end models may use low-pressure aluminum castings, plastic and mild steel fixings whereas the high-end models often are machined from solid alloy, use ceramic bearings, hardened stainless fixings and precision gearing. In the last 5 years, the belt drives used in the higher-end units are increasingly being used to replace the spur gears in the smaller units, the main advantage of which is to reduce mechanical backlash in the drive. They are often quieter too. The same goes for optical encoders. These sensors work out the absolute position of each axis and once a mount is aligned and remains switched on when it is manually moved, it still knows where it is pointing. There are two physical drawbacks of the GEM design: imaging through the meridian and leg clashes.

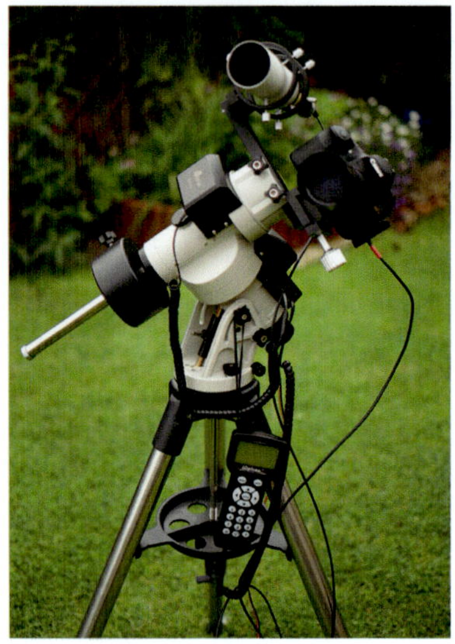

*fig.3 At a similar price level to the SkyWatcher AZ EQ5 is the iOptron iEQPRO 30. This also has a belt-driven worm drive and has permanent PEC too. There is often some variability in periodic error on budget mounts and it is advisable to check its performance as soon as possible after purchase. It is also noted that even with a saddle plate and separate guide scope, the counterweight is at its minimum limit. It is not uncommon for some mounts to presume heavier equipment and are consequently more difficult to balance with lightweight equipment and may require the purchase of smaller counterweights, or add weight to the imaging end.*

### Meridian Flips and Leg Clashes

During normal tracking, the main body of the GEM rotates around the RA axis. At the point at which the counterbalance shaft is horizontal and pointing east, the telescope will be pointing due south or north (at the meridian) depending on its declination setting. (In the Southern Hemisphere, the counterbalance shaft points west.) As the mount continues to track, the camera-end of the telescope continues to lower and may collide with the tripod legs when imaging at high altitude. To avoid this, the mount controller detects when the telescope crosses the meridian and stops tracking. At this point, to continue tracking, both RA and DEC axes need to flip 180 degrees. In an automatic sense, this is not without some risk of cable entanglement or collision, which is why most mounts simply stop moving and require manual hand-holding. Automatic flips are possible. In most cases, issuing a slew command to a spot past the meridian will cause the flip. Before doing so, some familiarity with your system is essential. With bulky telescopes, leg clashes may occur just before the telescope reaches the meridian, especially on mounts that are fitted directly to a tripod. A pier extension lifts the scope away from a leg obstruction and a permanent or mobile pier support allows even greater freedom of movement without fear of obstruction. In practice, most GEM mounts will continue to work without issue past the meridian for a few degrees and in some systems, the user sets a precise tracking limit before any touch condition can occur. There are other practical issues with meridian flips that need managing. The first is image centering; any alignment error or flexure requires a small adjustment to re-center the object. The second is a consequence of the flipped image which may require the guider software to reverse the RA guider polarity.

## Other GEM Considerations

Selecting a GEM model can be quite tricky and if wrong, an expensive mistake. Appearances are skin-deep and unlike a telescope, how do you evaluate and rate the claims of the internal mechanical performance? Budget is only a guide; every model, no matter how expensive, will have different user experiences. In the last few years, there have been a growing number of new models in the £700–£1,500 range replacing the old stalwart, the SkyWatcher NEQ6. The obvious purchasing consideration is the precision of the mount; periodic error, backlash and positioning accuracy. Any particular model's specification is a starting point. There is usually a range of values according to the manufacturing tolerances, with tighter control and less variation in the high-end mounts. There are several drive mechanisms to choose from; friction, belt and gear, each with their supporters. Some companies use stepper motors and others use DC brush-less motors with positional feedback. The implementation, however, is as important as the design architecture (just as there are awful carbon-fiber bicycle frames and fantastic steel ones). For instance, gear systems have less flexure but the mechanical interfaces have greater potential for backlash and periodic error, the opposite of toothed belt-drives.

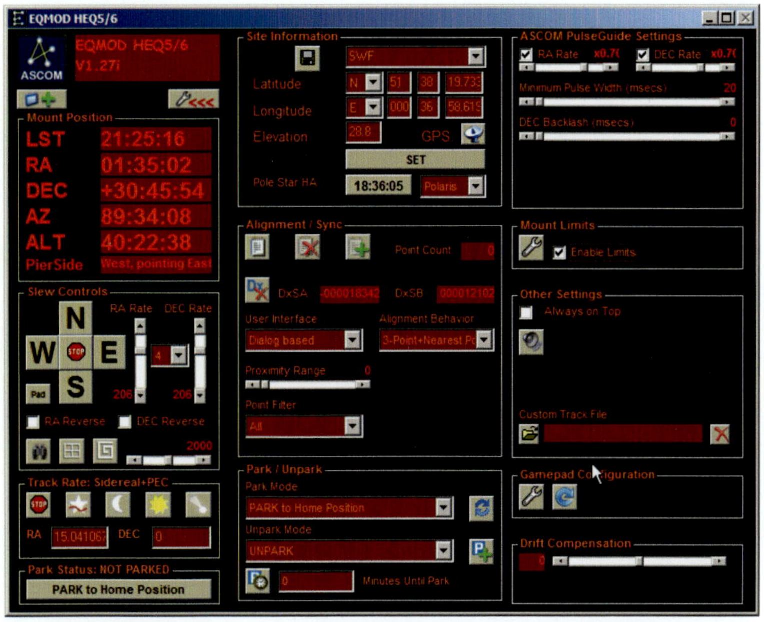


*fig.4* *Several years ago, a group of software engineers wrote a sophisticated freeware application for the popular SkyWatcher mounts as an alternative to using the handset. EQMOD ASCOM was the outcome and has a faithful following. More recently, SkyWatcher themselves have produced their own application for Windows and some other platforms.*

A few companies use friction drives; an interesting solution, similar to a rim-driven record turntable. My Fornax LighTrack has a simple friction drive and the announced AstroTrac 360 uses one too. In a GEM design, there is no actual coupling, as with a gear or belt system and it requires precision optical encoders on each axis to give position feedback, increasing the price. As a result, the motor control is more complicated and expensive but corrects for its own issues, including changes over time. (This simple metal to metal friction interface is not unique, it is the basis of the popular Crayford focuser mechanism too.)

Mounts with low periodic error are sought after and highly regarded. The chapter *Improving Tracking* goes into some detail about periodic error and how it affects imaging quality. From the various mounts I have owned and used, a budget mount is most likely to have a peak to peak periodic error of ±7–±30 arc seconds. Above that, it becomes increasingly difficult to remove either by built-in software corrections or guiding, prompting a re-think. It is to be expected that low-end mounts exhibit a wider performance range, both between models and also between production batches. Even when autoguiding, it is equally important that the mount does not exhibit sudden changes in periodic error, since these errors are difficult to measure and correct within a few guider cycles. It is an interesting time for the first-time buyer right now with new models being released each season. Internet forums are full of opinion and discussion, perfect to while away those cloudy evenings. The cost of a good mount is significant; robotic mounts are considerably more complex to make and operate than trackers. They have more payload capacity and their construction has to be rigid enough to carry two motor systems and the payload without flexing. Robotic mounts introduce a new dimension into the equation; software. This includes the low-level hardware code (firmware), the software for any handheld controller and the device driver for computer control, a third-party example of which is shown in fig.4. The software and hardware are of equal importance and it has been my experience that both can be equally troublesome. The worse offenders are typically the external device drivers, particularly

ASCOM drivers, which sometimes appear as an afterthought and rely on a third party to make the best of the mount's existing application interface.

An important part of this key purchase decision is who to buy it from. Dealer choice is an important consideration. The specialist dealers are often practicing astronomers with hands-on knowledge. These are more than box-shifters and give pragmatic advice depending on your needs. I'm a strong believer that if you value their service, they should have your business. In the UK at least, I have found the dealers to be refreshingly collaborative and an inquiry with one will lead to another that has stock or deals with a different brand that better suits your needs. A bargain is not a bargain if you make the wrong choice or you are unable to use a product to its full potential. As a result, the experience and backing of your dealer easily outweigh their premium over a discount warehouse or grey import deal.

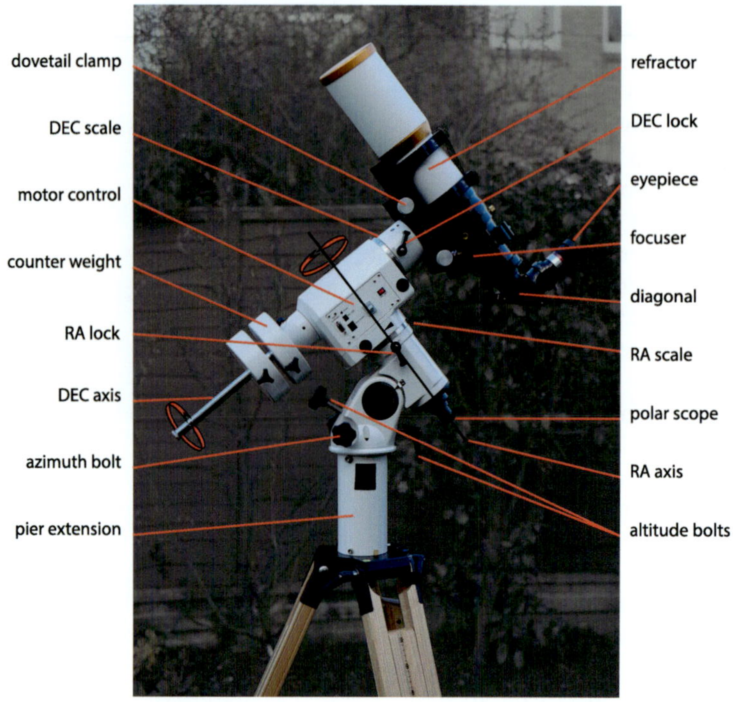


*fig.5 Anatomy of a GEM typical mount; in this case my first GEM purchase, the SkyWatcher NEQ6. This mount has since gone through a number of updates, improving the adjustment controls, drive mechanism and connectors. The short pier extension is to avoid leg clashes when using longer telescopes.*

On several occasions, dealer advice has made me rethink my requirements and avoided costly equipment overkill. Astrophotography is still testing the boundaries of robust economy and customer equipment-tuning and adjustment is part of the game. It is generally not a plug'n'play experience, as it is with computer equipment and digital cameras.

As you can imagine, many reported issues are "finger trouble" and, if there is a problem, it may be difficult to persuade an OEM to fix things, especially when the problem is intermittent or in the grey area of not living up to an expected performance, the most obvious of which is periodic error. Mount systems have complex interactions between software and hardware and establishing any issue's root cause is not trivial. While software updates can be delivered effortlessly, packaging up and posting a mount back to the vendor is a pain to all and which may not always be necessary.

I have included a few methods in the chapter *Improving Tracking* to reliably measure periodic error to help the dealer or manufacturer. In some cases, a manufacturer may offer to adjust or replace a rogue worm drive rather than the entire mount. Similarly, recognizing that economics encourage wider manufacturing machining and assembly tolerances, there is scope for after-market "tune-ups" and plenty of on-line videos on how to adjust and improve the assembly of popular mounts, some of which have been adopted by the original equipment manufacturer.

### Tripods and Piers

Cost-conscious designs often target the unexciting tripod for compromise. Light-weight models packaged with a mount may benefit from upgrading: there are several companies that make piers and tripods (in aluminum, carbon fiber, steel and wood) and supplied with adaptor plates to suit the popular mounts. The better models have firm clamps on the adjustable legs, leg bracing and tight leg-to-mounting plate interfaces. The heavier models may have screw-adjusted feet for accurate leveling. Several companies choose ash hardwood for the legs, as this has a natural ability to absorb vibration. The simplest test is to firmly grip and man-handle the tripod's mounting plate; the tripod should remain firm and while viewing, the effect of a small tap on the telescope should dissipate within a few seconds. For imaging purposes, the telescope height is of little concern and a tripod is more rigid when its leg extension is at a minimum, providing of course that the feet are wide enough apart so that it does not topple over during the process of attaching the telescope and counterweights. The earlier comments on providing a sound and stable footing for your telescope mount/tripod/pier are even more relevant with the higher image magnification and longer exposure times associated with deep-sky imaging. An image can be ruined by an inadvertent 1/1,000$^{th}$ degree movement.

Most mounts are designed with damp outdoor use in mind and use stainless steel fixings and aluminum throughout. Hardware is one of those things that succumb to cost-cutting; for instance, the popularity of after-market hardened stainless steel Alt/Az bolts for the original Sky-Watcher mounts tell their own story. There may be other occasions too when a little "finishing off" can improve a value-engineered product, to improve its performance and protect your investment. Some of these are quite intrusive, involving complete strip-downs and rebuilding. It can be rewarding to buy a used mount and try this. A word of caution though, it obviously invalidates any warranty and be realistic about your own practical abilities. We may not be as adept at taking things apart and putting things back together as we would like to think!

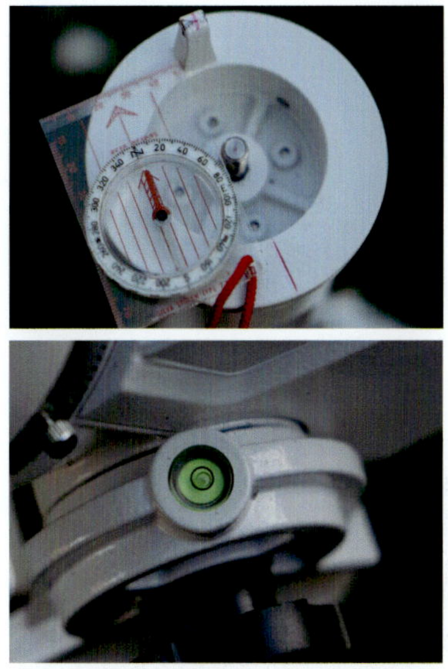

*fig.6* *When setting up a GEM, it helps later on to have a reasonable alignment, starting with the tripod or pier. Before the mount is fitted, the tripod is orientated so that it points to true north within a few degrees. The tripod is then leveled too, either by using the bubble level on the mount itself or a spirit level on the tripod platform.*

### Used Equipment

Buying a used mount may be a smart move. Not only will it be possible to buy a premium model at a budget price but it is often the case that the user may have already tuned the unit; including adjustments, accessorizing and running it in. It is not without risk, however. I would not buy one blind and would want to collect and check its general condition (which is often the case anyway since mounts are tricky to ship, especially if the original packaging is missing). At the same time, it is an opportunity to assess its likely treatment by inspecting the unit itself. In addition, a more general discussion on astrophotography with the seller reveals the reasons for the sale, usage, storage, example images and so on.

## Setting Up and Testing a GEM

Siting a temporary setup benefits from a little planning. For instance, a clear view of the celestial pole is handy for a quick mount alignment and although there are alternative methods, these take longer and are better suited for a permanent setup. Any installation should be situated on stable ground but additionally one should also consider the site, in regard to general safety, especially in a public place. An open space presents a wonderful vista but it is not a necessity for deep sky imaging. To avoid the worst of the light pollution and the degrading effect of the atmosphere, imaging ideally starts from 30° above the horizon. At 30° altitude, the optical path passes through twice as much atmosphere as it does pointing straight up to the zenith. This not only affects transparency and seeing but the angle introduces some refraction too. This is not usually seen, other than at high magnifications, for instance with planetary imaging, where the upper and lower part of a planet will have blue and orange/red coloring.

Bright lamps in the surrounding area can be another cause of grief: even though your scope may have a long dew shield, stray light from a bright light source will flare inside and affect the final image. Usefully, my local council switch off the street illumination after midnight, not out of consideration to astronomers but to save money.

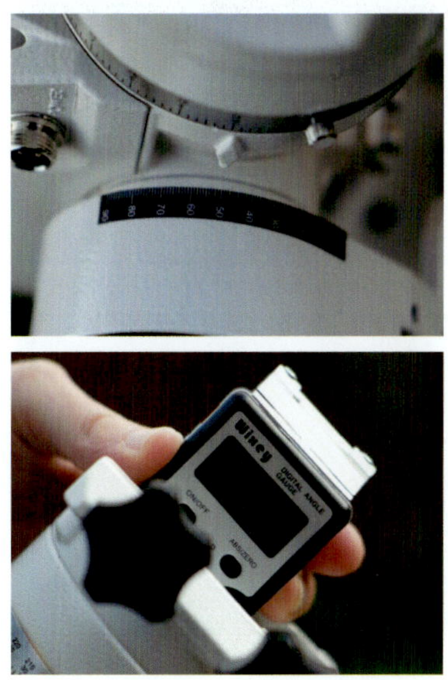


*fig.7* *The mount altitude is then set to the location's latitude. There may be a scale on the mount, or one can use a digital level (or smartphone app) resting on the dovetail clamp.*

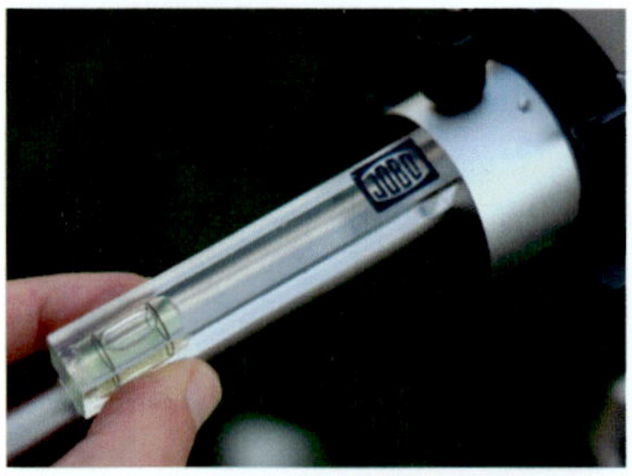

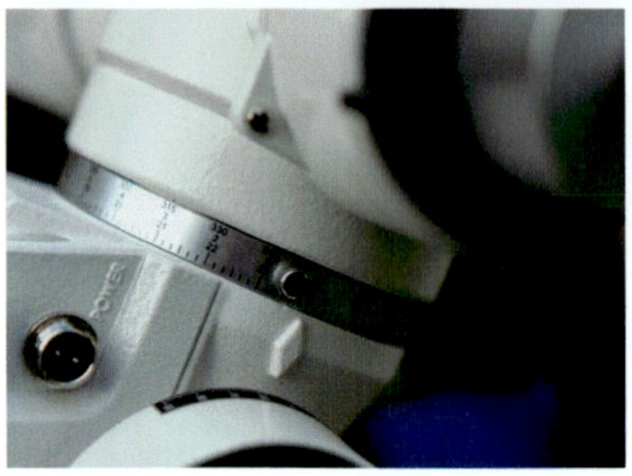

*fig.8 If your mount does not already do so, it is useful to establish an accurate home position, first by using the counterweight bar followed by the RA scale to rotate 90° (6 hours) so the bar is vertical.*

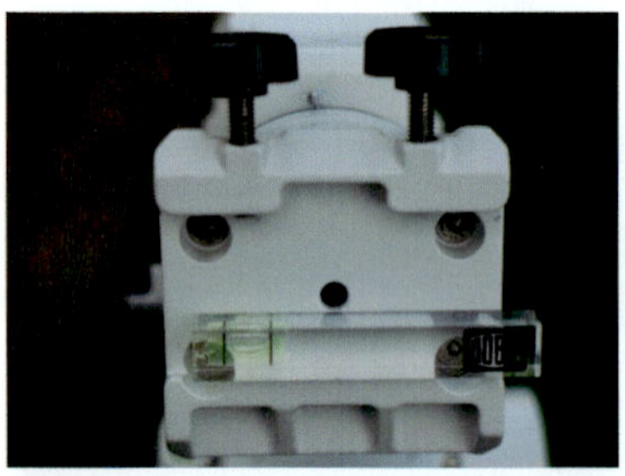

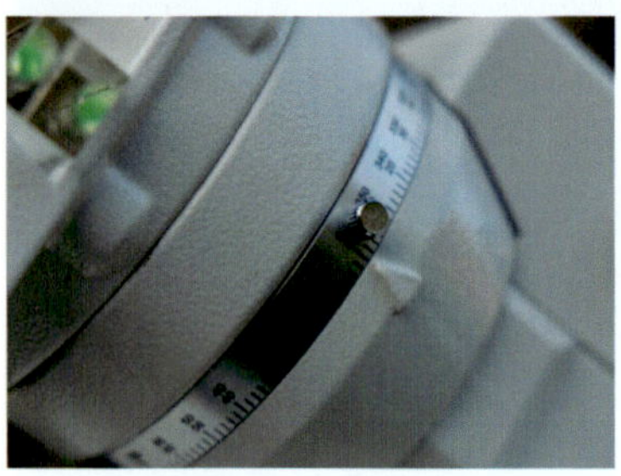

*fig.9 After setting the RA position, do the same for the DEC axis, using the dovetail clamp as a reference plane and the DEC scale to rotate 90° (or not, if you are using a side-by side saddle mount).*

### Tripod and Mount Alignment

In the case of an equatorial mount or a wedge-mounted fork, although the RA axis simply has to align with the celestial pole, there is some benefit from aligning the tripod with true north (fig.6) and leveling (fig.7) as it improves the accuracy of the first alignment slew. As described in *Setting Up a Simple Mount System*, leveling a tripod is easier if you think of it in terms of east to west and north to south. You only ever need to adjust two legs to level a tripod. Place one of the legs facing north and away from you and slightly extend all three legs. Place a spirit level across the two other legs east to west. Adjust one of these legs to level the mount. Turn the spirit level 90° and adjust the north leg to level north to south.

### Finding Homing

Most mounts have a home position. On some models, this is the assumed orientation at switch-on, or it may be a particular position defined by sensors on its RA and DEC axes. Either way, this position is used as the initial reference for star alignment and slewing. This is not to be confused with a park position, which is usually defined by the user, although on many mounts, including the SkyWatcher's, one of the defined park positions is also "home." On this mount, if one powers-up and down in the home position, a slew to an alignment star or target will be reasonably accurate (assuming the time of day and global position are known too). Setting the home position is a one-time setting but the act of setting up and dismantling a system will likely lose this reference. This is avoided if the home position is marked either by using the RA and DEC scales or on the body of the mount itself, allowing it to be easily rotated back into place, before powering up.

The most common home position is with the counterweight bar pointing vertically down and the telescope pointing straight up towards the celestial pole. To set this up accurately requires a spirit level and the following sequence:

1. define a position on the RA axis with the counterweight bar horizontal
2. rotate the RA axis by 90° and mark this position, with the counterweight bar pointing downwards (fig.8)
3. rotate the DEC axis so the dovetail jaws are horizontal (fig.9)
4. rotate the DEC axis by 90° (unless you are using a saddle plate to mount a camera and guide scope side by side) and mark this position.

Some mounts, like the SkyWatcher's have adjustable RA and DEC scales, others omit them entirely and may already have markings on the mount body that make steps 1–4 unnecessary. At worse, it may require a few slithers of electrical tape (which can easily be repositioned) or a fine permanent marker to define these references for the home orientation.

### System Assembly

With the smaller systems and given the opportunity, assemble the telescope and imaging components together in a well-lit space. Most telescopes are kept assembled to a dovetail plate for convenience usually via a clam-shell or tube rings that clamp the telescope. Handling a larger telescope is tricky and it helps to fit a handle between the tube rings on a refractor. I use an old sturdy leather belt, cut down to length so that it bridges the tube rings with a few inches to spare. After checking the optics are clean, pop the lens cap back on. If the telescope was last set up for visual use with a diagonal, remove it and insert extension tubes to achieve a similar focuser position with the imaging system. In the early days, I converted an inexpensive 2-inch Barlow lens into an extension tube, by simply unscrewing the optical element. I quickly abandoned this idea when I noticed the sag caused by the weight of the camera assembly and the play in both 2-inch clamps. Screw couplings are best. Several manufacturers now produce adjustable flatteners. These have a general optic that will work with similar telescopes with a built-in helicoid mechanism. This allows for precise spacing and has the advantage of non-intrusive adjustment.

With those refractors that have an accessory field flattener, these require precise spacing to the sensor to ensure consistent focus across the frame. If you are using a Canon or Nikon DSLR, you are in luck, since most telescope manufacturers sell an accessory bayonet coupler to the flattener which automatically sets the right spacing. Some telescopes, field-flatteners and focus tubes have a rotation device. You only need one and each is a potential source of flexure. After screwing the system together I double-check all the rotation features for play. Those that use three small fasteners to grip an internal circular dovetail are prone to work loose. I check these grub screws periodically and gently tighten them to secure the coupling. The all-nylon grub screws are a little too soft and I replaced mine with nylon-tipped stainless steel ones. These still clamp without biting into the dovetail and allow the assembly to rotate. Finally, once the camera is orientated at the right angle, I lock each rotation feature.

Next comes the autoguider system. With consumer cameras, this is likely to be a piggy-back scope or an adapted guide scope. If these are dedicated for the purpose you can save time by keeping them in their in-focus position using their focus-lock feature. It is useful to recall that precise focusing is not required for guide cameras and some applications prefer a slightly out of focus image, as it helps with the accurate measurement of the guide star. In the case of an off-axis-guider, these usually have a depth and focus adjustment. Both affect focus and so it is best, once set, to lock down. The depth adjustment inserts the pick-up mirror further into the optical field. Viewing from the telescope end, ensure it does not obscure the camera sensor in any way. I have mine so there is a tiny amount of space between the edge of the mirror and the long side of the sensor. I have a QHY camera and off-axis guider and although it is possible to directly couple, it is only possible to achieve focus with both if there is at least a 12-mm spacer in between. Fortunately, their filter wheel is 17-mm deep and so this camera "assembly" is a bolted combination of all three.

The next step is to assemble the telescope to the mount, preferably before fitting any wiring to the telescope; it is hard enough to carry the ungainly mass of a telescope without the additional trip hazard of trailing cables. When attaching the telescope to the mount there are a couple of tips to keep things safe and to avoid damage. Some mounts have sensitive drive systems and the assembly should be carried out with the drive clutches disengaged to prevent damage. During the assembly, the unbalanced system may also suddenly swing round. The trick here is to reduce the imbalance at any time so that you can easily support the telescope in any position. On the mount, loosen the clutches and swing the counterweight bar so it points downwards. For stability, slide and fix a counterweight onto the bar. Loosen the dovetail plate clamp and, cradling the scope, gently place or slide it into the dovetail clamp. With one hand holding the scope in place, quickly tighten the dovetail clamp and pop in the safety screw or tether. Hold the counterweight bar firmly and carefully assess the balance. If the assembly requires a second counterweight, now is the time to fit it and adjust both, so that the counterweight end just swings down of its own accord.

From here, it is safe to fit the various cables and wiring. *Remember, to avoid static damage, do not turn the power on (and that includes computers) before all the power and communication cables have been connected.* Extend the telescope's dew shield and wrap the dew heater tape around the telescope. This should be immediately behind the dew shield and as close as possible to the exposed optical elements. In damp conditions, I keep the lens cap on until the dew heater system has been on for a few minutes. Route the various cables from the computer to the cameras, focuser, filter wheel, dew heater and so on. If the connectors stick out too far and are at risk of catching on things, like tripod legs, consider changing the standard cables to those with right-angled connectors. Next, look out for potential cable snags; Velcro® cable ties are inexpensive and are an excellent way to keep the cabling from dangling, as is nylon mesh to bundle cables together (compare fig.10 with fig.11). To keep the balance consistent, route cables close to the DEC axis. (I attach a cable clip to one of the spare holes in the middle of my dovetail plate.) Set the focuser to the approximate focus position. (I set the focuser stepper motor's "home position" close to the in-focus position and to the nearest 10-mm marking on the engraved scale. In that way, it is easy to re-establish if I lose the reference position.) Once everything is assembled, you are ready to fine-tune the balance.

### Balancing

The general concept ensures that the telescope is balanced about the declination and right ascension axes so that the mount's motors are not put under undue strain in any orientation. Final balancing should be completed with the full setup, including cameras, guide scopes and cabling. For simple setups with in-line cameras, this is a simple two-axis check:

1. Tighten the DEC clutch and slacken the RA clutch and swing the counterweight bar to the horizontal. Without letting go, slide the counter weights back and forth until it is balanced. If the bearings are stiff, gently move the assembly in each direction and aim for a similar resistance to movement. Some mounts have electronic balancing and give a balance indication by monitoring the motor current in each direction.

2. To check the scope's balance about the DEC axis, with the counterweight bar horizontal, tighten the RA clutch. Support the telescope horizontally, slacken the dovetail clamps and carefully ease the dovetail back and forth to adjust the fore-aft balance point (fig.12). Remember to do this without the metal lens cap. Carefully tighten the dovetail clamps and if you have not already done so, screw in a safety stop or hook a safety cord around the tube rings and extended dovetail plate to prevent any accidental slippage.

Well, that is the theory. There are invariably a few complications; for instance, with heavy cameras, the focus travel also affects the DEC balance and ideally, the focuser should be at the focus position for balancing. It speeds things up for next time to mark the balance position on the dovetail plate against a marker placed on the dovetail clamp (I use a slither of white electrician's tape). If the scope has a large off-axis mass, for instance, a Newtonian design or a heavy off-axis guide scope, it may require additional balancing around the DEC axis:

3. With the counterweight bar still clamped horizontally, rotate the telescope to point straight up and balance about the DEC axis. In the case of a dual mounting bar arrangement (fig.13) slide the mounting bar along the dovetail clamp. A Newtonian or a lopsided assembly may require more ingenuity.

Balancing on this third axis can be quite tricky: a long scope may foul the tripod legs and a Newtonian scope has to be rotated in its mounting rings, without shifting it longitudinally, to place the focuser and camera in line with the DEC axis. (If you fit a third mounting ring, butted to the front of one of the main rings, you can use loosen the main rings and use this third ring as a fore-aft reference.) Other methods include an oversize

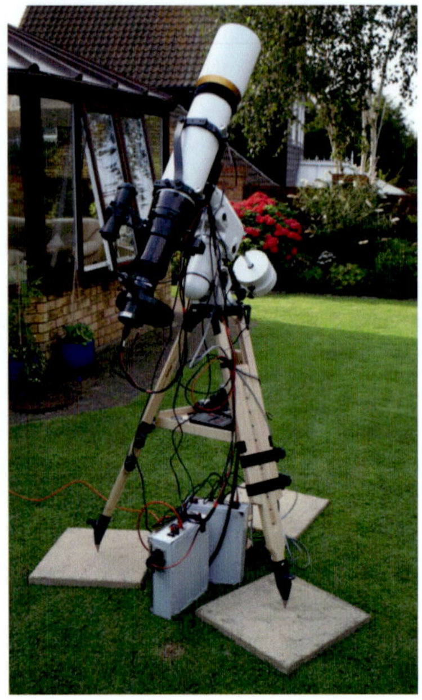

*fig.10 In the early days I wired up all my systems independently and, although I had two battery boxes with some of the modules housed in with them, the cabling was a mess and caught on handles and flopped about, causing tracking issues.*

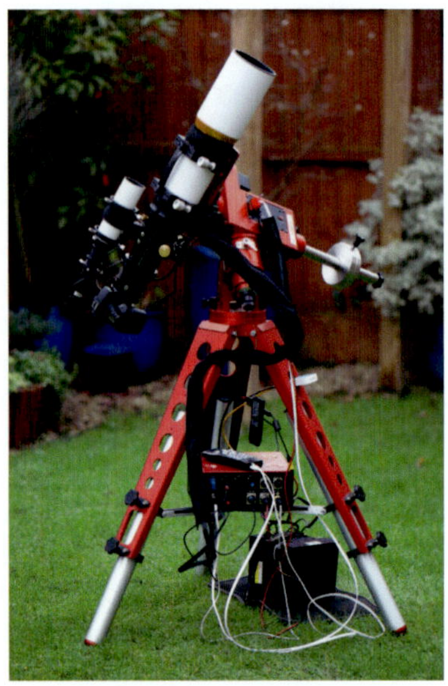

*fig.11 A later system uses a nylon mesh cable sleeve, to keep all the cables together and form a short cable loom. This makes assembly more reliable, repeatable and quicker too. The nylon mesh bundle is less likely to snag as the mount rotates.*

dovetail plate to which a weight is attached to one side. Some mounts are more forgiving than others. Those mounts that use an all belt-drive, rather than a worm or direct drive, benefit from careful balancing. Finally, just as you achieve perfect balance, I'll mention for those mass-produced models that use a traditional geared drive, it is sometimes suggested to introduce a small imbalance, to reduce declination backlash!

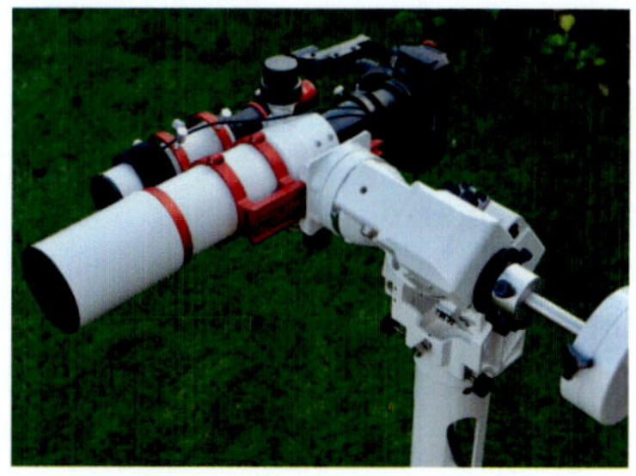

*fig.12* *When all the elements of an imaging system are in line, balancing the system is a matter of sliding the counterweight along the bar and the entire imaging system along the dovetail clamp. In the position shown, with the clutches disengaged, the finer points of balancing require judgement in so much that a small push on each axis and in each direction has the same travel. (It is normal to do this with all the cabling in place and they have been omitted here for clarity.)*

### Mount Limits

Everything is almost ready for connecting up to the computer but there are a few last one-time things to make note of: now is the time to define your local horizon (which can be a nominal altitude or a series of Alt/Az coordinates) and more importantly, to check for leg clashes especially when imaging near the meridian at high declinations. In each case, you need to slew the mount about and my advice is to stand by the mount and use the 4-way controller on the mount's handset.

Defining the imaging horizon is useful in more established setups and when you want to plan your imaging sessions with some precision. Otherwise, a simple 30–40 degree limit is usually sufficient. In the instance of the complex horizon – some computer programs accept a text file with a series of altitude/azimuth angles that define the local horizon. Others have sliders which set the horizon altitude for each compass point to the same effect. Another method of working out the local horizon is to make a panoramic image at the imaging site using a camera mounted on a level tripod. The programs differ in the detail but after merging the dozen or so images into a panorama, crop the image so that it forms a full circle. If you apply a square grid (for instance a Photoshop view option) it allows you to work out the horizon altitude in degrees. Even if the local horizon extends below 30° altitude, it makes sense to limit imaging to 30° and above.

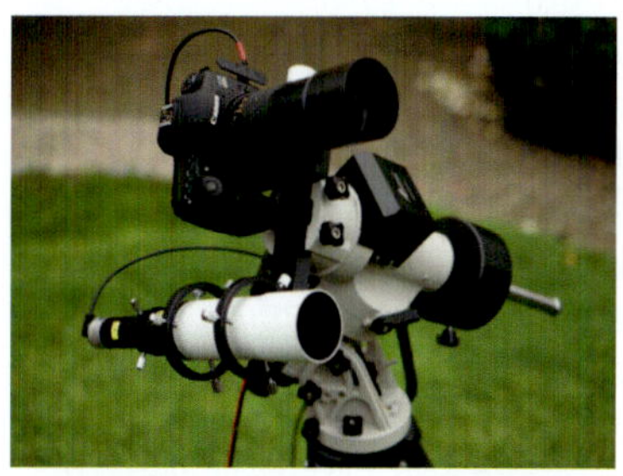

*fig.13* *Balancing is more complicated when the guider and imaging systems are mounted side by side. Here, after balancing the overall weight and the fore-aft balance along the camera dovetail plate, the system requires balancing on a third axis, by sliding the saddle plate side to side in the mount's dovetail clamp, with the optics pointing straight up. Finally, when the system is balanced, the entire camera/guider system should rotate evenly around the DEC axis with a little push and at any initial angle.*

Most mounts continue to track past the meridian for 10° or more unless software settings instruct otherwise. This is normally the point when leg clashes occur. Wide-bodied filter wheels and exposed electrical connectors make matters worse. Leg clashes potentially damage sensitive equipment and it is important to know the safe movement limits. To establish these, mount your worst offending scope (normally the longest one) and from the home position, rotate it so that counterweight bar is horizontal. Now rotate the scope to point straight up and check if the scope can swing past the legs without obstruction. The trick is to repeat this at slightly different RA positions until the scope just clears the legs. Note this RA value, either from the RA readout from the handset or from the mount setting rings. Depending on the model, enter this value into the mount's handset or take note for later inclusion into the PC control software.

Lastly, although we have not mentioned fork-mounted telescopes a great deal, they often have limited clearance for bulky cameras at high declinations. The fork arms are simply not wide or long enough to allow a camera to swing through unimpeded. Most mount control programs have a maximum declination setting to avoid clashes of this kind, above which it will not be possible to image. The better handsets have safeguards that request the user to confirm a second time that a slew is safe when it is near the limit, as do the better planetarium applications.

## Mount Initialization

Most mounts require some form of initialization before they can be used. To aim at the right part of the sky, they need an accurate sense of time and place. On top of that, they need to align with one or more points in the sky to fine-tune their bearings.

Most mounts require them to be set to the home position before switching on (as described earlier) and a few, which have an accurate position sensor on both axes do not, but require "homing" after powering up, so that the mount can find this reference position. The manner in which mounts determine time and place varies, added to which, many mounts do not have battery back-up and forget this information when the power is turned off. Those with built-in or accessory GPS sensors can acquire the information directly, usually through a utility command on the handset. If these options are not available, the data is entered manually. My SynScan handset falls into this last category. The site location in longitude and latitude is required, as is the time zone, elevation, date, time of day and daylight saving status. This is most easily accomplished by referencing the data from a smartphone application. (If you have a PC connected it is also possible to populate this data automatically from the computer via a series of ASCOM software commands. The book's website has a utility program to do just that.)

### Polar Alignment

Polar alignment is the next step in the initialization and for a GEM it follows similar lines to that of the simpler tracking mounts, described in the chapter *Setting Up a Simple Mount System*. The polar scope will likely not have perfect centering in its delivery condition and will first need centering on a distant target or star. Some GEMs have a polar scope permanently mounted within, or on a short rigid arm. Some of the internal ones have a static orientation or rotate with the RA axis. For those in the Northern Hemisphere, having entered the site location and time, the mount's handset may display a Polaris hour angle position to align on the polar scope clock-like reticle. Not all mounts have a provision for a traditional polar scope and some of the reticle designs are admittedly better than others. For those using a PC in their imaging setup, the popularity of the PoleMaster electronic polar scope has prompted a range of adaptors for use with any mount (or clamped to the dovetail plate). Some mounts additionally use their star alignment procedure to also align the mount's RA axis to the celestial pole (see below). This is useful if you are in the Southern Hemisphere or Polaris is not visible.

### Star Alignment

After polar alignment and before imaging, it is usual to do an initial alignment to the stars. This alignment accommodates all the inaccuracies of the polar alignment, mechanical tolerances, site and time errors as well as telescope mounting too. At its simplest level, the handset asks the user to select a prominent star, slews to where it thinks the star is and prompts the user to use the 4-way slew buttons to center it in the eyepiece/viewfinder. Once centered, a button press on the handset enables the mount to work out an offset that it can apply to all other slews. More sophisticated algorithms request one or two more bright stars. With more reference points, the handset can null out the effect of polar misalignment and any telescope cone-angle errors too, caused by the telescope being mounted at a slight angle to the dovetail plate. This alignment procedure is not mandatory in all circumstances but advisable if you are running the mount without PC assistance. After 2-or 3-star alignment, there is an option on some mounts, which do not have a built-in polar scope, to use the alignment data to effect polar alignment. The implementations differ between mounts but in general, the handset will prompt you to slew to another star and request the user to center the star, first using the slew buttons and then using the Alt/Az adjustment bolts on the mount. The star and polar alignment steps are repeated for greater accuracy. It is possible to achieve 1 arc minute accuracy in two or three cycles. More sophisticated mounts achieve this automatically through PC utilities.

After the initial alignment, there are a few other things that are worth setting up. These parameters are usually stored in permanent memory and are a one-time-only affair. These include autoguider rate, home and park positions, tracking speed, backlash setting, flipping modes, PEC training and so on. Some of these are accessible to an external PC too through the mount's application interface or ASCOM driver. We are almost ready to slew to an object and start imaging and, if you are using PC control, you may need to additionally instruct the mount to accept external commands, for example, the "PC Direct Mode" on SynScan handsets.

### Next Steps

The next four chapters, in short succession, discuss complete imaging systems, with PC control of the mount, guider and focuser.

# Acquisition Systems

*We are never far from a microprocessor. Astrophotography is no different and there are many different uses of computers to create sophisticated acquisition systems.*

Wherever we look, computers and software pervade modern life. Astrophotography is not exempt and even if the user is simply using traditional photographic equipment, they will undoubtedly be using a computer and image processing software to enhance their images. Astronomy software options are expanding and becoming ever more sophisticated, at a rate to rival that of the digital camera revolution in the last decade. This chapter is an overview of what software can achieve, some of the terminology and the considerations and choices open to you. Image processing software alone moves into a different ball-game as the demands of calibration and manipulation are outside the scope of traditional applications. At the same time, extended imaging sessions favor computer control and automation, which is the subject of later chapters.

## Image Acquisition

When one wishes to move beyond a simple cable release or intervalometer, computer control of the imaging rig is the next logical step. It brings with it a whole host of opportunities and challenges at the same time. There are some days when I can sit back and relax and watch it all work it out for itself and there are others, when I wish I had taken up oil painting. It, of course, presumes one has a computer, typically running Windows but potentially OSX or Linux. (At the time of writing, while there are many utility applications for mobile devices, image acquisition is not one of them.) The applications and utilities cover a wide range of functions:

- polar alignment utilities (useful)
- camera control and acquisition (essential)
- planetarium and mount control (useful)
- guiding (essential)
- focus control (essential)
- filter wheel control (useful)
- mount alignment (useful)
- weather monitoring (nice to have)
- notifications (nice to have)
- remote control (nice to have)

This list looks intimidating at first and is a mixture of things that operate certain equipment, utilities that make life more convenient and those that improve imaging performance. For the budget conscious, there are many affordable commercial applications and free software applications that cover all the essential needs and a few that we didn't know we needed. To give an idea of the range, fig.1 shows a list of popular software applications, pricing and general capabilities at the time of writing. The choice is extensive, but at the same time, rather confusing and it is not a practical proposition to evaluate them all. I recommend using a few of the simpler applications to start with and stick with them for a while, to fully understand their foibles. If you chop and change about too frequently, the result is a perpetual, chaotic learning curve. After a while, one can make a more informed purchase decision from one of the more advanced applications.

Most software applications generally support one of the two activities; image acquisition and processing, a few do both. Some of these will be introduced in later chapters when their time is due. First, we need to step back and consider computers.

### Computing Hardware and Platforms

Alternative operating systems are a favorite punch-bag for Internet forums. I use both Macs and PCs daily but prefer Apple Mac OSX in general, especially for image processing and publishing. It is not something to get hung up over, as both operating systems have their good and bad points. The reality is, however, that there are simply more astronomy programs available for the Windows platform. Having said that, one only needs a working system so a choice of say 10 rather than 3 planetarium applications is not a big deal. A more important issue though is hardware support; in OSX, the astrophotographer is reliant on the application (or operating system) directly supporting your hardware. That can also apply to Windows applications but usefully, most hardware manufacturers support ASCOM, a vendor-independent group of plug and play device drivers that provide extensive hardware and software interoperability. ASCOM runs on Windows and although Mac software may improve with time, presently the image capture, focuser and planetarium applications do not support all available astronomy hardware. It is more likely that Indie, running on Linux will overtake Apple for astronomy applications. I initially assembled a system around Apple OSX, including planetariums (Equinox Pro, Starry Night Pro and SkySafari), image capture (Nebulosity), autoguiding (PHD2) and image processing (Nebulosity and Photoshop). I produced several pleasing images with this combination and a MacBook Pro with a 9-hour battery life. As my technique improved, however, I became increasingly aware of poor alignment and focusing and I eventually switched operating systems to facilitate better control. In recent years, the runaway success of the Raspberry Pi has not gone unnoticed by the astrophotography community and the low cost and independence is appealing. At the time or writing, however, the number of supported devices and general ease of use has some way to go to match the established platforms. While interesting, ultimately we require a system to support all our hardware and above all else, be reliable and fully supported by the OEM.

Operating systems are also generally migrating from 32- to 64-bit. The latest version of OSX requires 64-bit applications. Emulators and virtual OS environments may appear to be the answer but in practice, on a Mac, it is faster and more reliable to simply boot into a Windows BootCamp partition. In practice, I continue to use Windows 7 Pro x64 as well as Windows 10 Pro x64. The latest version of Windows 10 has an automatic update feature that cannot be entirely disabled. System updates during December 2017 caused chaos for many astronomers and subsequent updates have been problematic for individual applications in Win 7 and Win 10. For a *standalone* PC, there is less risk from vulnerabilities and I will continue to use Windows 7 as long as it is viable, even though Microsoft have dropped support.

Laptops are a popular choice both for convenience and portability. Dew and computers do not mix and if I do use my MacBook Pro outside, I keep it off the ground and use a hard-shell case and silicone keyboard cover. Another alternative is to use an inexpensive Windows laptop or one of the miniature media PCs. Battery life is an obvious concern for remote locations and an external lithium battery pack is an effective, if expensive, means to supplement the computer's internal battery, to deliver a full night's use. The demands placed on the computer hardware are not as extreme as that needed for modern PC games or video processing and it is possible that a retired PC from an upgrade may be ideal for the purpose. While Netbooks appear to be sufficiently powerful, my 1.6 GHz, 1 GB netbook had marginal processing resources and insufficient screen resolution to display an application. The shared resources also caused issues during image capture from an 8 megapixel CCD camera, causing stripes in the image.

| Title | Price ($) | PC Mac | Mount Control | Auto Focus | Guide | Advanced Control |
|---|---|---|---|---|---|---|
| **Packages** | multi-purpose packages | | | | | |
| Maxim DL | 400-600 | PC | Yes | Yes | Yes | Yes + API |
| AstroArt | 165 | PC | Yes | Yes | Yes | science focus |
| The Sky X (Pro) | 349 | PC/Mac | Yes | Yes | Yes | Yes + API |
| Nebulosity | 80 | PC/Mac | No | No | No* | No |
| Images Plus | 239 | PC | No | Yes | No* | majors on processing |
| **Acquisition** | acquisition programs | | | | | |
| Astro Photography Tool | 20 | PC | Yes | Yes | No* | various utilities |
| Backyard EOS Backyard Nikon | 50 | PC | No | No | No* | majors on DSLRs |
| N.I.N.A | Free | PC | Yes | Yes | No* | + sky atlas |
| Sequence Generator Pro | 99 | PC | Yes | Yes | No* | Yes + API |
| **Utilities** | plate solving and guider utilities | | | | | |
| PHD2 | free | PC/Mac | N/A | N/A | *links with acquisition programs (above) | |
| MetaGuide | free | PC | N/A | N/A | *links with acquisition programs (above) | |
| Astrometry.net | free | PC/Mac | N/A | N/A | N/A | all-sky plate solver |
| ASTAP | free | PC | N/A | N/A | N/A | plate solver/stacker |
| Pinpoint | 199 | PC | N/A | N/A | N/A | plate solver |

*fig.1 A wide range of image acquisition software and utilities, prices and capabilities. Some are multi-functional, including planetariums, image processing, modeling and scientific functions. Others are aimed squarely at more modest needs and budgets and are more relevant to the aims of this book.*

## Connectivity

Connectivity is key; Mac laptops are increasingly abandoning the traditional mixture of display and USB ports in favor of USB-C. Astronomy equipment generally uses USB 2.0 (with the exception of the faster CMOS cameras that require USB 3.0) and RS232 serial. A few devices use WiFi, Bluetooth, FireWire or Ethernet interfaces

USB has a limited range; the USB 2.0 cable limit is about 5 m and USB 3.0, even less, at 3 m. The various astrophotography forums bear witness to the fickle nature of USB communications. It is common for a member to condemn the imaging hardware or software, only to find out it is a cable or hub issue. It is false economy to use inexpensive USB hardware. I use high-quality brands and industrial-strength hubs, like the ones from StarTech, that work in colder conditions and have built-in power regulators. The USB cable limitation also dictates the placement of the computer. That is not always desirable and it is possible to extend the range of a USB cable. There are a number of repeaters, extenders and hubs, some of which work better than others. For a few years, I used a system that extended USB over CAT 5 Ethernet cable, up to 300 feet. At one end of the cable was a transmitter, connected to the computer's USB port and at the other, a 4-way USB 2.0 port. It was utterly reliable and I only replaced this system, at similar cost, when I decided to go wire-free with a miniature PC module at the mount, entirely operated over WiFi using remote desktop software.

Good ol' RS232, operating at 9,600 bits per second serial is commonplace in astrophotography hardware and while it is all but forgotten by the computer industry, it is reliable and convenient for low-speed device control. Crucially, it works well over long cable lengths. Few modern computers include serial ports and we rely on USB/serial converter cables. Each module that has a serial interface, requires its own USB/serial adaptor. Quite a few modules use serial-based communications but have a USB socket. These use the same integrated circuits as the adaptor, but in this case the chips are soldered directly onto the module's circuit board. In either case, the PC's device manager treats these devices as a virtual communication port (VCP), assigning each a unique COM number. I have 6 COM ports in my system, interfacing to mount, focuser, observatory, power control and weather systems (fig.2). Just as with USB hub designs, there are a few different chip-sets used within the adaptors, some of which have been found to be more robust than others. In practice, I find those based on the FTDI chip-set are reliable and I notice that most of my modules use that chip-set too. COM ports are assigned dynamically and unless everything is permanently connected, the assignments can change, messing up connections. I open each USB serial port in Device Manager and manually set the COM port number in the advanced properties. This should be remembered for the next time you power up.

In addition to USB, WiFi/Ethernet is essential. The Internet is the most convenient means to set an accurate clock time, update the software and allow remote operation. Backup is essential and a high-capacity external hard drive is essential to store the gigabytes of image and calibration data that astrophotography quickly acquires. After each night's imaging, I copy the image data over to an external drive and keep that safe. It is useful to keep the original data since, as one's processing skills improve, another go at processing an old set of image files often produces a better result.

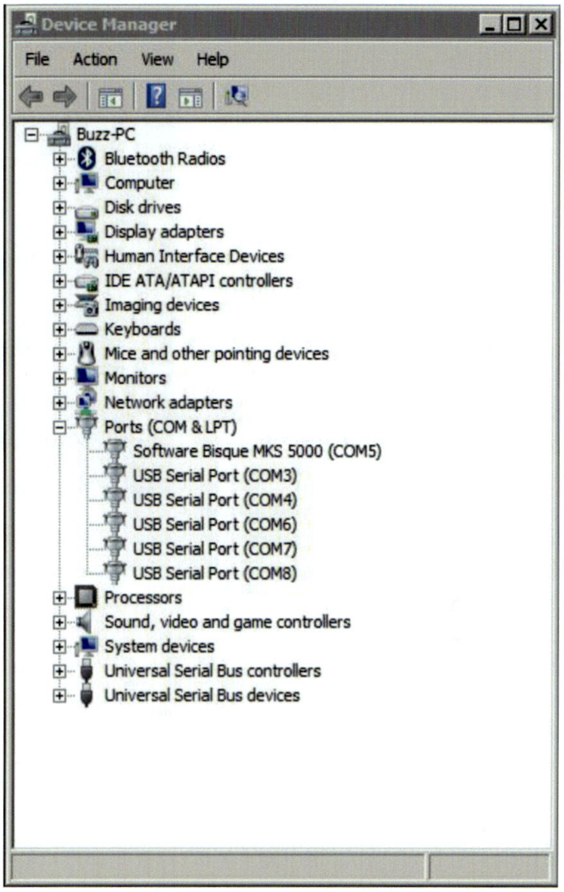


*fig.2 RS232 is not dead. There are 6 serial ports in my observatory system, including the mount, focuser, weather monitor, roof controller, power control and environment sensors. Each uses a USB to RS232 adaptor or chip in the module.*

## ASCOM

Throughout this book there are numerous mentions of ASCOM. The ASCOM initiative is a consortium of software developers and manufacturers who came together to find a way of standardizing the software interface for astronomy devices. The problem is that every device has a different way of operating and, as they proliferate, it becomes an impossible task for the application developers to write code for each one. ASCOM defines a standard set of methods and properties (subroutines and variables if you are as old as me) some of which have to be implemented and others that are optional. Most hardware has some form of interface, with a defined set of commands. This might be a serial interface, with a menu of commands and responses, or more sophisticated, as in the case of camera interfacing. This interface is defined and made public for developers to write their own device driver. These protocols are called the device's API, or APplication Interface. A device's ASCOM driver uniquely interfaces to the API but outwardly, shows a set of standardized commands and responses. In the case of the ASCOM driver that I wrote for my roll-off roof, it has just a few serial commands: open, close, status and so on. Any application can control my roll-off roof by calling on my driver, without having to know how my roof operates. The latest ASCOM platform, utilities and some drivers are found on the ascom-standards.org website. Others are obtained from the support pages for the device. While not perfect, ASCOM is one of the reasons that most astrophotographers choose Windows to run their observatories. In some cases, where a particular piece of equipment is very common, the application provider may well write their own driver to make full use of the device's API. Advancing technology sometimes leaves the ASCOM protocols behind and it takes a while for standards to catch up with new features. It is not uncommon for users to incorrectly blame image acquisition software for ASCOM driver implementation issues.

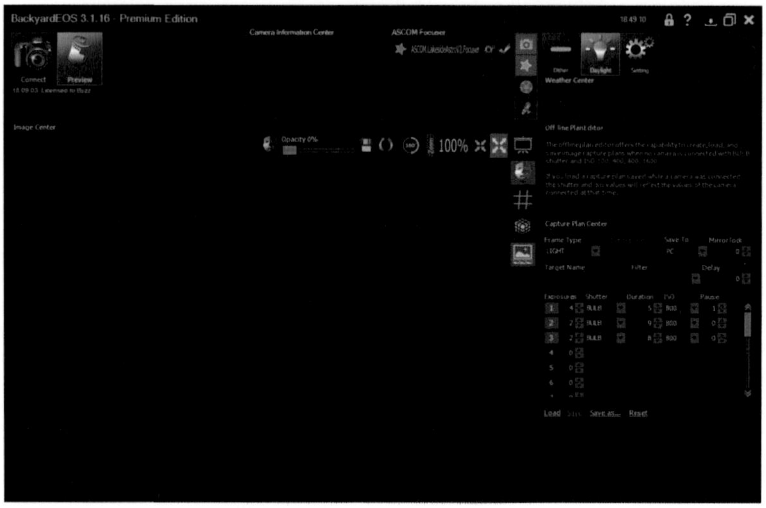


*fig.3* *Backyard EOS (BYE) and its Nikon version are popular acquisition programs for those who use their DSLR for imaging. It can replay sequences of exposures and provide previews to help with focus and framing.*

### Software Choices

Image acquisition software offers a dizzy range of capabilities: some do a single task very well; others take on several of the major roles of telescope control, acquisition and image processing. There are several major integrated packages including Prism, Maxim DL and TheSkyX. The fully featured versions are not budget options ($500–$600) but they are able to display, control, capture, autofocus, guide, align and process images (to varying degrees). These are overkill for our purposes and it is possible to find several applications that, together, perform the same task for less outlay. Deciding which applications to choose is a difficult and personal process. You will undoubtedly get there after a few false starts and as your experience grows you will likely move away from the simpler applications. Thankfully, many companies offer a fully-featured trial period with which to evaluate their product. This may give adequate time to check for hardware compatibility and basic performance (providing you have clear skies). It is tempting to continually change applications and workflows, as a result of forum suggestions and early experiences. Something similar occurred in photography, with the allure of magical film and developer combinations. In both cases, the recommendation is the same; it is better to stick to one system for a while and become familiar with it, before making an informed decision to change to something else.

Choosing an application is not easy, especially as there is a healthy degree of feature overlap between titles. What follows is a walk through some of the utilities and applications that will be useful to you, in the likely order in which you encounter them:

### Camera Control and Acquisition

A plug-in intervalometer works well with a standard camera on a single-axis mount and for landscape work, it may be all that you need. As the imaging time extends, acquisition becomes more complicated or you wish to control other functions, remote control is the answer. Canon and Nikon publish an Application Interface or API. These permit camera control and image download over USB (and increasingly WiFi). The thing that sets the acquisition applications apart is their hardware compatibility and automation. The actual function is straightforward but there are many potential interfaces and protocols, which change with time. Most applications will directly control a Canon or Nikon camera via its USB port, setting their ISO and exposure time, triggering the exposure and downloading their RAW files directly to the computer (rather than the memory card). Some can even move the focus position of a DSLR lens.

Other major brands are less well served. These same applications increasingly support dedicated CCD/CMOS cameras too, allowing the photographer to use the same software after camera upgrades. For beginners, Nebulosity is hard to beat for ease of use. Other low-cost alternatives have matured into fully-featured applications that include sequencing, filter wheel and focuser control, automatic centering and other utilities. There is no such thing as a free lunch and the original uncluttered simplicity and ergonomics suffer at the hand of a growing list of features (especially on a small laptop screen). There also comes a point when, like video recorders of the last century, half the features are never used and clog up the interface. I always check an application works with my hardware and assess the quality of its support from the web page/forum too before purchase. Bad news travels fast and it is instructive to see how long it takes to fix a bug.

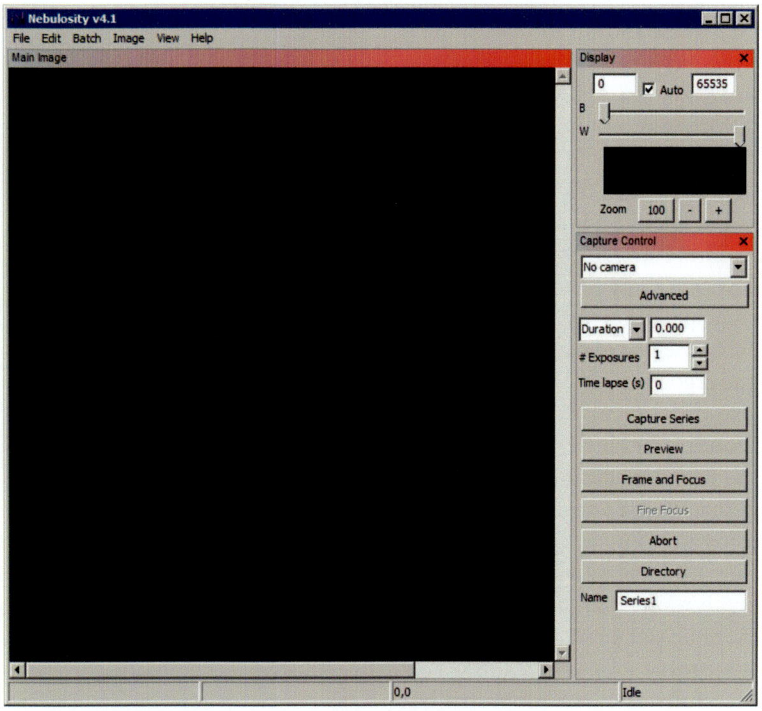


*fig.4 Nebulosity has a simpler, less intimidating interface and is an ideal choice for the beginner using any digital camera. It also includes essential image calibration and processing tools, all for a small outlay. It has a hands-on approach, with manual focusing and target centering. It is available for Mac and PC versions.*

Nebulosity 4 is available for both Mac and Windows platforms and has both acquisition and image-processing tools. It has a very simple interface and is my top pick for the beginner. For Windows, NINA is a relatively new open-source acquisition application, Backyard EOS and BackyardNikon have a loyal following, as does the "Swiss-army knife" Astro Photography Tool (APT). I typically use Sequence Generator Pro in my observatory. It is similarly priced to Nebulosity 4 and although it does not process images, it features advanced device control and automation, to support all-night unattended imaging.

## Polar Alignment

There are perhaps more polar alignment utilities than there are decent imaging nights in a UK year. It can be a bit of a crusade. I can see the appeal if one has an observatory and can dedicate an entire night to polar alignment but for a mobile dual-axis mount setup, I find a polar scope and autoguiding are sufficient. For those polar scopes that have a clock face, a smartphone application is often the most convenient method to set the precise position on the reticle. For single-axis mounts, polar misalignment may cause drift in declination, which cannot be corrected with guiding. Some time ago, I invested in the QHY PoleMaster camera-based polar scope. It has adaptors for numerous mounts and achieves sub 30 arc second alignment in about 5 minutes. Its supporting software supports Macs and PCs.

## Planetarium and Mount Control

Things become more interesting when you want to image deep-sky objects. These dim objects are difficult to locate in the sky and will generally be invisible on a camera LCD screen. The answer is some form of planetarium, preferably linked to a telescope mount. A planetarium is an electronic map of the sky, including stars, galaxies, comets and nebulae. The heavens are on the move and a planetarium dynamically adjusts to your geographic location, date and time. When these are on a smartphone, equipped with GPS, compass and a gyro, the better ones dynamically adjust to the phone's orientation.

At this point, you know roughly where to point your camera, usually with respect to some prominent star or constellation or the coordinates in RA/DEC or ALT/AZ. If, however, you have not been able to contain your excitement and already have bought a dual-axis motorized mount, you will know also know that the handset or planetarium will be able to point the mount in the right direction via serial, USB, Ethernet, WiFi or Bluetooth interfaces.

There are many applications covering these functions, with the principal difference between them being their hardware compatibility and ergonomics. The heavyweights also do acquisition, with a corresponding heavyweight price. Some are very graphical, others less so, with a lower demand on computer resources. For imaging purposes, they all have sufficient data and precision to plan and point most mounts to the target object. I own the pretty and educational PC/Mac fully featured programs but often use a simpler iPad application SkySafari, for image planning or simply enter the target object into a planning utility. Four free planetariums are Stellarium, CdC , HNSKY and C2A. Once the imaging sequence is running, however, the planetarium is largely redundant. For this purpose, I rate the planetarium applications by their ease of navigating around the sky; searching for an object, displaying its information, zooming in and displaying the camera's field of view for any particular time in relation to the horizon and meridian. These are basic requirements but there is an amazing difference in the usability of available programs.

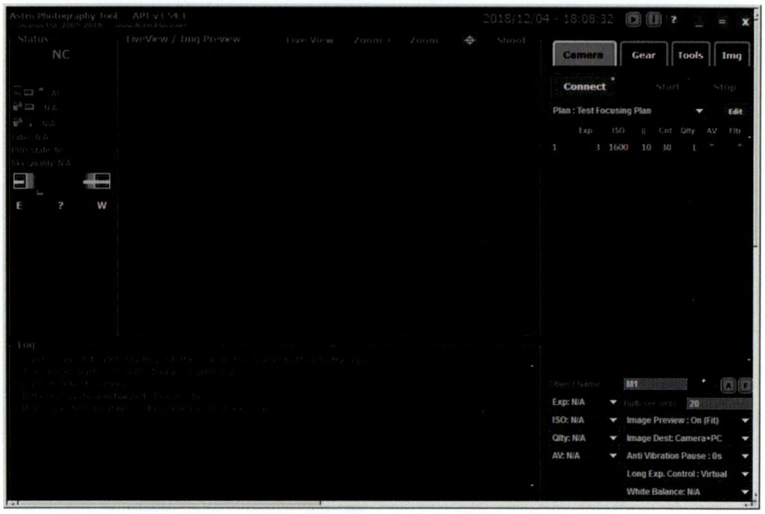


*fig.5* *APT has many more controls and utilities than BYE and concentrates on image acquisition sequences. It interfaces to many device types and other applications to extend its functionality. It can populate targets with coordinates, command a mount to slew and center on the target and run autofocus routines. This optional dark interface is useful in dim ambient settings.*

If the pictorial aspects of a planetarium are not required, one may simply use a planning tool to provide the location of or identify, promising targets. AstroPlanner and Skytools 3 offer a database selection approach; selecting objects using a set of user-entered parameters, including position, size, brightness and so on. These can interface to the mount or alignment programs as well as generate a target data file that is read by the image acquisition program. These assume, of course, that you wish to place the object in the center of the image, which is not always the case.

Most mount manufacturers define an interface control protocol that allows PC/Mac/mobile control through a serial or USB port. Physically, this either couples directly to the scope or via the mount control handset. Some of these drivers are fully-fledged applications too,

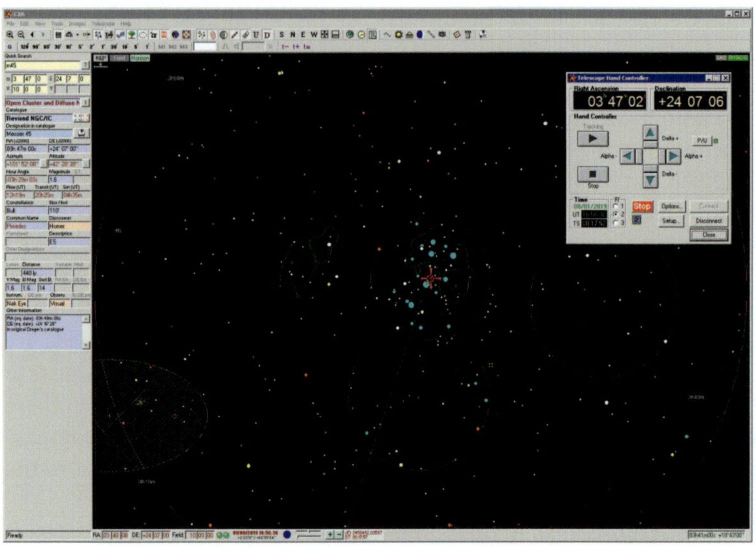


*fig.6* *A screen grab from the free planetarium C2A. Like others, it offers an object search facility and gives the boundaries of common objects as well as telescope control. Stellarium and Cartes du Ciel (CDC) are two more popular choices.*

emulating handset controls and settings on a computer. In the case of the popular SkyWatcher EQ mounts, an independent free utility, EQMOD, largely replaces the Synscan handset and allows direct PC-to-mount control, including a host of useful features and utilities.

### Autoguiding

Even with excellent polar alignment, guiding improves tracking during extended exposures. Although there are some standalone guider systems that do not require a PC, many choose one of the few standalone applications, of which PHD2 is far and away the most popular with still cameras and MetaGuide, with video cameras. In general, these connect to a dedicated small guide camera and after locating a suitable guide star, repeatedly take short exposures. They quickly compute the exact center of the selected guide star and detect any slight deviation, to an accuracy of about $1/10^{th}$ of a pixel. The application then issues a corrective guide command, either through the ST4 hardware interface or a software command, resulting in a tiny adjustment in the mount's tracking. These movements may be as small as a $1/20{,}000^{th}$ degree. When choosing the application the compatibility and ease of use are the primary considerations; ST4 is slowly being upstaged on the more expensive mounts by software guide commands like pulse-guide. On the other hand, single-axis mounts are not as intelligent and they rely on ST4's simplicity. Most guide cameras and some imaging cameras have an ST4 output. Failing that, it is possible to find a USB-driven ST4 interface and ASCOM driver. Guiding is a challenging subject in its own right and is explained in much greater detail later on.

### Focusing

After guiding, accurate focusing makes a big difference to image quality. While manual focus will suffice for a while, there are a number of reasons why software-assisted focusing is a good thing and required if high-quality imaging is on the agenda.

- as the image magnification increases, the merest touch of the focus knob causes the star to wobble around, making it difficult to judge the best focus
- in the case of the longer focal lengths, the focus position drifts as the telescope cools down, preventing focusing being a one-time activity
- a computer is a better judge of image focus than the human eye
- a small amount of defocus exacerbates any optical defects

Focusing software helps at a number of levels. At its simplest, it can acquire an image and give you a numerical evaluation of the focus with either a manual or motor-driven movement. If the focuser is motorized and is under PC control, it can additionally move the focus position, with minimal vibration and find the best focus via a number of image acquisitions, assessments and moves. Some applications, like Nebulosity, do not have focus control but can take repeated exposures and give a numerical readout of the peak intensity and star size. When these are maximized and minimized respectively, this is the point of best focus. Some other packages like APT can do the same, directly controlling EOS lenses and analyzing focus mask images better than the human eye. The more advanced software packages have an autofocus feature that integrates with the camera control. These find the best focus position by progressively checking the appearance of a star at different focus positions to work out the best setting. Some others will characterize the thermal trend of a telescope's focus and automatically adjust with ambient conditions.

### Mount Alignment

Mount alignment requires a sense of time, location and an initial physical orientation. Locating an almost invisible object is challenging and there are many software alignment utilities. To begin with, an accurate sense of time and place is essential. PCs do not generally have good clock accuracy and they rely on periodic Internet updates. In practice, my computer is only turned on for an imaging session and its clock is usually out by several seconds. Each second is equivalent to 15 arc seconds of pointing error and as my imaging computers are connected to the Internet, I use the free utility, Dimension4, to explicitly synchronize the PC to a time server using the network time protocol (NTP).

The sky orientation also depends on your geographic location and is most conveniently obtained from a GPS sensor or a utility on a smartphone will locate one's latitude and longitude for the mount's handset or application driver. From this point, alignment concerns itself with accommodating the initial mount alignment and mechanical errors in the system. Some mounts have a reference orientation, others require synchronization with 1–3 prominent stars using a utility on their handset. This alignment will get you in the ballpark, such that your target will be in the sensor's field of view. Some PC applications or features within an imaging application will go one better. These use a utility program called a plate-solver.

This analyzes an image and compares it with a star catalog, computing the image scale, orientation and precise center coordinates to fractions of an arc second. AstroTortilla, PlateSolve 2, ASTAP and Astrometry.net are several free plate-solvers. They are typically called from within the image acquisition program as part of an accurate centering process. This process compares the true position with the desired one and then issues a slew command to adjust for the pointing error.

### Weather Monitoring

Weather monitoring gives peace of mind and reduces the risk of equipment damage during unattended operation. At its simplest level, a simple rain detector module (triggered by raindrops bridging two conductors) closes a relay, causing an alarm to sound. Prevention is better than cure and for all-night, unattended imaging, I use a cloud detector. This device measures and compares the ambient temperature with the infrared sky temperature. This comparison is interpreted as a cloud cover measure and can either close a relay or be read by software to take action.

Weather prediction does not have to involve costly hardware, there are now various Internet weather feeds that provide weather and predictive rainfall for your locality. One such is OpenWeatherMap, for which an ASCOM driver was released with ASCOM platform 6.2. The Dark Sky application is equally effective on the web, smartphone or watch. One of its features is an imminent poor weather warning, with sufficient time to put things away. Be careful with plastic covers. I know from experience that they can act like a sail in windy conditions and I once had a 50-kg system blow over in a freak wind gust.

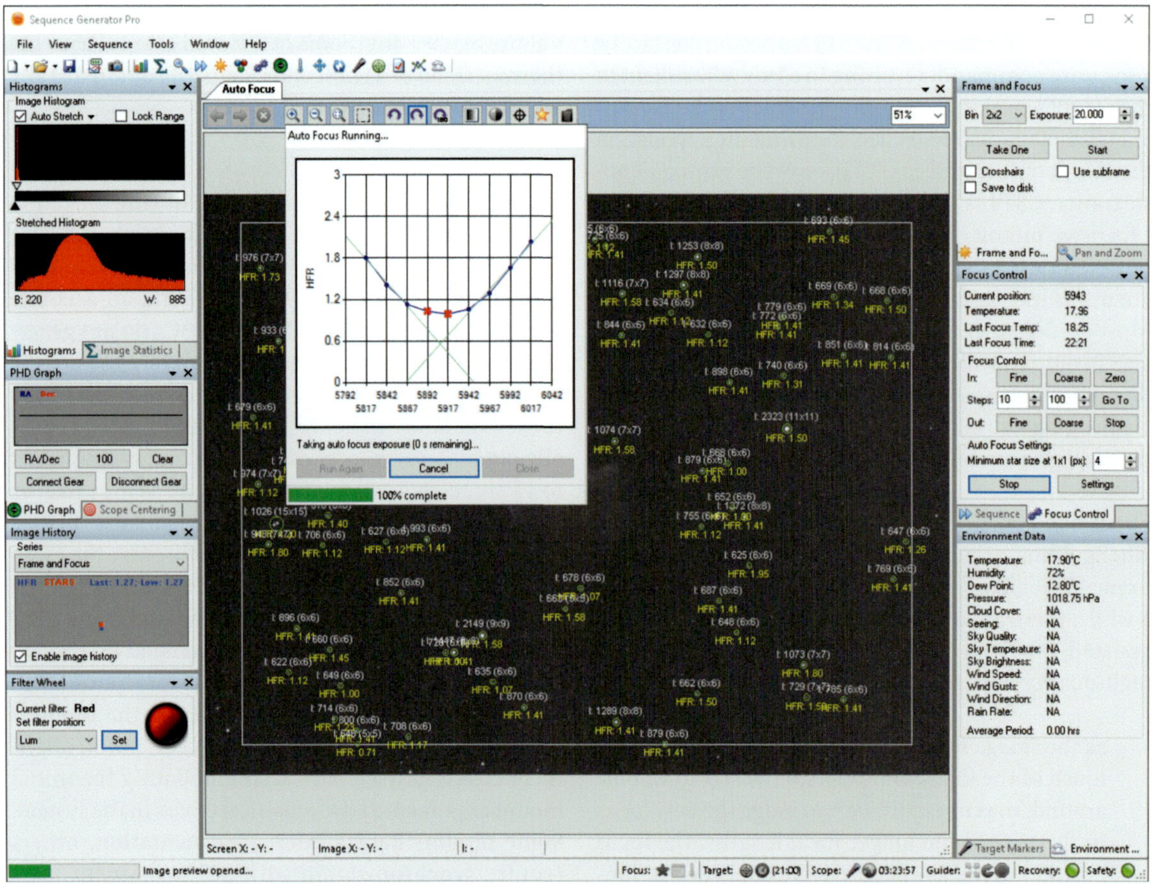


*fig.7 I use Sequence Generator Pro extensively as the acquisition program for my observatory. It allows full automation and unattended operation. It works well with simpler equipment too and although it is a little more expensive than the entry-level applications, if the intention is to progress to more serious imaging, it may be worth skipping these from the start. Its developers are responsive to feedback and for still-image work, is hard to beat. Other applications are better suited to planetary video work or EAA, (Electronically Assisted Astronomy), which sits between hard-core astrophotography and visual astronomy.*

# Setting Up an Acquisition System

*Put your feet up and let the computer take control. Well, in theory anyway.*

While we may have already experienced the advantages of PC camera control and assisted focusing in wide-field imaging on a simple tracking mount, full computer-controlled imaging is a watershed moment. This benefit comes from combining a dual-axis mount and PC-assisted imaging, discussed in the previous chapters. This combination allows one to control the imaging sequence entirely from a computer. As the complexity of the system increases, however, so do the number of things that can go wrong and it will take a while before you can rely on a system to perform without manual intervention. For my permanent installation, I have now reached the point where I'm ready to start in a matter of minutes and before that, with practice, I could assemble and align my portable system in the back yard and be imaging in 15–20 minutes.

In this chapter, we put aside the simpler Backyard EOS and Nebulosity applications and examine the setup and operation of popular applications such as Astro Photography Tool and Sequence Generator Pro, which permit more automated control. The following chapters look at the specialist topics of autoguiding, autofocus and remote operation.

As mentioned before, computerized imaging requires a mount that allows external PC control. I should clarify; that is a computer running Linux, Windows or Mac OSX. While tablets and smartphones can connect to a mount and slew it to a target, they are presently unable to do the other aspects of robotic control such as image acquisition, focusing, guiding, centering and so on. Although the mount and modules may communicate via serial cables, the camera system will almost certainly be USB 2.0 or 3.0, mandating a computer or powered hub close to the mount.

### PC Setup

The demands of image acquisition are not that taxing and it does not require a top-spec machine in the field. In fact, a high spec machine will generally have greater power consumption and may well suffer from shorter battery life. Many choose an inexpensive laptop when they start out and later on, choose a brick or stick computer if they are moving on to remote control. My recommendation would be to load the machine with Windows 7 Pro or Windows 10 Pro. There are a number of reasons for this; firstly there is more software available for Windows than any other platform and secondly, the pro version gives you more control over forced updates and allows remote control too. Laptops are not ideal for image processing and if possible, dedicate this machine to the task of acquisition, which then allows you to strip out all applications and processes that are not required, including mail, social media and so on. The best route is to do a clean OS install and then wait a few hours for the operating system to update itself. In Windows 10, Microsoft can force updates and restarts on what it thinks are critical updates. Since my computer is single purpose, is not used for e-mail or browsing and is working behind a firewall, these updates are not critical to me. Before I was wise to it, these unplanned and unannounced installations caused my observatory system to reset in the middle of the night. The update policies are continually changing and currently, in Windows 10, you can only delay updates through the settings menus. Some system updates in recent years caused issues for several astronomy applications but were eventually rolled back. In frustration, I reverted to Windows 7, though I know at some time, an application or device will force me to upgrade again. Some of the more knowledgeable users have discovered a method that requires approval to load and update the system, by modifying the Windows Update property in the Group Policy Editor. This may change again and so an Internet search on a tech website will offer the best up-to-date advice on this.

When it comes to loading and configuring applications, I have a preferred order that seems reliable:

1. disable fancy Windows themes, animations and screen-savers
2. set the desktop to dark grey or black, to minimize light output
3. enable .NET 3.5 framework in Windows
4. install ASCOM platform
5. install NOVAS (ASCOM)
6. install device drivers
7. install ASCOM device drivers
8. set up document and catalog folders

9. install autoguider software
10. install planetarium(s)
11. install acquisition software
12. install plate solvers and catalogs
13. install utility applications and drivers
14. install a Network Time Protocol utility
15. simplify and organize desktop and pin essential applications to tool-bar.

This sequence may not load every device driver in all cases and it may be necessary to plug in all the astro devices to trigger the operating system to prompt for the final device drivers. In particular, with multiple serial devices, it is useful to connect them in a fixed configuration. This helps to prevent multiple instances occurring in the device manager but it does not always prevent COM port assignments from changing. When these do, it causes the stored ASCOM COM port connections to fail and, if several use the same adaptor hardware, there is no way of telling which is which, other than by trial and error. The solution is to disconnect all the serial devices and open Windows device manager. Now, insert a cable, one at a time and fix its COM port number manually (and make a note of which is which for later). There are a few other essential operating system settings. Possibly the most important is power management. The default power management settings do not work for astrophotography and several may need modification to preserve battery capacity and to prevent unintended system shutdowns. Here are a few suggestions, which are found in the advanced/additional power settings in the Power & Sleep menu:

Additional Power Settings menu:
- put the computer to sleep > never

Advanced Settings menu:
- turn off hard disk after >5 minutes if solid-state, 20 mins if not
- Wireless Adaptor Settings > low or medium Power Saving
- Sleep > all options to never/off
- USB Selective Suspend setting > Disabled
- Power Buttons and Lid > lid close > do nothing
- PCI Express > all options to off
- System cooling policy > passive
- Processor Power Management > 70% max on battery
- Display > turn off after 2 minutes, 50% brightness, disable adaptive brightness
- Graphics Power Settings > Maximize Battery Life

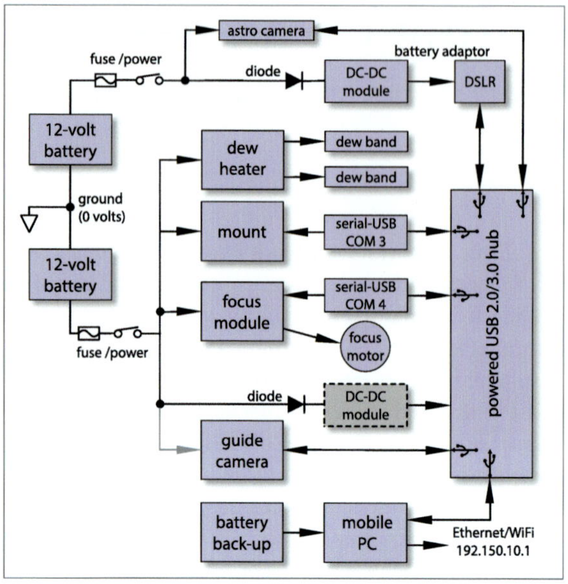


*fig.1 A typical portable acquisition system connection diagram, including optional power for guide camera and alternative power options for imaging camera and USB hub.*

## Good Connections

There are a number of recommendations for the general connectivity of what is now becoming a more complex system. My background is in electronic engineering and these precautions give your system the best chance of performing reliably. At some point, it will make sense to use two power sources, which are designated "clean" and "dirty." Telescope mounts, motor controllers and dew heaters create electrical interference and are considered "dirty." Conversely, to get the very best out of a camera, it helps if they have their own clean power

> In the case of SkyWatcher mounts with SynScan handsets, the serial port is via an RJ connector on the handset itself. It is also possible to remove the handset and its umbilical connector and apply serial commands directly to the mount electronics from the utility PC application EQMOD ASCOM. There is a catch, however; the connection between the handset and mount uses TTL Serial, with signal levels of 0–5 volts and not the customary RS232 signal voltages of ±12 volts. Direct RS232 connection will damage the mount's electronics and one must use the USB-TTL Serial version of the (FTDi) adaptor cable.

supply. DC power supply outputs are floating and it is good practice to tie their 0-volt terminals together. When I constructed my own power and USB distribution box I brought all the 0-volt connections to a single binding post, called a star-point earth. This reduces the chance of electrical interference and discharge issues from floating outputs and earth loops. For good measure, I also fit ferrite clamps to the power cables to reduce radio frequency interference.

One of the worse culprits for electrical interference are amateur dew heaters controllers. These modules, usually designed for motor or LED lighting controllers modulate a 12-volt supply to the dew heater band. The dubious ones do this by simply turning the power on and off in rapid succession. When this is applied to an inductive load, a high voltage spike is generated on each occurrence of the power turned off (you can often detect the interference with an AM radio). The better modules have a voltage clamp on the output as well as capacitive filtering. Some amateur mounts have plastic housings for their electronics and are more sensitive to electrical interference than those that have metal enclosures. In one case a pulsed dew heater caused a borrowed mount to slew erratically and I changed to a 15-watt linear controller, using an adjustable voltage regulator circuit and sold as a fan controller. The only downside to linear controllers is that they dissipate the energy that is not sent to the dew heater and require a small heat sink to keep them cool.

The way in which you connect the USB devices makes a difference too. Here are a few golden rules:

- keep all cables as short as practical
- power all hubs
- use high-quality hubs and cables
- try to avoid connecting hubs in a daisy chain
- identify each cable with a marker at both ends
- choose the hub version (USB 2.0 or 3.0) according to the camera interface
- video cameras (or fast frame rate still cameras) are best connected directly to a PC USB socket

As mentioned before, and for good reason, inexpensive USB hubs are designed for domestic environments and many struggle in cold conditions. After the significant investment in shiny astronomy equipment, it makes no sense to compromise it with false economy. I use industrial USB 2.0/3.0 7-way hubs, which have a wider operating temperature range, metal enclosure and conveniently accept a 7–12-volt power input. Lending these to friends has fixed their connection issues.

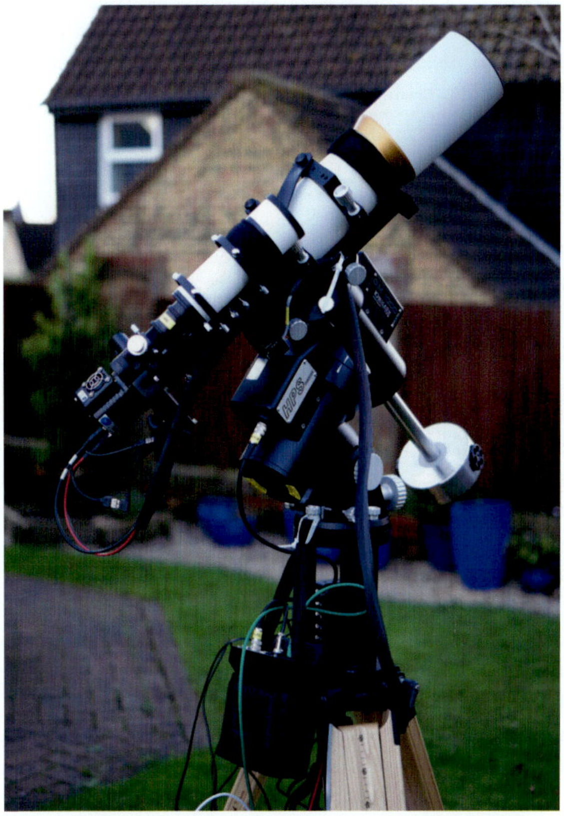

*fig.2 Like many mounts for the beginner, this premium model does not have thru-mount cabling and to ensure good tracking it is essential to bundle the various cables into a loom and carefully secure it so that its drag is minimized.*

Fig.1 shows a typical configuration for a simple system. This layout is annotated with the COM port assignments as well as the static IPv4 address that is assigned to the PC in the broadband modem. It will come in handy later on! Here the camera is connected to the PC via a USB hub. If you are taking video images a USB 2.0 hub may throttle the data stream and cause lost frames. In this case, connect the camera directly to one of the other USB ports on the PC.

### Configuration

Practical sessions are more productive if you spend some time setting up the acquisition system beforehand. This requires knowledge of your equipment devices and their specifications but it allows effortless connection as well as enhanced features such as planning, automatic centering, focusing and guiding. In time, an additional telescope or camera will allow you to image with several configurations. The better acquisition programs allow one to save a configuration or equipment profile and then apply it to an imaging

| optics | mount | cameras | guider | focuser | filters | plate solve | planetarium |
|---|---|---|---|---|---|---|---|
| **Telescope:**<br>type<br>focal length<br>aperture<br><br>**Field Flattener:**<br>spacing<br>reduction<br><br>**Guide Scope:**<br>focal length | **Connections:**<br>driver name<br>COM port or IP address<br>baud rate<br>polling rate<br>GPS port<br><br>**Location:**<br>time<br>time zone daylight<br>savings? epoch<br>longitude latitude<br>altitude<br>horizon limits<br>meridian limits | **Hardware:**<br>driver name<br>CCD temp setting<br>download time line<br>order<br>gain<br><br>**Image:**<br>pixel size<br>binning<br>pixel count<br>angular resolution<br>field of view (FOV)<br>read noise<br>dark noise | **Hardware:**<br>driver name<br>line order<br>binning<br><br>**Image:**<br>pixel size<br>noise reduction<br>angular resolution<br>field of view<br>dark cal. file<br><br>**Guiding:**<br>calibration time<br>exposure time<br>settling criteria<br>backlash setting<br>aggressiveness<br>flip RA? | **Hardware:**<br>driver name<br>COM port<br>baud rate<br>home position<br><br>**Calibration:**<br>approx focus posn.<br>slope or aperture<br>backlash setting<br>focus exposure<br>focus binning<br>find-star setting<br>microns per step | **Hardware:**<br>driver name<br>COM port<br>#filters<br>filter names<br>focus offsets<br>reversible?<br>changeover time | **Environment:**<br>catalog path<br>epoch<br>timeout setting<br>pixel scale default<br>exposure default<br>binning star<br>count<br><br>**Other:**<br>hot pixel rejection<br>magnitude range<br>expansion % | **Location:**<br>longitude<br>latitude time<br>altitude time<br>zone horizon<br>epoch<br><br>**Display:**<br>horizon setting<br>object filters<br>grid settings<br>cardinal settings<br>catalog choice<br><br>**Other:**<br>telescope connections<br>imaging connections<br>plate solve connections<br>camera FOV setting |

*fig.3 Setting up an imaging system requires information on the connected devices and their characteristics. These are the most common ones and it helps if you document your components systematically for easy retrieval later on. Many of these are required by the image acquisition application to work effectively and it takes a little exploring to find out where to input the values.*

sequence, saving setup time and improving reliability through consistency. Fig.3 shows some of the key parameters that may come in useful. I document my equipment parameters, in a simple spreadsheet on a cloud server, for instant retrieval.

A number of these are computed from others, for instance, the field of view and imaging scale from focal length, sensor dimensions and pixel size. An Apple Numbers spreadsheet does this for me and a template is available on the book's website. In the case of the more advanced Sequence Generator Pro application, it has the ability to store multiple instances of three types of information: user profiles (including locality), equipment profiles (the device and operational settings) and image sequences (the target and exposure plan). In practice, the first two are pre-populated and then they are associated with an image sequence. Astrophotography Tool is more hands-on and remembers one set of settings. Having completed one configuration or equipment profile, the others are usually minor modifications to it, arising from a change of optics or sensor. The PHD2 guiding software has its own equivalent equipment profiles to store the key parameters for the guide camera, optics and guiding setup.

Some of the parameters cannot be read off a specification sheet and require measurement or a record of some settings that work (e.g. autofocus settings). With that in mind, Sequence Generator Pro has the ability to update an equipment profile from a working session, so that it picks up on any fine tuning to the various settings and options during or after an imaging sequence. In practice, one initially goes through each of the equipment tabs and populates the required information, which in turn set up software connections between applications and device drivers (fig.4). It is a good idea to then do a quick try-out, including targeting, imaging, guiding and autofocusing before saving the configuration for later use.

Apart from the obvious benefit of being able to control the mount's position and running an image sequence, robotic imaging benefits from three major control systems; autofocus, autoguiding and auto-centering. These are powerful tools that ensure quality results from extended imaging sequences. These all require some care to get them working reliably, especially the first two, whose specialist chapters follow this one. The last, auto-centering, is a truly remarkable feature that is quickly found to be indispensable (and free). This feature overcomes the practical problem that a mount rarely points accurately to the intended target coordinates, due to a number of factors including initial alignment errors and mechanical flexure. This last one is especially noticeable when, even after accurately centering on the target, after a meridian flip, in which the telescope swaps sides to continue tracking, the target is no longer centered. The auto-centering process is deceptively simple; after the mount has slewed to the target coordinates, the true position is determined and the mount's position is adjusted to correct the pointing error. In practice, things like gear backlash can restrict a small movement and procedure may require a second or third iteration to home in on the target.

In applications like Sequence Generator Pro, auto-centering can automatically occur at the beginning of an imaging sequence or after a meridian flip, to ensure the target is initially set correctly and remains so. The remarkable gizmo in the middle of this process is the algorithm that works out where the telescope is actually pointing. This process is called plate solving.

### Plate Solving

One of the outcomes of the huge body of work that maps the millions of stars is an algorithm that is able to detect and correlate stars in an image with a star catalog database. In practice, a short exposure is taken (about 10 seconds) and the plate solver goes to work, using the telescope's reported position and approximate image scale (arc seconds per pixel). After an initial image analysis discards hot pixels, it identifies the centroid pixel positions of the major stars in the image. The plate solver then uses some very clever math to describe multiple star position using geometric patterns and correlates this with a star catalog of the region for a match.

The output from a plate solver is the image scale, angle and the precise RA/DEC coordinates of the image center to within fractions of an arc second. This analysis usually completes in about 10 seconds or so, depending on how far out the initial assumptions were. The scale assumptions are usually acquired from the acquisition application's equipment profile and assumed target from the sequence data or telescope mount. This type of data is also stored in the FITS image file's metadata. It is still possible to plate solve when the image scale, orientation and approximate RA/DEC position are unknown, using an all-sky plate solver, though this may take several minutes to complete. Being lazy, however, is a false economy as false matches are more likely when the software has no initial hint.

The output of the plate solver is compared to the intended image target and any automatic adjustment is made to the telescope position. In practice, this is done by the acquisition program issuing a small slew command to remove the error, or in a more proactive sense, sending the true position to the mount using a sync command and then issuing the initial slew command once more. In yet more sophisticated systems, the sensor angle information is also used to instruct the user to rotate the camera body to the required orientation manually, or in premium systems, using a motorized rotator.

For the purposes of image centering, there is no reason to purchase a commercial application and the free packages, including PlateSolve 2, AstroTortilla, ASTAP and All Sky Plate Solver with all-sky backup from astrometry.net work very well indeed. These applications are usually set up within the acquisition software so that it can call them as utilities to establish the pointing error and start imaging once the error has been reduced below a defined threshold. I choose a value of around 15 pixels for normal imaging. Each of the plate solver applications has its favored star catalogs. There are a couple of things to watch out for when installing these. For instance, the General Star Catalog (GSC) is used by planetariums, for displaying stars, as well as some plate solvers. These both require the same astrometry data but in different compression formats. To avoid confusion, put one version in a planetarium catalog folder in My Documents and the other in the plate-solve program folder. As part of the setting up process, the plate-solver applications require you to provide the path to their catalog. Some planetariums have an associated catalog compiler that converts the otherwise disparate formats into a single version dedicated for the purpose. The free planetarium C2A and others have extensive catalog management support and compilers for their planetarium, as do the premium applications.

### Applications

At the time of writing, most astronomy applications and utilities are 32-bit and do not require a 64-bit operating system (though they will run in one). There are a few 64-bit applications and these will not run in a 32-bit version of Windows. This will increasingly be the case over the coming years and several mainstream applications will need to move over. Windows 7/10 have some tools to run Windows XP compatible software but I found a few utilities, such as PERecorder, stubbornly refuse to run in a 64-bit environment. If you do upgrade operating systems, many astronomy

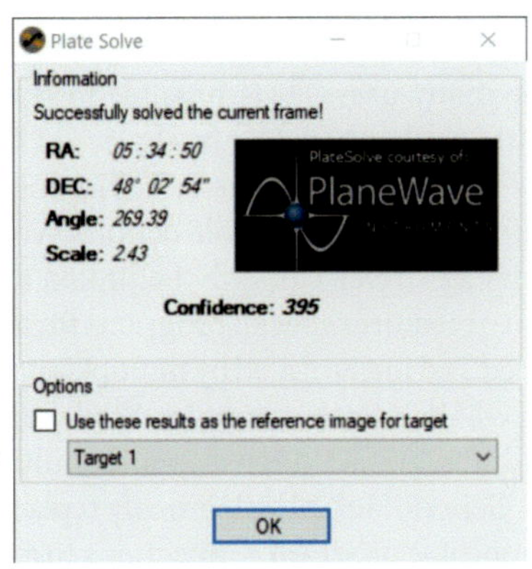


*fig.4 A typical outcome of a plate-solve application gives the coordinates of the image center in RA/DEC, sensor angle and pixel scale along with a confidence weighting.*

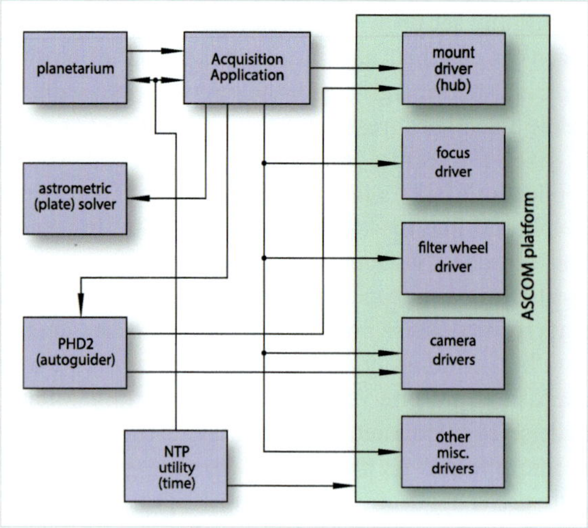


*fig.5 After the physical connections, the software links between applications and devices follow. The physical device drivers also have device simulators which can be helpful to try things out, without breaking anything.*

programs store configuration and setting files that can be copied over to the new installation. This may not be obvious and it is worth checking the normally hidden AppData directory in your user folder. You may need to check their contents with a text file editor and change any paths from "/Program Files/" to "/Program Files (x86)." A number of programs have a finite number of activations and you must deactivate or de-register them before formatting the drive upon which they run. The most notable example is Adobe Photoshop prior to their Creative Cloud® versions.

### ASCOM connectivity

Finally, there is the linking of the programs through their ASCOM interfaces (fig.5). This is a confusing subject for many users. It is possible to daisy-chain programs through others to a final piece of hardware. Fig.5 shows an example of the ASCOM connectivity between programs. The default condition is that an ASCOM device driver can only be linked to a single application. It requires special programs for a device to be controlled and interrogated by multiple applications. Those ASCOM drivers that accept multiple connections are called hubs. ASCOM developed a multi-purpose hub called Generic Hub, which broadly replaces POTH to satisfy multiple program connections to a mount. It has expanded since then to encompass other roles too. Many modern telescope drivers automatically act as a hub for mount control and the latest observing conditions class allows multiple connections too by design.

The ASCOM configuration is a one-time-only setup but requires a little care, noting the daisies in the chain sometimes have to be in a certain order.

### Image File Formats

File formats have been touched upon earlier and we all know by now to stay clear of 8-bit JPEG formats; for serious astrophotography, the only time we use this is to upload the final image to the web or to have it printed. We also know the high-quality option on any digital photographic camera is the RAW file, an "unprocessed" file. These are usually 12- or 14-bit depth images stored in a 16-bit file format. Unfortunately, these files do not contain unadulterated image data (for example, they mess about with dark levels) but we have to make the best of it.

Dedicated CCD cameras output raw data. When used with a dedicated image capture program these are commonly stored in a special scientific FITS format. The "Flexible Image Transport System" is an established open format extensively used in the scientific community and works up to 64-bit depth. Just as with the other image file formats, the file contains more than just the image. It has a header that contains useful information, which allows details of image capture to be stored for later use as well as processing details. In astronomy, this includes things like place, time, exposure, equipment, sensor temperature and celestial coordinates.

A typical imaging night may capture many individual files and the FITS header is often used to automatically sort and order the files into separate objects, equipment and filter selection by the image processing programs. During the software setup, it is a good idea to find the part of your image capture program that defines custom fields in the FITS header and check it is adding all the useful information. This saves time during batch processing, as it helps to group like-images together automatically.

The benefits of storing in FITS are not limited to astro-cameras; the better acquisition applications have the option to store DSLR or mirrorless camera RAW files in their native RAW format or in FITS. During image processing, the software assumes a FITS file is monochrome unless told specifically it has to de-mosaic (de-Bayer) the image. There is more than one way to de-mosaic a color sensor file and the standard methods used by photographic programs are not optimized for astrophotography. Given the choice, now that you are becoming a serious imager, use an acquisition program and save your camera files as FITS.

# Improving Tracking

*Welcome to the imperfect world of autoguiding and periodic error correction.*

As telescopes get longer and as exposure times lengthen too (often as a result of slower optics) the key to quality imaging is improved tracking. Even with a perfect mechanical mount (which almost exist but are ludicrously expensive), perfect tracking is not guaranteed. Atmospheric refraction, minute errors in polar alignment and general flexure in the metalwork conspire to introduce small tracking errors.

In the short exposures we have been using up to now, these tracking errors have not been intrusive. The higher magnifications and longer exposure times make the effect of tracking errors more obvious in the final image. To some extent, many error mechanisms are reproducible and as a consequence, it is theoretically possible to measure the tracking errors and construct a mathematical model that corrects the tracking error for any mount position, equipment combination and environmental condition. Such models are used by 10 Micron, Software Bisque and others to improve pointing and tracking accuracy. These systems, however, are beyond our budget and the solutions adopted by others include Periodic Error Correction (PEC) and autoguiding. This chapter explains both and becomes quite technical, as there really is no short-cut to good performance other than through sheer dumb luck.

Although PEC is the first line of defense and goes a long way to make tracking acceptable at modest focal lengths, it first requires a knowledge of guiding and so it helps to return to PEC once we have explained the fun and games of autoguiding.

## Autoguiding

Not all mounts support PEC. For those that do not, or to counteract all the other things that conspire to degrade tracking, autoguiding is the usual solution. The concept of autoguiding is simple; a secondary imaging system takes continuous short exposures of a star and analyzes its precise position in each download. If the

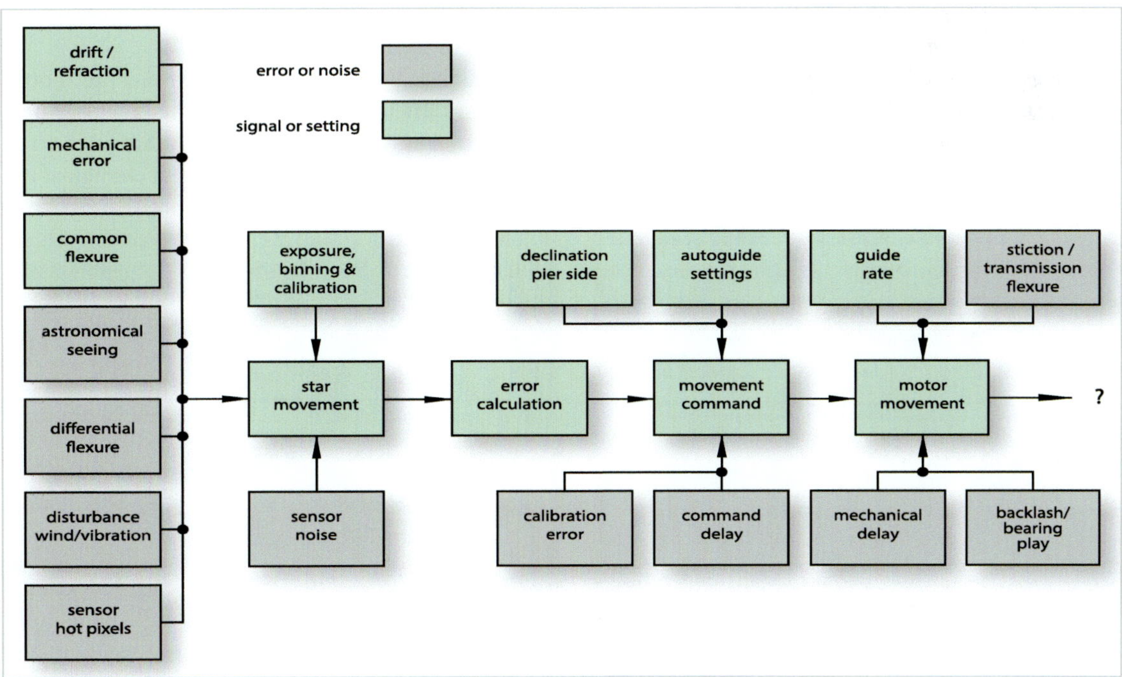


*fig.1 It is a wonder that autoguiding works at all when one realizes all the factors that affect the measurement and correction of tracking errors. If that was not bad enough, some of these change from night to night and for no apparent reason.*

star moves from its initial position, even by fractions of a pixel, the system issues movement commands to the mount, to correct for any error. As is often the case in astrophotography, success or failure is in the detail and, although it can be trivial, it usually takes several attempts to arrive at the best settings.

There are many reasons for this; in engineering terms, autoguiding is a giant feedback loop and, depending on the system behavior within the loop, it is possible to make things worse rather than better. There are also many external factors that mess things up too, the most obvious being the effect of seeing conditions. Fig.1 gives you some idea of the challenges and operation of an autoguiding system. Getting it to work is all about finding the right compromise, which is often a case of "less is more."

With all these things going on, a sense of perspective is very instructive; a polar misalignment of 5 arc minutes (achievable with a polar scope) causes a maximum drift rate of 0.02 arc seconds per second. In comparison, a mount with a typical peak to peak tracking error of 20 arc seconds might have a maximum rate of tracking error change of 0.2 arc seconds per second. That means the tracking error measured in subsequent seconds will be at worse, 0.2 arc seconds higher or lower. Here's the kicker, at the same time atmospheric seeing conditions can easily be 2 arc seconds and subsequent measurements of a star (assuming perfect tracking) will vary up to ±2 arc seconds between measurements. These numbers are approximations but each is of a different magnitude and dominated by the effect of astronomical seeing.

You can see this pictorially in the two graphs in fig.2 and fig.3. These tracking graphs from PHD2 autoguider software show the apparent tracking error with short and long exposures (with the guider output disabled). These clearly show that long exposures average out the effect of seeing but there is no such thing as a free lunch; long guider exposures increase the risk that the system cannot keep up with a rapidly changing tracking error.

It is very tempting to react to the tracking error reported by consecutive short exposures. In this case, the software reacts to seeing more than tracking error and the system is sent into a frenzy of unnecessary activity, increasing rather than decreasing the real tracking error. To add to the fun, when you try this for real, it will then be apparent that the guiding conditions at low altitude may be different to those higher up, due to the improving seeing conditions.

## Home Truths

The key takeaway is that autoguiding is not trivial. By its very nature, it is measuring an apparent tracking error, which itself is almost certainly inaccurate and then making a mount correction after the event, by which time the mount's tracking error has changed. This system is always playing catch-up and it is susceptible to being too slow to correct an error, or too reactive, making it potentially unstable. As said before, it is all about compromises, finding the right settings that balance the exposure time, aggression and filtering options to distinguish between real tracking errors and the effect of seeing.

This is not the whole story, however, the mount itself plays an important role in the tracking performance. The various mechanical characteristics, including backlash, stiction, the rate of change of PE,

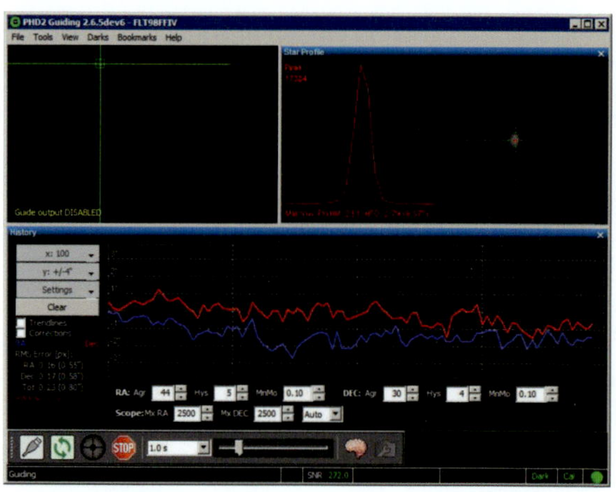

*fig.2 This tracking measurement over 100 seconds uses 1-second exposures to measure the center of the star. The two traces, for DEC and RA, constantly change from sample to sample.*

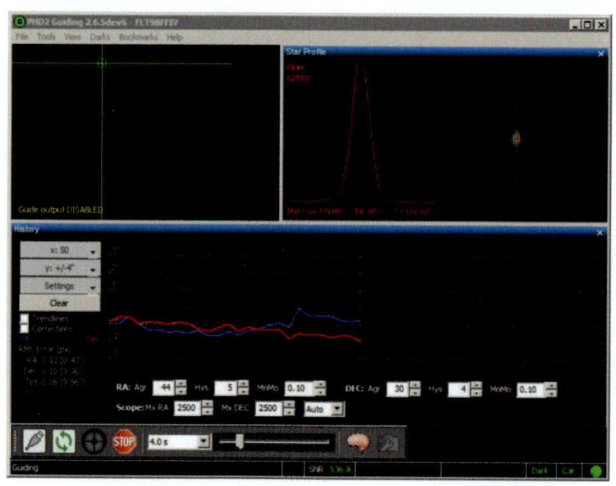

*fig.3 As fig.2, over 100 seconds but using 4-second exposures. The two traces, have much less sample to sample variation. More significantly, the tracking error is easier to detect.*

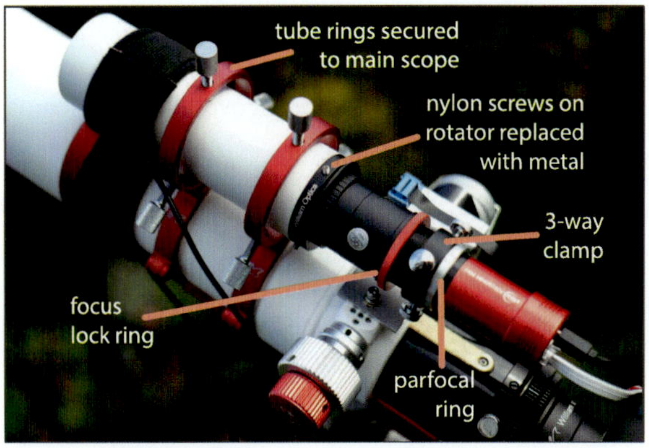


*fig.4 It is essential to make the guide scope assembly as rigid as possible. Compliant fixings are useless. This configuration works well and once focused, it needs no further adjustment.*

alignment and balance all play their part. Even a well-behaved system will change its tracking performance between sessions, depending on the star brightness, focus, collimation, the mount orientation, wind and so on. As a result, it is usually necessary to monitor and tweak the guider settings for each session and target. In my observatory system, my apparent tracking performance is typically about 0.4 arc seconds, but it can go as high as 1.1 or, in favorable conditions, 0.3. The key word here is "apparent," as we have seen the tracking error as measured by PHD2, or any other software for that matter is fooled by external error states caused by signal noise and seeing conditions. The ultimate measure of tracking error is the star size and shape in the image but again, this can also be affected by seeing.

## Practicalities

All of this is very interesting but it does not necessarily help with getting started. What follows are a set of recommendations, starting with general principles, single-axis mounts and moving on to the unique issues that occur with guiding dual-axis mounts. If you are not yet at the stage of having a dual-axis mount, one can skip that last bit and concentrate on getting good polar alignment and RA guiding.

The first golden rule is to make the underlying tracking as good as it can be. This includes:

- sure-footing of tripod or pier on solid ground
- accurate polar alignment
- don't overload the mount (typically up to half the specified payload)
- ensure the mount's mechanism is clean and lubricated
- balance the equipment about all axes (there are a few exceptions, which we will discuss later on)
- make the mount, telescope and imaging camera as rigid as possible, paying attention to the focus tube and camera mounting.

After the mount and imaging camera system, attention switches to the guide camera and optics. When starting out, the likelihood is that you will use a separate short focal-length lens fitted with a CMOS or CCD guide camera. Guide cameras typically have a small sensor and additionally have a guider output (ST4) interface. These sensitive cameras have fast download speeds and typically have a 1.25-inch diameter barrel and a C-mount thread to cover a range of mounting options. To help, they often have the ability to download a subframe around the guide star, rather than waste time with the entire image frame. The most popular CCD unit is arguably the Starlight Xpress Lodestar, but there are CMOS options that are a third of the price. The CMOS units have extremely fast download speeds, allowing rapid exposure sequences (or video) and often double up as planetary imaging cameras.

Guide scopes are usually short, small aperture refractors or even old prime camera lenses. It is not necessary to have the same focal length as the imaging system; the guider software has the ability to measure a star centroid to a 1/10$^{th}$ pixel. A 220-mm focal length lens will guide up to 600 mm and possibly a bit longer. More importantly, it is essential to make the guider system as rigid as possible as the rig moves slowly across the sky. In particular, any flexure or sag between the imaging and guiding system, commonly referred to as differential flexure, cause the imaging and guide cameras to see different tracking errors. Although the autoguider program will not see a tracking error, the image will. It will always be present if you use a separate guide scope but it may not be significant with shorter imaging instruments. This is not OCD, an arc second is only 1/3,600$^{th}$ degree. The guide scope mount and the camera attachment are the most likely weak spots. Single screw eyepiece fixings are usually a false economy. I avoid them where possible and screw my cameras into their adaptors and the focus tube. On my 50-mm guide scope, I use a three-fixing compression 1.25-inch eyepiece adaptor (fig.4). Metal to metal interfaces are best and I avoid auto-centering adaptors that use compliant rubber. In the case of dual-purpose visual/imaging guide scopes, discard flimsy brackets and mount the tube rings to a Vixen plate or directly to the main refractor's tube rings. On many guide scopes

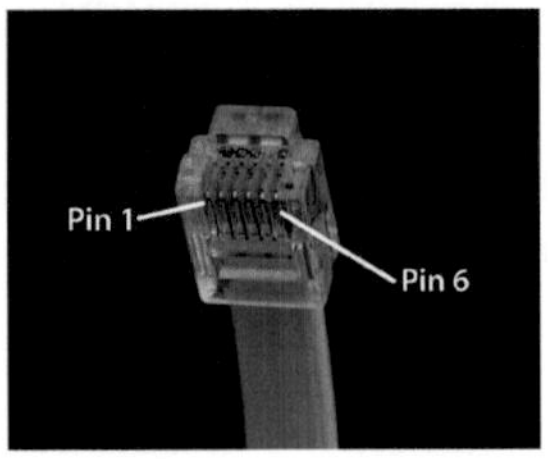


| Pin | Connection |
|---|---|
| 1/6 | Right (RA+) |
| 2/5 | Up (DEC+) |
| 3/4 | Down (DEC-) |
| 4/3 | Left (RA-) |
| 5/2 | Common |
| 6/1 | No Conn. |

*fig.5 So much for adherence to standards. The ST4 "standard" pinout has a number of implementations, some of which are confusingly in the reverse pin order. Single-axis mounts may use other connectors, e.g. 3.5-mm audio plugs as they only require 3 electrical contacts.*

themselves, a source of flexure are the soft nylon grub screws that clamp the camera rotation ring. Replace these with metal ones (fig.4). I initially bought one with an attractive double helicoid focuser (like a camera lens). This makes focusing very easy but, in practice, the barrel floated about as the focus helicoid had too much play, was filled with gloopy lubricant and the lock screw did nothing to stop it moving about. The older unit has a less seductive, simpler threaded focus tube, but it uses a lock ring to secure the focus position. This makes the assembly completely rigid (fig.4) and more suitable for guiding.

The mount and its software interface (if any) determines the guider interface choices. Many mounts have an ST4 guide port, permitting a direct correction to a guide camera. This hardware interface is based on 2-way (E-W) or 4-way (E-W, N-S) switched inputs. The duration of the switch closure determines the amount of adjustment at the selecting guide rate. Most mounts and cameras use an RJ-12 connector but the pinout may not always be consistent. The two most common configurations are shown in fig.5. If something is clearly wrong, it often just requires one of the connectors to be flipped over on the customary 6-way flat cable. I have a box of RJ connectors in 4, 6 and 8-way styles and an inexpensive crimping tool, for just this situation.

More advanced mounts additionally have some form of software guider interface, typically using the ASCOM interface. These are generically called pulse guiding. This eliminates the guiding cable and its potential to fail or snag. If the software implementation has been thought through, pulse guiding combines the PEC and guider corrections intelligently, rather than have the independent corrections fight one-another. For that reason, pulse guiding has the potential for improved guiding though, in reality, it may not be noticeable.

### Single-Axis Mount Guiding

While my Fornax LighTrack II has a guider input, its native PE is so low that the benefit over the added complexity and cost of a guider system is small. Like other simple trackers, its guider input only works for the RA axis and the DEC axis is fixed. On those trackers that use a worm gear system, however, the benefit will be noticeable on deep-sky images. The considerations for single-axis mount guiding also apply to the RA guiding principles on a dual-axis mount.

The RA and DEC axis behave quite differently. In the case of the RA axis, the motor is rotating the telescope at the rate of 15 arc seconds per second. Even during guiding, the motor continues to move and never changes direction; guide inputs merely momentarily speed up or slow down the motor. The great news is that since the motor is always moving in the same direction, the gearing system is also in constant engagement in the same direction and, even if there was some slack in the gearing, it should never come into play (literally!). One of the reasons for this is how the guider input affects the motor. One of the guide parameters is the guide rate. This rate is a fraction of the tracking rate and typically is in the range of 0.25x–0.75x. In those units with a handset, it is one of the available settings. If the guide rate is 0.5x, a common default setting for many mounts, the actual tracking rate flips to either 0.5 or 1.5x the sidereal rate, for the duration of the guide "pulse." This subtle change in tracking speed catches up with the tracking error. In my permanent installation, guide pulses last 10–150 ms, every few seconds or so.

### RA Guiding Challenges

RA guiding quality is influenced by the mechanical properties of the mount. There are several things to watch out for:

- worm gear engagement
- mechanism lubrication and cleanliness
- balance

The majority of trackers and mounts use worm drives. The advantage is that the worm can easily turn the gear but the gear cannot turn the worm, protecting the rest of the delicate drive mechanism and with no possibility of slippage. If the worm is linked by spur gears, as long as the gears are constantly rotating in one direction, any balance or backlash in the transmission system is largely irrelevant. If the worm is not engaged properly, or a sprung-loaded worm does not have the right tension, backlash can occur between it and the worm gear. (The

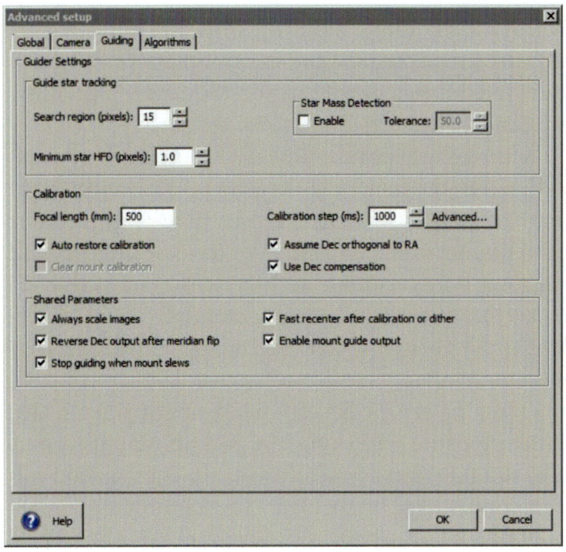


*fig.6 The calibration information in the guiding tab in PHD2 allows it to convert tracking errors from pixels to arc seconds.*

worm shaft itself runs in bearings and may also have some lateral play.) Although the worm is constantly rotating, the balance around the RA axis affects the gear engagement. It is normal to start with a telescope that is balanced around each of the rotation axes. If the RA drive has a little backlash it sometimes helps to deliberately introduce a slight imbalance to add some resistance against the direction of movement. This ensures the gears are always touching on the same surface. As a result, some users initially set a small imbalance towards the East (in the Northern Hemisphere). This may help as long as you remember to change the balance when the mount flips sides (assuming it is an equatorial mount).

Since worm gear engagement is crucial to tracking performance, some manufacturers provide instructions or adjusters for tuning the mechanism. This includes the potential to clean the gears and apply fresh lubricant, as it is easy for particles from the manufacturing process or environment to get into the mechanism. However innocuous, their effect can cause a rapid unpredictable change in tracking that is difficult to guide out. Lubricants become sticky over time and the preference is to clean and service with a synthetic PTFE-based lubricant. Before you start dismantling your mount, however, please check with your equipment dealer that your actions will not affect its warranty. There is almost certainly a YouTube video that explains the tuning process for your particular mount. Some of these are quite adventurous and extend to replacing the OEM worm with a precision version or changing the spur gears to a belt drive.

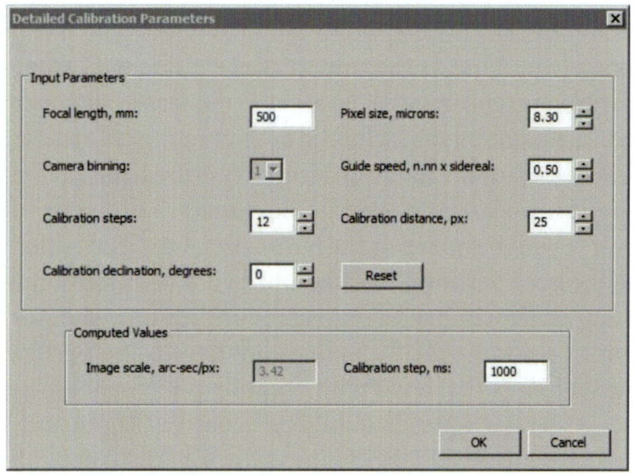


*fig.7 If the calibration parameters are not auto-populated, type them in. This sets the calibration pulse size and also enables meaningful tracking data during guiding.*

## Guider Calibration

After polar aligning your mount and connecting your equipment, we move on to guider calibration. This works out the complex relationship between the mount and guide camera. (If the periodic error of your mount is poor, in extreme circumstances it interferes with the guider calibration. In these cases, if it has a PEC feature, once you calibrate the guider, it is useful to repeat the calibration after you have trained your mount and enabled its PEC feature.)

Assuming PHD2, as the most popular guiding application, we need to provide some essential information to get the most out of it. First, select the camera, guide interface ("Mount") and if required, in the case of mounts with an ASCOM interface, the mount driver too ("Aux Mount").

Clicking on the "brain" brings up four tabs: global, camera, guiding and algorithms. PHD2 works best with the following information in the camera and guiding tabs:

- guide camera pixel size (camera tab)
- guide scope focal length (guiding/calculate tab)
- guide rate (guiding/calculate tab)

The focal length is typed into the calculate tab (fig.6). Depending on whether the camera and mount have an ASCOM connection, the other parameters are potentially auto-populated from the drivers or otherwise, need manual input. These four parameters determine the step size required during calibration as well as converting the tracking graph scale from pixels to arc seconds. PHD2 has the ability to be fired up by another imaging application and controlled via its server interface. Here, we are going to take over control ourselves. First, is the subject of the guide star; some calibrate at the same declination as the target but more consistent results are achieved by calibrating on a star at low declination. Using a 1-second exposure time, hit the loop button and focus the guide camera. Auto-select a star (from the menu, or the button that has been added to the latest version). The magnified view of the optional star profile window helps enormously, as does the star width readout. The focus does not have to be perfect (it seldom is that good with cheap two-element guide scopes) but the profile should be as narrow as possible, with the lowest HFD readout and displays a classic peak ending in a point like the profile in fig.2. If this has a flat top when focused, it is saturated and you should reduce the exposure time or choose a dimmer star. If you now shift-click the green guiding icon, PHD2 will calibrate itself. (With a single axis-mount, disable DEC guiding. To do this, select DEC guide mode to "off" for the guiding algorithm in the algorithm tab.)

The calibration routine moves the mount West and then East by about 25 pixels, working out the relationship between the guider pulse length and the star movement. Once the calibration completes, it saves the data and immediately starts to guide the mount. PHD2 will remember the calibration while it is active (or recall it, if you enable the auto restore feature in the guiding tab). This is a handy time-saver and avoids recalibrating for each session, providing the guide scope and its camera are kept exactly in the same orientation between sessions. PHD2 also has some additional utilities that measure polar alignment and mount characteristics.

## RA Guiding Parameters

The right guide settings are often elusive. They are interactive and while there is some logic to setting them, even then, a little experimentation may improve things without fully understanding why. The key ones are exposure time, aggression, minimum move and guide rate. In addition, depending on the guider algorithm, we can add hysteresis and slope weight.

Setting the exposure time is a balancing act. It typically lies in the range of 2–4 seconds. If it is shorter, the star intensity is weak and sensor noise may interfere with the measurement and, as we have already seen, the outcome is less accurate as the system is more reactive to seeing. It does, however, potentially react to real tracking errors without delay. Longer exposures average out the effect of seeing in a general blur, improving the tracking error calculation but potentially too late. If the exposure is too long, the peak star intensity may also saturate the sensor, affecting the star position calculation and is to be avoided. (The star profile in fig.3 is very close to saturation and any more exposure would cause PHD2 to show a red "SAT" warning flag on the status bar at the bottom of the window.)

The aggression parameter sets how much of the tracking error is guided out on each command. Considering that the tracking error measurement is inaccurate, settings are usually in the 25–75% range. It is usual that longer exposure times can cope with more aggression and conversely with short times. A good starting point for this pair of settings are 2 seconds at 50% and take it from there.

The last of the three settings is "minimum move." This is an interesting control. The setting itself defines the minimum move in fractions of a pixel. In combination with the guide rate, it defines the minimum adjustment that is issued to the mount. If it set too high, it locks in a tracking error that will never be corrected. If it is set too low, the mount may not react, as some models have a practical lower limit of adjustment. In these cases, it is better to give a single substantial guider input, rather than a dozen micro-adjustments that get

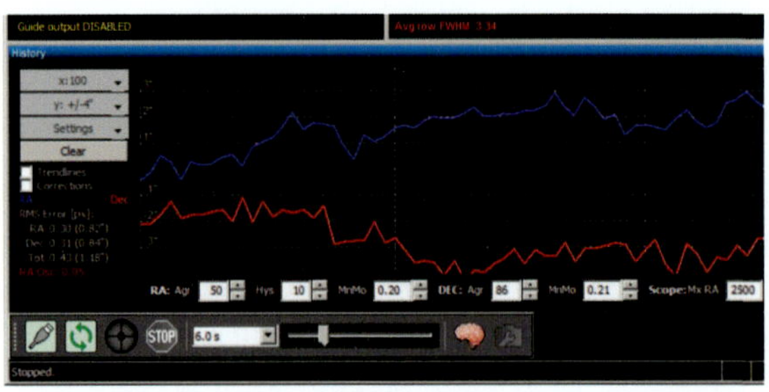

*fig.8 The unguided tracking of a Fornax LighTrack II over 6 minutes is remarkable. Apart from a slight drift on both axes, there is very little ripple on either trace (<± 1 arc second). It does not need heavy-handed guiding.*

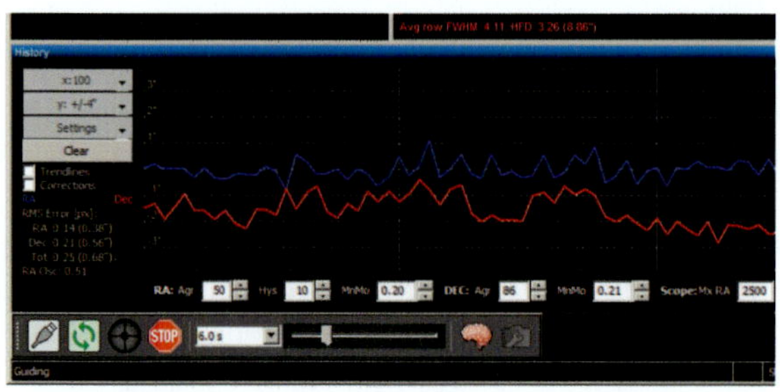

*fig.9 With 50% aggression, the RA RMS tracking error is <0.4 arc seconds! The alternating ripple indicates slight over-correction. Backing off aggression and increasing the min move may improve things further.*

lost. For example, my small guide scope with a focal length of 220 mm, fitted with a popular guide camera has a pixel scale of about 7.5 arc seconds per pixel. PHD2's default 0.2-pixel minimum move is equivalent to 1.5 arc seconds. At a guide rate of 50% (7.5 arc seconds per second), it requires 200 ms pulse to correct a 1.5 arc second error. A higher guide rate requires a shorter pulse width to make the same tracking adjustment. My old Meade LX200 mount would not react to anything under 50 ms, whereas my other mounts respond to pulse widths as low as 10 ms. A number of mount manufacturers recommend a min move value that equates to about half an imaging pixel. As the guider may have a different scale (as a result of different optics and sensor), a guideline minimum move (in pixels) is calculated:

*min move (pixels) = 0.5•imaging scale / guider scale*

PHD2 offers a number of different guide algorithms. These alternatives use the same tracking error measurements but calculate the guide pulses in different ways. The requirements for RA and DEC are different and the algorithms attempt to isolate the slowly changing tracking error from the rapidly changing seeing, using a variety of low-pass filtering options. A long exposure time is, in effect, a low-pass filter, as is taking account of the last few tracking errors and combining them to make an intelligent prediction of the error. It is worth pointing out that these are imperfect solutions; while any low-pass filter will reduce the effect of seeing, at the same time it delays the tracking correction. Some of the more advanced ones are non-linear. They realize that the underlying tracking error has a maximum rate of change, so it makes no sense to issue guider commands that, in general, exceed that. This last statement is the clue to better guiding. The trick is to realize that the tracking error generally changes very slowly and thrashing the mount around, chasing seeing, is the wrong way to go.

### Guiding Analysis

In PHD2, one can only change the guide algorithm while guiding is inactive. To get started, I recommend the hysteresis option, which combines recent corrections and the current error, with the hysteresis value set to 0 or 5%. This setting allows one to get a feeling of how the system responds to the autoguider. After polar aligning the mount, choose a star at low declination, disable the guide output, set PHD2 going with a 5-second exposure time and watch what happens for 10 minutes or so. The 5-second exposure filters out the worse of the seeing and it should be apparent how well the mount tracks without any assistance (but if possible, with PEC enabled). The tracking will then show the residual alignment and mechanical issues and, most importantly, their rate of change (fig.8). At one time it required quite a bit of experience to work out an effective guiding strategy. The latest version of PHD2 has a guiding assistant that does some of the work for you. With the guider running, start the Guiding Assistant. This disables the guider output and starts measuring the tracking error. As it does so, various statistics are calculated and updated on screen. It takes several minutes for the values to stabilize. On the right-hand side are a series of recommendations on the principal guiding parameters, including minimum move and exposure time. It does not mention aggression. The best value for this setting considers the accuracy of the tracking error measurement. If the "apparent" error is mostly seeing, it hardly makes sense to try and entirely remove it with a 100% setting. The higher the confidence in the real tracking error, the safer it is to use a higher setting. A higher setting is more compatible with a longer guider exposure time. If the tracking graph has alternating W<>E excursions, it is usually a sign of over-correction and requires a smaller aggression setting (fig.9) shows this on RA trace on a single-axis Fornax mount).

When you consider the fickle nature of seeing conditions and all the other error modes, guiding performance with any group of settings may well change during a session. It is all too easy to be heavy-handed with multiple parameter changes. In obvious cases, it will be required to tame a particular issue but when the system is basically tracking, the advice on the openphdguiding.org website tutorials suggests making small changes and taking your time to make an evaluation.

### Other Settings

There are further settings scattered among the dialog screens. For noisy sensors, PHD2 can bin exposures, which lowers the sensor noise content to improve the star position calculation accuracy. A further setting, labeled Star Mass Detection, looks for a sudden change in star intensity, usually associated with the software accidentally locking onto an adjacent star. This triggers an alert, which can abort acquisition in some systems. This sounds useful but, in common with others, I disabled the feature as the false triggers, caused by vapor trails were more annoying and potentially lost too many exposures.

### Common Symptoms and Cures

The following are some of the more obvious errors which one can detect from the guider graph and potential ways of fixing them:

- alternating positive and negative swings – reduce aggression and/or lengthen the exposure
- error diverges, stays the same or very slowly converges back to zero over multiple corrections – increase aggression and/or shorten exposure
- alternating ripple with frequent small adjustments in either direction – increase minimum move
- multiple moves in a slow oscillation – reduce hysteresis and aggression, shorten exposure time

### Dual Axis Mounts

If things were not challenging enough, on dual-axis mounts we have the DEC axis to contend with. Tracking errors on the DEC axis are thankfully free of periodic error but will have drift, caused by polar misalignment, errors arising from general flexure of the imaging system and increasingly towards the horizon, a creeping error caused by atmospheric refraction. In a rigid, well-aligned system, the amount of DEC correction is minimal. That is the good news. On the other hand, the DEC mechanism is basically stationary but is required to move small amounts in either direction, at the mercy of the uncertainty of mechanical play (backlash) and stiction. On those mounts that use a series of gears to reduce the stepper motor rotation to the worm, this can be significant and is the reason that mount suppliers are increasingly switching to belt-driven worms and upgrade kits to convert existing mounts. These toothed belts rely on multiple touch points (unlike a gear that has single tooth engagement) to even out mechanical tolerances. They do not have backlash per se but they are less rigid and so, depending on the opposing forces in the drive mechanism, they may not fully deploy a reverse movement. Again, before experimenting with guider settings, it is good practice to check the mount's gear engagement and adjustment. If there is some backlash on the worm, adjustment or a small deliberate fore-aft telescope imbalance can help.

In general, a well-aligned mount will track better in DEC than in RA and the guiding can be less aggressive as result. A good starting point is to take your established RA guiding settings and back them off a little for DEC. Coping with backlash is the principal hurdle to overcome.

To begin with, gear backlash is only an issue if the guide pulses change in direction. If you have an alignment error, the drift will be in one direction and, in theory, one only needs to send guider pulses either North or South. This idea is capitalized on by several unique DEC guiding principles:

- higher minimum move
- deliberate polar misalignment
- resist switch guiding algorithm
- backlash compensation algorithm

In the first case, the minimum move is set to a higher level so that it reduces the number of guide pulses and statistically favors one direction, to account for an underlying drift. This works quite well if the backlash is small. The second idea is interesting and goes against conventional wisdom, as the mount is deliberately misaligned to the pole by several arc minutes. This causes a drift in DEC (and potentially RA too) and so the guide pulses favor one direction over the other, reducing the incidence of gear reversals. This idea is often coupled with the third idea. The resist switch guiding algorithm is semi-intelligent. If it is issuing guider pulses in one direction, it will not switch polarity unless it sees three successive requests to do so. If there is a general drift on the DEC axis, this algorithm mostly ignores apparent tracking errors (due to seeing) in the opposite direction to the drift. Putting in a minimum move equivalent to about half an imaging pixel helps too.

The last option is just that. If all else fails, it is possible to use this feature. In essence, if the guider commands change polarity, the guide program adds an additional movement to overcome the slack in the system. This, of course, requires one to measure the backlash in the first place. That can be difficult since it may not be consistent at different DEC angles. Even when you have a reasonable measure, it needs backing off. It is better to remove 90% of the backlash rather than potentially overshoot and cause the system to oscillate. At the back of your mind, just remember that it is often the case that over 50% of the apparent tracking error is not real and mostly due to seeing conditions, in which case, there is no need to thrash a mount about and make matters worse.

## Periodic Error Correction

PEC works by applying small real-time corrections to the RA tracking rate to compensate for the repeating principal mechanical errors in the gear system. These are usually associated with a worm gear system. As the worm turns (usually about once every 5–10 minutes) tiny mechanical errors translate into a slight

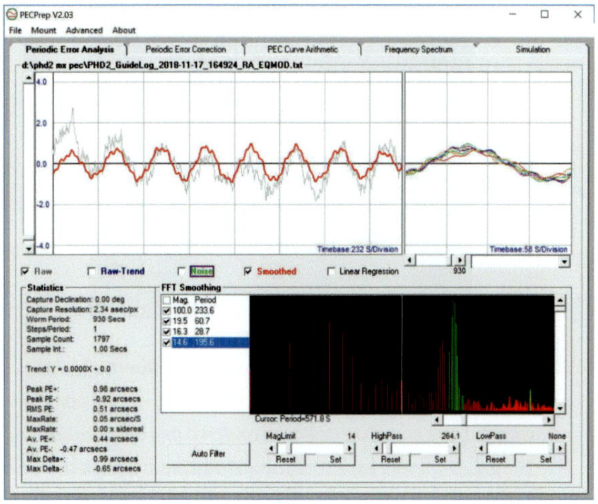


*fig.10 The PECPrep utility reads an autoguider log file and its various filters isolate periodic errors. Here, my Paramount MX has a clear periodic error of about ±1 arc second.*

acceleration or deceleration of the tracking rate. A full revolution of the worm returns the error to zero and in-between, there is a cyclical or periodic error (PE). The mean error is often reported in arc seconds and an excellent mount will have +/- 1 arc second or less. Budget mounts are more typically in the range +/-7 to +/-30. Fortunately, the cyclical nature of the tracking error helps with its correction. It is possible to make rapid measurements of a single star's position over several turns of the worm and characterize the error profile. Special software applications then calculate the required opposing tracking correction, as the worm rotates. In practice, PEC reduces a mount's tracking error to about $1/10^{th}$ of its initial value. In extreme cases, if a mount's tracking error changes at a fast rate, PEC will be playing catchup and is generally less successful.

It is also important to note that PEC only concerns itself with the worm error. The software routines that analyze the star position deliberately filter out other tracking errors, caused by polar alignment and other gearing defects. After PEC, the tracking error is usually dominated by other mechanical and alignment issues. Precision gears are expensive to manufacture and budget mounts show a wide variation in performance, confirmed by consumer reviews and personal experience. There are notable exceptions, the AstroTrac and Fornax tracking mounts employ a unique drive system that reduces PE to a level expected in a premium mount.

The saying "you cannot make a silk purse from a sow's ear" also applies to PEC. A recent mount purchase featured "permanent PEC," in which the mount monitored the guider input for a period of time and calculated its own error correction. The native periodic error of this mount (and two replacements) was extreme and unusable for imaging. Even with its PEC feature enabled and after autoguiding, stars were clearly elongated in images taken with a modest focal length refractor.

PEC measurement and execution require two things; the ability to know the precise worm angle and a look-up table of corrections for different angles. In the first case, some mounts have a sensor mounted to the worm axis that detects a "home" position, from which it can work out the other worm angles. In the second case, the mount, mount's handset or the application software has the ability to store and replay the corrections as the worm turns. To calculate these corrections, we essentially use autoguider software to measure the tracking error (or corrections) over several worm rotations.

This is usually done with the autoguider output disabled so that it simply measures the star position every few seconds over half an hour. While it is doing that, it writes a log file that is used by a utility such as PECPrep from the EQMOD project to extract the error associated with the worm rotation period (fig.10).

In practice, we want to isolate the error from the worm gear as best we can. To do that requires good polar alignment and seeing conditions. We cannot just choose any star, after carefully aligning the mount to the Celestial Pole, point the telescope South, and select a star near the Celestial horizon (DEC = 0).

Next, we need to measure the tracking error. To do this, disable the correction output or remove the ST4 cable and start PHD2 guiding with a 1-second exposure. Let this run for 15–30 minutes. In this way, PHD2 is being used as a tracking measurement tool and does not interfere with the mount's movements. During this operation, since the guider output is disabled, the automatically generated log file records the tracking error over the period. Depending on the mount model, it is usual to use an external application to extract the tracking data and filter out those errors that are not associated with the worm to form an error correction table. The plotted error versus worm angle resembles a lumpy wave. A table version is either uploaded to the mount or its application software for use in real time. For the popular SkyWatcher mounts, although the handsets are unable to store PEC, an independent free utility EQMOD ASCOM and its accompanying PECPrep utility analyze the log file, extract the PEC table and then play it back to the mount. Other manufacturers use different methods and these features may be self-contained or require third-party applications. While the details of the software operation vary from mount to mount, there are a few common best practices:

- accurately align the mount axis to the pole
- choose a bright star near the meridian
- choose a star within 20 degrees of celestial horizon
- choose short exposures (1 second or less) with no delay in-between exposures
- take measurements over several worm cycles, the more the better
- take note of the mount position in regard to which side the telescope is on (i.e. side of pier); this information is often required to determine the polarity of the correction; if the polarity is wrong, you will soon know, as the periodic error will be twice as bad!

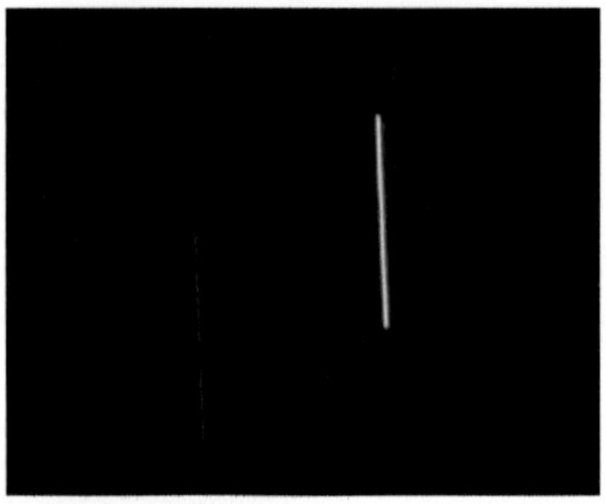

*fig.11 A direct way to assess periodic error is to deliberately shift the mount's polar alignment by about 3° to the East or West and aim the camera due South and at a low declination. Here a 20-minute exposure shows the drift in DEC. If there is any periodic error, the line becomes a wave, here the Fornax unit star trail is almost perfectly straight. (The PE magnitude is calculated using the pixel scale in arc seconds/pixel.)*

### Assessing Mount Tracking Performance

I have returned two mid-range mounts whose periodic error was appalling and a premium mount whose system design was incapable of working with an autoguider. A mount purchase represents a significant outlay and it is a good idea to run a few tests to check that it is performing as expected. The most obvious measurement is its periodic error (before PEC). The previous paragraphs indicate one such method, using an autoguider to calculate the tracking error over 15 minutes or so. This may either be by screen-grabbing the guider graph (fig.11), or analyzing the autoguider log file with a utility such as PECPrep (fig.10).

It is an unfortunate truth, however, that some suppliers use customer inexperience to deny a reported issue and question the validity of their complaint. Certainly, to be sure of any PE measure, one must be careful to ensure the imaging scale is correct. This requires double-checking the pixel pitch of the guide camera and the focal length of the lens in the guider setup tabs. Mistakes do occur; I discovered that PHD2 requests the sensor's pixel pitch directly as a parameter from the guide camera and overrides any manual input. I quickly realized that the stated pixel pitch was incorrect for the sensor and informed the camera manufacturer. 48 hours later, I had a camera firmware update with corrected parameters. Unfortunately, the corrected pixel pitch value made the reported PE even worse.

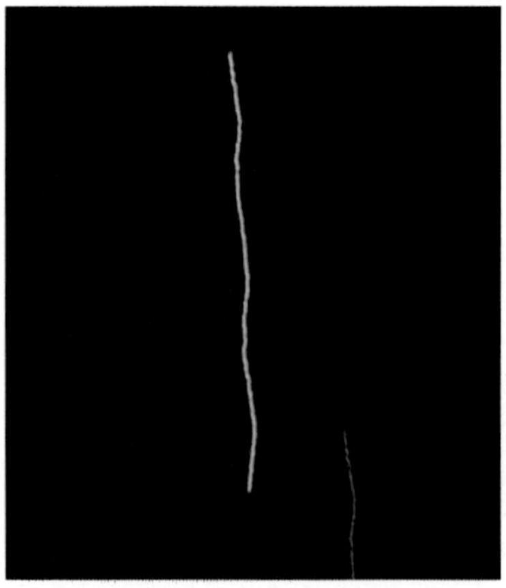

*fig.12 As fig.11, this time with the SkyWatcher AZ-EQ5 worm-driven mount over 15 minutes. The wobble in the star trails is caused by periodic error, which is approximately 7 pixels in magnitude, or ±7 arc seconds peak to peak. This is a better than average performance for an entry-level mount. This mount is also capable of correcting most of this error by using its in-built PEC utility, which monitors an autoguider input over a worm cycle.*

A simpler and more direct method does not involve any software and is less prone to misinterpretation or disregard. It requires the mount to be polar aligned and an imaging system mounted on it. You need to know the pixel size of the imaging camera's sensor and the focal length of the telescope. If you are using an APS-C camera with a pixel size of about 4 microns, a focal length of 300–500 mm is ideal. With these, you need to work out the pixel scale, in arc seconds/pixel, where the pixel size and focal length are expressed in meters:

*scale = 3,600•arctan(pixel size/focal length)*

### The Camera Never Lies

The PE assessment is very simple; first, set the camera to its lowest ISO speed and put it into"B" mode. Then, deliberately misalign a polar-aligned mount by about 3 degrees East or West using the azimuth bolts. It does not have to be exact and, in practice, it is about the same as moving Polaris to the edge of the field of view of a typical polar scope. Now, aim the camera due south, near the celestial horizon (DEC = 0). Turn off any autoguiding system or periodic error correction and start the mount tracking. Finally, take a few exposures of about 15–30 minutes each. (In regions of strong light pollution, it may be necessary to use a special LP filter, or simply aim at a nearby bright star and use an ND filter, to prevent the background washing out.)

During each exposure, there will be a few rotations of the mount's worm drive and in each image there will be extensive drift along the DEC axis. This causes obvious star trails over the exposure period. Any periodic error shows up as a wiggle along the line. The example in fig.11 shows an almost perfect result, where there is no detectable wobble and backs up a claim of sub-arc second PE. The example in fig.12 is from a SkyWatcher AZ-EQ5 mount, using the same lens and camera. The test was carried out with a Canon EOS 60Da and a William Optics Zenithstar 73, giving a pixel scale of 2 arc seconds/pixel. The assessment is carried out in Photoshop or Affinity Photo, on a magnified portion of a star trail and using the ruler or measurement tool, set to pixels. The wobble has a magnitude of 7 pixels, which corresponds to ±7 arc second peak to peak periodic error. This is a good result and since there are no obvious rapid movements in RA, there is a good chance of reducing the error with PEC and guiding. Even without PEC, the guider performance of the AZ-EQ5 mount (fig.13) shows that the belt drive system works well and, even without PEC, keeps up with tracking errors.

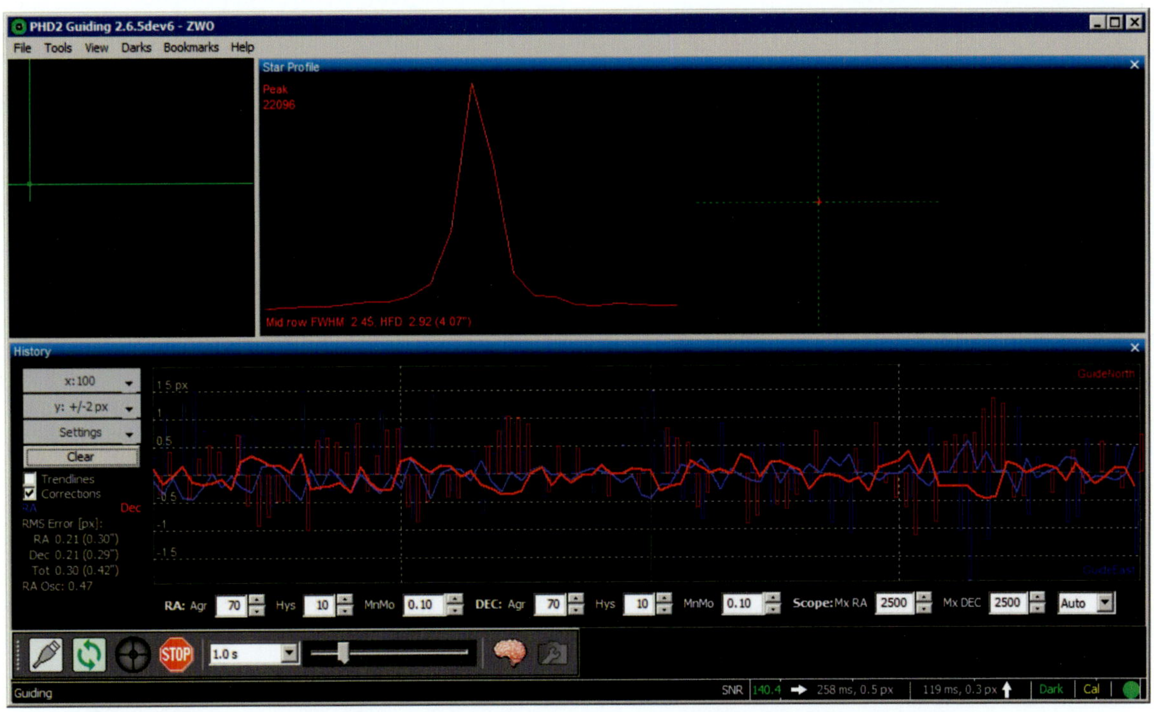


*fig.13 This SkyWatcher AZ-EQ5 guides well. With very little experimentation, a respectable RMS error of less than 0.5 arc seconds is achieved using default PHD2 parameters. The vertical red and blue lines represent the guider corrections. On the red DEC trace, successive corrections in the same direction suggest a small amount of backlash, but nothing substantial. In practice, the guider exposure at 1 second is probably too short and is picking up a disproportionate amount of seeing. Further tuning might include trying 2- or 3-second exposures and alternative DEC guider algorithms, like "resist switching."*

# Improving Focus (part 2)

*Using a computer to automatically determine the best focus position improves image quality over extended periods and enables unattended operation too.*

We take focus for granted on modern consumer cameras. Their powerful algorithms evaluate the image in real time and not only focus at a point but are able to track a subject and even predict the focus position of a moving object. In astrophotography, although we are working at infinity, it is by no means a fix-and-forget process. The extended imaging time during the night requires repeated monitoring and focus correction; in most cases, carried out in-between exposures. These corrections are necessary to compensate for several complications:

- thermal expansion or contraction of the telescope assembly
- flexure and mirror movements (in the case of reflectors)
- focus mechanism defects
- focus shift with wavelength

As we have come to appreciate, there are many practical challenges that affect the apparent resolution of an image, with focus accuracy being just one of them. They all add up, however, and poor focus is especially noticeable on star fields, globular clusters and galaxy dust lanes. In practice, there is a certain amount of latitude, as the conventional critical focus zone does not account for the net effect of a small amount of defocus. Others have proposed a more holistic approach that factors in seeing, diffraction and the focus quality impact. In the equation below, $D$ is the aperture diameter in meters, $f$ is the f/ratio and $Q$ is focus tolerance as a percentage or the seeing in arc seconds. The critical focus zone in microns is approximately:

$$\textit{focus zone} = 2.2 \bullet \textit{seeing(arc seconds)} \bullet \sqrt{Q} \bullet D \bullet f^2$$

For example, assuming average seeing conditions of 2 arc seconds and a 20% deterioration from a focus error with a modest short refractor, the critical focus zone is about 50 microns. At the optimum focus, the diffraction-limited star diameter is theoretically about 1.6 arc seconds, but seeing ruins this. At the limit of the focus zone, from 5 steps of the focus motor, the star diameter bloats to about 2.2 arc seconds.

Autofocus requires a motorized focus mechanism; these impart little vibration into the system (unlike manual handling) and as such, the imaging software is able to make an immediate short exposure and focus assessment. Apart from being convenient, automatic focusing avoids human judgment and enables unattended remote operation of the focus mechanism. I for one do not wish to donate my bodily fluids to hungry insects and as soon as I was able, I used a computer to automate focusing. There are several different strategies and in practice, you may use several, depending on circumstances:

- manual autofocus
- triggered autofocus
- automatic filter offsets
- automatic temperature compensation

**Manual Autofocus**

Of the three amateur means for determining focus, the most popular is to image one or more stars and measure their diameters at different focus positions. Another method measures the relative intersections of diffraction spikes generated from imaging a bright star through a focus mask and another novel method introduces astigmatism into a star image and uses the fact that the star is perfectly round, rather than elongated, at the point of focus.

In most cases though, the autofocus algorithm uses the imaging camera to take a short exposure. It then selects one or more stars and measures their diameter, moves the focuser, repeats and makes intelligent adjustments. Some algorithms require the image to be at approximate focus before running, others can work it out for themselves. These usually take 7 or more exposures at different positions (either side of the focus position) and work out the position that has the minimum star diameter. Poor seeing conditions and image noise add some uncertainty into the measurement and some algorithms average several readings at one focus position or evaluate multiple stars. At one time there was a pioneering free autofocus plug-in, FocusMax, which neatly integrated with some high-end applications. This software learned the relationship between the change in star diameter with focus position for a particular telescope, which then allowed it to focus with just a few samples at different focus positions. Even so, it could take several minutes or more for these programs to complete. In good seeing conditions, however, a focus mask system can autofocus relatively quickly and equally usefully, characterize an optical system, including the precise relative focus positions for individual imaging filters or the focus shift at different ambient temperatures. Manual autofocus is better than dead reckoning, but to maintain the focus accuracy throughout the imaging session, it requires repeat operation and sleep deprivation. It is possible to set automatic prompts to repeat manual autofocus but a better method is for the acquisition software to automatically trigger autofocus events for itself.

### Triggered Autofocus

A major benefit of triggered autofocus is one of non-intervention and although optimized mask systems achieve high accuracy and their software has a basic autofocus routine, it is not trivial to integrate their function into an acquisition system. It requires an electromechanical means to place and remove the mask over the optics as well as control software to synchronize its operation with the acquisition and guiding software. More conveniently, the better acquisition applications offer a range of autofocus triggers (as in fig.1) which anticipate the conditions which require focus fine tuning:

- ambient temperature change
- filter change
- mirror shift (after slewing or meridian flip)
- time or exposure count interval
- half-flux radius (star size)

These are used singly or in combination. With a refractor, I typically choose 0.5–1°C temperature and an interval. With a large reflector telescope, which warms up and cools down more slowly, I choose 0.5°C temperature, mirror shift and intervals of 1 hour.

### Automatic Filter Offsets

In both of the above setups, I choose not to use a filter change to trigger an autofocus sequence. That is not to say that a filter change does not need a different focus, it is just that the focus shift is a constant and if the focus positions for the different filters are already worked out , it is possible to simply make the required motor adjustment in a single action, without taking more measurements. This is made possible by either the acquisition application or the filter wheel's ASCOM driver properties storing focus offsets for each filter position. The acquisition software simply reads the values and applies the relative move immediately after the filter change.

### Automatic Temperature Compensation

As the temperature of the optics or the tube/truss change, so does the focus position. A temperature controlled autofocus trigger is always playing catchup to some extent. To address this, some autofocus modules monitor the ambient temperature and have a mode to constantly change the focuser position in real time. Several image acquisition applications also have this feature built-in and require a temperature source,

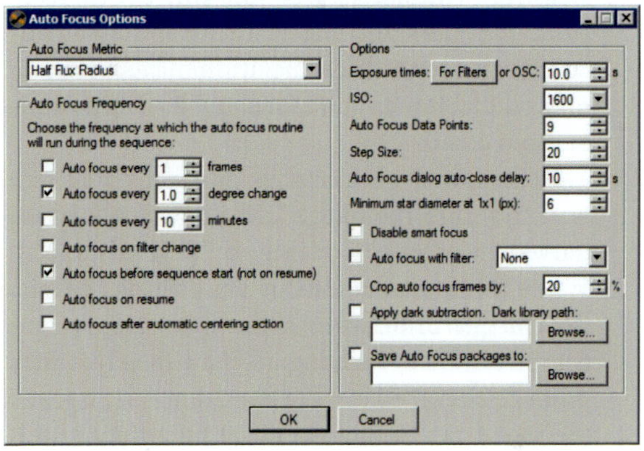


*fig.1 Sequence Generator Pro had an extensive array of autofocus triggers and settings. These permit unattended imaging with the knowledge that, one way or another, the focus will not drift during an imaging session or a rogue autofocus run will not permeate the entire night.*

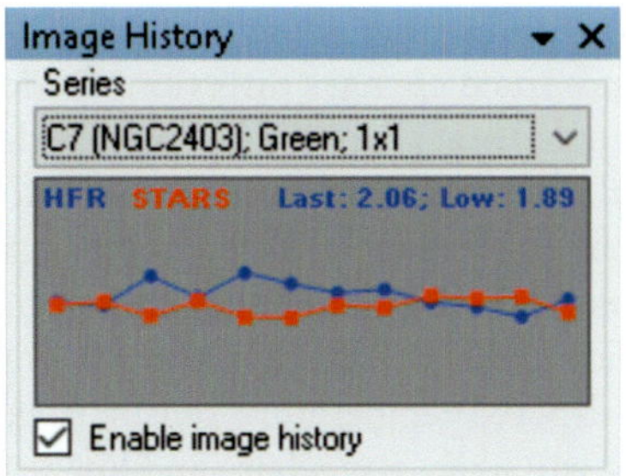


*fig.2 The image history monitor in SGP records the star size and number for each frame. Here, the conditions are stable but if the HFR size starts to increase at the same time as the number of stars declines, it is a sign of drifting autofocus or thin cloud.*

either from the focuser module or a weather monitor. To use such a method, the user has to first calibrate their system. This is best done by recording the focus positions for a number of ambient conditions and then calculating the number of steps per °C. (If you wish to use a focuser module to make the corrections and have multiple systems, you will need to enter its unique calibration each time you swap over.) The module (or acquisition software) then adjusts the focus position as the temperature changes. In this scheme, no time is wasted with lengthy autofocus routines after an exposure. It all sounds wonderful but it is has some practical limitations that one should consider:

1. Temperature sensors react quickly to ambient conditions but lumps of glass and aluminum do not. Long after the temperature sensor has stabilized, the focus position will continue to shift. The glass and metal also stabilize at different rates so the focus position may shift in an unpredictable manner, especially during periods of rapid cooling.
2. Most inexpensive focus tubes (and some premium ones too) shift the image laterally when they are moved in or out (typically by several pixels). While traditional autofocus algorithms perform their magic in-between exposures, temperature compensation is working all the time. This may produce elongated stars or even double images in the worst-case scenario. What about autoguiding I hear you say? Well, if the autoguider uses a separate scope, it is ignorant of the shift and if it is using an off-axis tube, it may eventually correct the error but there potentially will be a ghost of the temporary deviation in the image.
3. All focus mechanisms and gearboxes on motor systems suffer from some backlash. If the temperature trend changes direction, the focus tube may not move at all until the change is significant. Traditional backlash control is unlikely to help as it deliberately overshoots the focuser tube movement and retreats by a set amount, with the potential for significant image shift and temporary blur.
4. If an exposed temperature sensor is exposed to dew, it will likely give a false and low reading, accompanied by an incorrect focuser position.
5. Adjustments are relative; they do not take account of other mechanical error states, such as focus tube slippage (most likely on Crayford focusers) and mirror movement. A major benefit of autofocus routines, however, is that they make an absolute measurement and usually eliminate many of these error states.

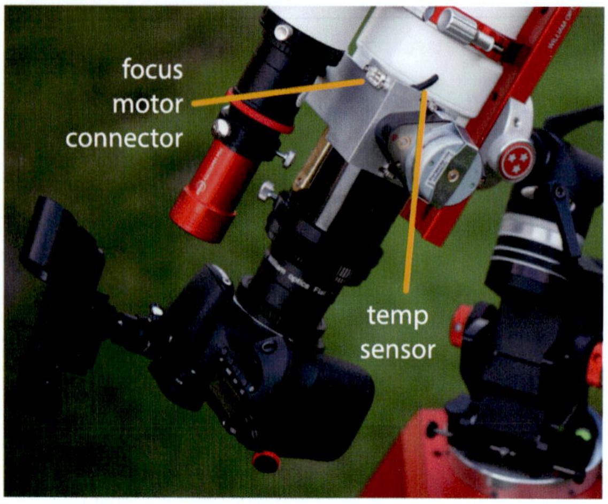


*fig.3 Typical of several systems, the Lakeside focus motors have a temp sensor mounted by the motor connector. Some others have it inside the focus control module, making it more susceptible to local heating. The temperature is available as an ASCOM property to any acquisition application.*

It is for these reasons I prefer to trigger an ordinary autofocus run for every 0.5–1°C change and when imaging with the larger scopes, include a standard interval too, to prevent erratic temperature movements and slow thermal responses causing delayed focus drift.

### Setting Up for Autofocus

There are a couple of things you can do to improve the reliability and convenience of autofocus. For starters, I have a single focus motor control module that I share between 4 telescopes. It is useful to have each set up at approximate focus when I start imaging. For that reason, I set a default focus tube extension for each telescope for a focuser position of 5,000. When I swap systems, I return to position 5,000 before disconnecting the control lead. In that way, I can record the focus position for each system and return to it instantly. Following on from that, I also note the motor step size in microns, by measuring the focus tube extension for say a 2,000 step move. This is useful for setting the measurement step size in the autofocus algorithm. This is often different for each setup. In my systems, a single step moves the focus tube from 3–5 μm (0.003–0.005 mm).

A motor gearbox and any gearing in the focuser tube will have mechanical play. Reversing the focus tube direction requires a number of steps to take up

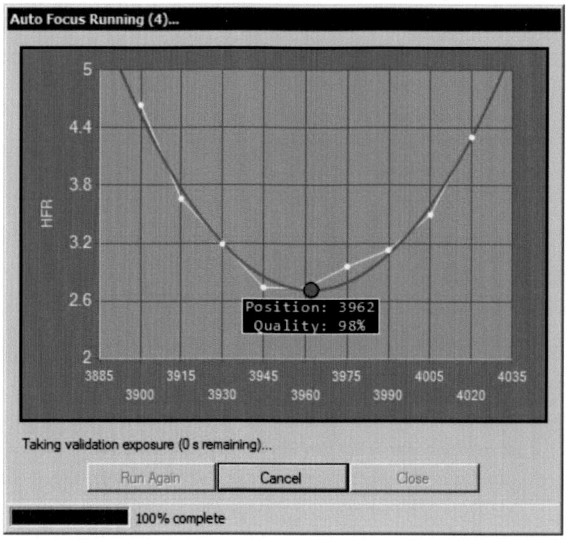


*fig.4 The majority of the autofocus algorithms evaluate the star diameters at various focuser positions. When plotted, they form a V-curve, with linear "arms" which intersect at the best focus position. The individual exposures are typically under 10 seconds and may confine themselves to a cropped central region or a single designated focus star. The linear relationship breaks down near focus and the latest versions of SGP uses a curve-fit to improve focus accuracy.*

this play or backlash. In my system, it takes about 50 steps to change direction. To overcome this, the motor always approaches the focus point from one direction and in this example, the focus control module is programmed to overshoot and wind back 100 steps after moving outwards. This guarantees the focuser always moves inwards on its last move, overcoming any likely backlash and as it is against gravity, keeps the focus mechanism gears engaged.

For those imagers who use separate filters and a monochrome sensor, it is normally required to autofocus after a filter change or, as mentioned earlier, make a pre-determined adjustment. This will account for variations in the system focal length at different wavelengths; simple refractors may not focus all wavelengths at the same point and in addition, is the effect of inserting a different filter. Although filter sets are often advertised as parfocal (implying they have the same focus) this is only an approximation. Introducing a filter in front of the sensor changes the physical distance required to achieve focus. The reason for this is that the physical and optical thickness are not the same thing. The optical thickness is the physical thickness multiplied by the refractive index (glass is about 1.5). The glass substrate has dispersion too (its refractive index varies with color) and so there are also small differences in the effective optical path length for each color. These small focus shifts become more noticeable at fast aperture ratios and at high magnifications. The good thing is that these shifts are repeatable and after doing a once-only calibration to accurately measure their respective positions, I simply move the focuser by the required number of steps when I change the active filter. I perform the calibration for each of my optical setups and store their values in the acquisition application's equipment profile.

### Autofocus Process

After a user command or trigger, the autofocus process measures the star diameters for a range of focus positions either side of the current position. The step sizes and the number of steps determine the range of the autofocus measurement. Too small and it potentially will miss the focus position, too large and the software may struggle to identify the star size reliably (the reduced star intensity is more difficult to measure and with telescopes with a central obstruction, a severely out of focus star resembles a donut). For a f/5.6 system, I typically set 9 or 11 steps over a 1.5-mm range. At f/4, I halve the step size.

The focuser usually moves outwards first (with gravity) by half the range and works its way inwards to the other end of the range. To speed things up, binning the image increases the intensity and speeds up each image download. One can also select a fast download speed and use subframes to avoid unnecessary download and evaluation time. Even so, it normally takes a few minutes to complete the evaluation. The outcome is usually a "V" or "U" curve, like the one in fig.4, showing the star diameter at different positions. It is normal for the bottom of the curve to flatten out as optical diffraction prevents a star having a zero diameter. If a telescope is out of alignment, the star shape is increasingly distorted when it is out of focus and, in the case of my reflector telescope, when it is even slightly out of collimation, the arms of the "V" do not have the same slope.

At the end of the process, the software calculates the position of the lowest star-size value, either by intersecting the slopes of the "V" or a weighted average of the 2 or 3 best points. The focus motor is then driven back to this point. Here, the key word is "back" and implies a reversal of the focus mechanism direction. This is where backlash may occur and why it is a good idea to have backlash compensation set in your system, either within the control module or the acquisition software. My main acquisition software has a backlash compensation option but I choose to set it within the control module and disable all other options to avoid any possible conflict. The outcome is a deliberate overshoot of the "back" command and then a reversal of the overshoot to the final focus resting place.

Some acquisition programs monitor the star sizes after each image download. This information is a useful diagnostic during acquisition. In the case of Sequence Generator Pro, it plots the number of stars and star widths (HFR) for each successive exposure (fig.2). An increasing HFR value may be the result of faint cloud, mist and drifting focus (as well as worsening seeing conditions).

As you might expect, there are a few process alternatives. Some autofocus routines require the telescope to slew to a fairly bright star, which permits short exposures (under a second) and rapid evaluation. Others, like the one in SGP, measure multiple stars in the target image, which are often fainter. In this case, I set an autofocus exposure of 5–10 seconds, through a clear filter, to ensure that it can detect enough stars. Focusing through colored filters requires longer exposures and becomes excessive in the case of narrowband filters and the results are more likely to be influenced by tracking issues too. For this reason, I usually set up my system to switch to a clear filter for autofocus and then switch back to the required imaging filter and apply the pre-defined focus offset, before the next exposure begins.

### Focus Defects

When starting out with a new telescope it is a good idea to check out the general alignment as soon as possible. To do this, after aligning the telescope to the celestial pole, carefully focus the telescope and take a simple short exposure of a star field with an APS-C sized sensor (or larger). Now, examine each of the corners of the image at 100% magnification and check the star shapes. In a perfect system, the stars will be pin-sharp into the corners and perfectly circular at the same time. More likely, the stars will be slightly fuzzy or elongated. In the case of a curved focus plane, the stars in each corner will be slightly elongated, radiating from the center. If this is the case it may be worth adjusting the camera distance to the field flattener and evaluating again with the spacing ±1 mm and repeating to find the best setting.

If the sensor is tilted (either due to a focus flexure, camera sensor alignment or optical collimation issues) the stars will have variable focus across the image and at the same time, slight elongation along the same line, rather than radially. To remedy tilt, some astro cameras have tilt-adjusters on their faceplate, as do some focus mechanisms (and optical cells). While it is expected that the user will have to adjust the mirror collimation on a reflector telescope, alignment issues on refractors will almost certainly need expert help. As refractors have faster aperture ratio's, the issues become more noticeable. Some manufacturers declare that they align their sensors, others do not but provide the means to adjust them. One of the construction projects in *The Astrophotography Manual* shows how to use a small laser module to assess the sensor tilt.

*fig.5 It is a good idea to check star shapes in the corners of a few trial exposures. This is the top left, with elongated stars pointing towards the image center. The other three were similar, indicating a curved focus field. If the corners are different, it is more likely an image tilt problem, caused by the optical axis not being orthogonal to the sensor.*

Tilt can also come from the other parts of the assembly. It is relatively easy to spot a problem but difficult to make an assessment. For that, there are image analysis tools (for instance, CCDInspector) that measure and compare tilt and field curvature. It is worth noting star elongation can also be caused by tracking issues. In this case, all the stars are elongated by a similar amount and can be confused with tilt. If the mount's periodic error/tracking is under control, this should not be the case with short exposures. Small amounts of defocus also make lens aberrations easier to spot, including obvious colored fringes and coma in the outer regions. (As noted before, in some cases it is easier to manually focus a DSLR image by minimizing these aberrations in the image margins than judging the size of a central star.) In fact, optical alignment is commonly judged by evaluating out-of-focus stars in the center and outer fields. If you are interested, *Star Testing Astronomical Telescopes* by Harold Suiter is the go-to book on the subject. It is amazing how sensitive a star is to minute misalignments or errors in an optical system.

# Problem Solving

*In astrophotography, a measure of success is how well we overcome setback and failure!*

The past few chapters have concentrated on PC-controlled (robotic) imaging. Astrophotography is not a plug'n'play hobby. If the physics of capturing a few photons every minute were not enough of a challenge, the diversity of individual setups and the unforgiving nature of astrophotography place additional random obstacles in one's path. With the introduction of computers and software, problems pop up at the most unexpected moments and are not always easy to diagnose correctly. While it may be convenient to classify issues into hardware and software pigeon-holes, systems are more complex, with all kinds of interactions. For instance, it is very common for a user to blame their acquisition software for crashing or behaving erratically when it is entirely due to a poor-quality USB hub. Even experienced imagers jump to the wrong conclusions, as something that appears obvious turns out to be a bizarre coincidence.

This chapter concentrates on issues associated with image capture. These turn up randomly and always seem to be the time-critical ones, consuming precious clear skies. While I can only highlight the most common problems and their likely causes, it is useful show you how to isolate an issue and ask for help. Here are my top 5 issues, mostly involving USB, or as astrophotographers call it, the Unreliable Serial Bus.

1. USB device disconnects – usually due to Windows putting the USB port to sleep. Fixed by disabling USB sleep in Windows' advanced power settings. While in the advanced power settings, it is a good idea to disable all hardware sleep modes.
2. USB device random behavior – often due to poor USB lead quality (especially when it is also being used to power a device) and/or insufficient USB hub power. Fixed by powering all hubs and using high-quality cables, as short as feasible.
3. USB device communication errors – often caused by an unreliable hub in cold conditions. Fixed by using a high-quality, rugged hub. (I have found the industrial versions to be reliable and my loaner unit has fixed several friends' issues.)
4. WiFi dropouts – there are many causes, proximity being the main one but also caused by interference from poor quality USB 3.0 hubs and cables that operate at the same frequency.
5. USB connected digital camera refuses to operate remotely – check whether the acquisition application expects the camera set to "M" or "Bulb" mode and the correct USB connection mode in the camera menu.

### Mechanical Issues.

Mechanical issues manifest themselves in three areas: pointing, tracking and focusing. On the one hand, many mechanical issues may appear "common sense" since they involve moving bits of metal that one can measure and touch. Certainly, if you can detect movement, flexure or play in a camera/telescope/mount/tripod assembly by hand, it will cause an alignment, focusing or tracking issue at some point. While there are obvious issues caused by poor tolerances, alignment errors and rigidity, the minute mechanical issues that cause mayhem with tracking are less obvious. For instance, even on premium mounts, if the mount's bearing pre-load is too high, a guider pulse on one axis appears to affect the other axis (diagnosed by a cross-test).

### Software and System Diagnosis

It requires megabytes of code to run an operating system, application and device driver. It only takes a few bytes to be wrong to cause an issue and that is before the hardware is taken into account. These complex systems are difficult to fault find and it is a common trap to confuse a symptom with a cause and follow the wrong path. The first step is to carefully observe; noting the conditions in which the problem occurs; when, how frequently, how consistently and if there have been any coincidental changes in the system. Documenting the conditions is a key step that helps to converge on the cause. It is important to know the conditions in which the problem does not occur as well as the one when it does. Similarly, validating a likely cause usually requires the problem to be turned on, off and on again. This might require swapping a USB hub or lead or undoing a recent software update, noting that the action of plugging/unplugging a connector may fix the issue by making a better connection. Even then, something as simple as a software update may appear to be the cause when all it has done is expose a hardware robustness issue. Infrequent issues are especially awkward, as one is never quite sure whether the issue has gone away or not. One last thing about problems and causes; engineers talk about special and common causes. If there is a sudden problem in a fixed configuration that has otherwise been behaving itself for months, the issue is unlikely to be a design issue or software. A sudden problem is often caused by a sudden change, like a hardware failure, software corruption or an "environment" change.

Usefully, many software applications and ASCOM device drivers automatically record logs. These text files are time-stamped lists of actions, responses and issues. These are invaluable to software developers as they enable them to identify an application bug or point to a hardware/driver problem. When you install the ASCOM platform onto a PC, it additionally loads a series of simulators for each of the hardware devices. These are useful to "play imaging" with a new application of check logical behavior, ruling out actual hardware issues. With all this, just remember that the most frequent culprit is likely you, the operator, and with the complexity of ever-evolving technology it is not surprising. Reliability improves with consistency and making use of equipment profiles and configurations helps enormously.

By recommending refractor telescopes in the beginning, we side-step a number of optical issues caused by poor optical element alignment (collimation) in reflector designs. Each reflector design has a unique method for aligning their mirrors to ensure they are spaced and angled correctly, to avoid misshapen or uneven star appearance.

| problem | likely cause | fix |
|---|---|---|
| initial alignment is completely wrong | accidentally did not align on Polaris or time/ location/daylight saving setting is incorrect (we have all done it some time) | confirm polar alignment and location and time settings in mount and applications |
| pointing accuracy after alignment is poor | poor mount homing accuracy, excessive flexure, telescope alignment, incorrect location and time (daylight saving?) | in addition to the above, check the epoch (J2000/JNow) and telescope mounting |
| excessive drift in DEC | poor polar alignment | check for stable platform and flexure |
| excessive drift in RA | poor polar alignment, wrong tracking rate | check alignment and sidereal tracking rate in mount |
| during guiding, periodic large RA errors | contamination in the mount gearing, cable snags, wind gusts | clean and lubricate the mount gears, especially the worm |
| during guiding, overshoot in RA | overcorrection of error or too sensitive to atmospheric seeing | lower aggression and lengthen guider exposure time |
| during guiding, DEC refuses to move for a period | gear play and stiction are most likely culprits / wrong guider algorithm | use guider backlash compensation /adjust worm engagement / gearing |
| during guiding, guider graph is good but image has elongated stars | most often occurs when a separate guidescope is used and it and the telescope flex differentially | make the guider assembly and coupling as rigid as possible, check for guider focus tube slop. |
| small focuser moves have no effect | usually caused by gear play in motor gearbox/focuser. | use backlash compensation or manually approach from one direction |
| images have radiating stars in corners | field flattener / spacing is missing or incorrect | add field flattener and adjust spacing to sensor to optimize corner stars |
| big faint blobs on image | dust on filter / flattener /mirror or dew | carefully clean optical surfaces |
| small darker blobs on image | dust or dew on sensor cover window | carefully clean optical surfaces |
| inconsistent focus across image | usually due to sensor tilt, from the camera construction or focus tube sag | it's the camera. if problem rotates with its angle, otherwise check for focus tube alignment and flexure. |
| changing image color from center to edge, especially with camera lenses | most likely caused by combination of fast aperture lens (f/4 or larger) with a dichroic filter (e.g. light-pollution filter) | try imaging without a filter or stop the lens down to f/4 |

# Eastern Veil Nebula (NGC6992)

*Getting the most from a color camera in light-polluted conditions.*

**Equipment:**
*Full-spectrum modified Fuji X-M1, intervalometer*
*73mm f/5.9 refractor (430-mm $f_L$)*
*field flattener, electronic focus*
*50-mm guide scope, guide camera*
*Fornax LighTrack II mount, counterweight kit*
*Light-Pollution (IDAS), Dual-Narrowband filters*

**Software:**
*PHD2*
*DeepSkyStacker, Nebulosity, Affinity Photo*

**Exposure:**
*IDAS filter 80 x 120 seconds at ISO 400*
*Duo-Narrowband filter 36 x 300 seconds at ISO 400*
*Dark, Flat and Bias calibration exposures to suit*

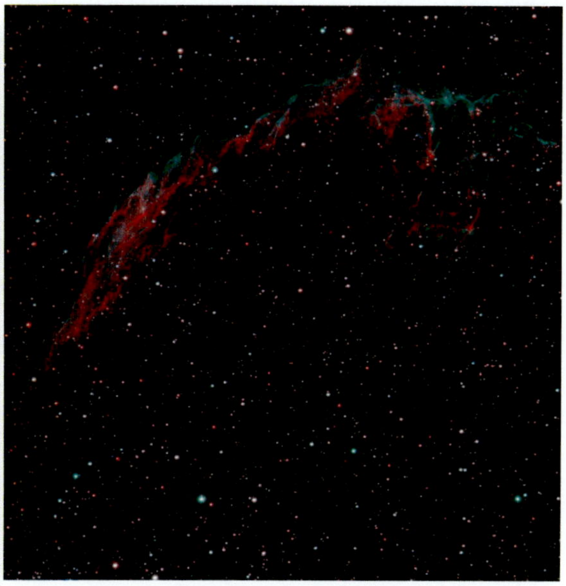

This assignment is an experimental departure, exploring the potential use of a cooled color CMOS astro camera. I normally use a mono CCD sensor in a dedicated astro camera. At some time new technology will either replace it or offer something different. Presently, CMOS sensors of a reasonable size are solely color CFA models (with a few exceptions). Yes, there are smaller monochrome sensors but these are more suited for guiding or planetary imaging. Color cameras are more challenging to use in heavily light-polluted locations and while it would be good to use one for the last assignment, it occurred to test the water first and avoid an unnecessary purchase. To that end, I took a used 16 MPixel Fuji X-M1 APS-C camera and had its UV/IR filter removed. This camera serves a dual purpose; it can take IR landscape images (fitted with a red or IR filter) or with its extended red sensitivity, pressed into deep-sky service to test the effectiveness of specialist light-pollution filters. It does not, however, have any internal sensor cooling which places a practical limit on individual exposure durations.

## Acquisition

The target, the Veil Nebula, is a popular one, with strong deep red and blue-green emissions. It is located in a heavily-populated area of the Milky Way and it has a dense starfield throughout. Coupled to a 350-mm focal length refractor, the sensor almost covers the full extent of the supernova remnant, whose regions are often imaged separately. Over several nights, the nebula was photographed from my back yard, initially through an IDAS light-pollution filter and then through one of the latest duo narrowband filters, designed for color cameras and which only transmit blue-green and deep red wavelengths. (These 2-inch filters screw to the 48-mm thread on the front end of the field flattener.) This imaging assembly is too ungainly to balance on a simple ball and socket head and it required a small counterbalance system (fig.1). After careful balancing around both axes and then polar-aligned with the polar scope, I waited for the optics to acclimatize and set the camera to manual focus, daylight white balance and Bulb mode. Using an iPhone as an aiming device, fitted to the camera's hot-shoe, it took a few trial exposures to center and rotate the camera to cover the nebula. Even so, there were some framing differences between imaging sessions. I should have anticipated this and, lesson learned, in future I will compose images with more "breathing space" around the target, to allow a generous crop of the final registered image stack.

Focusing was not easy, even using the bright star Vega with x5 LCD screen magnification, the diffraction spikes from the Bahtinov mask were difficult to discern. In the end, it was easier to find focus using the magnified image and nudge the focus position, 10 units at a time. (Later Fuji camera models have a x10 magnifier as well as a live video feed transmitted on their HDMI connector.) Fortunately, the ambient temperature was stable and only required a single focus confirmation after the meridian flip. Three hours total were taken with each filter; those through the duo-narrowband filter required more than double the IDAS exposures to bring each frame's star brightness to about the same level. Image acquisition concluded with a full set of bias, flat and dark calibration files. For guiding, I used a small laptop running Windows 7 and PHD2, connected to a ZWO guide camera. After calibrating, I used 2-second exposures to guide the RA axis.

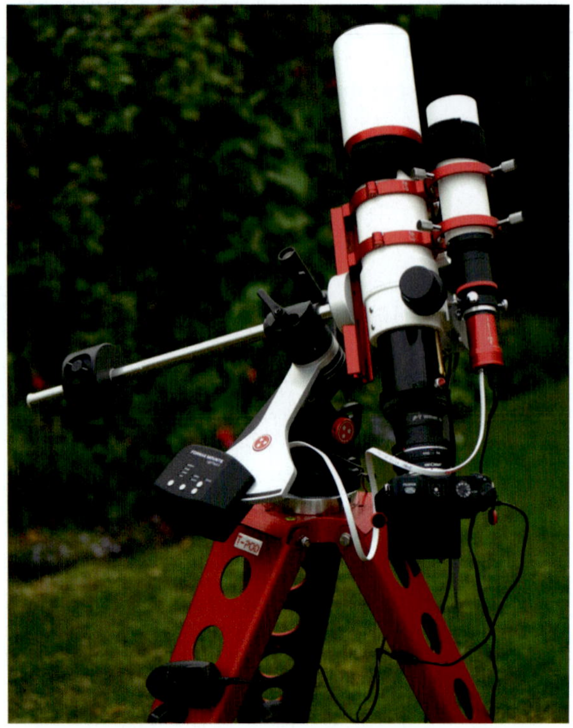

*fig.1 The basis of the imaging setup, using a short refractor, guide-scope and a modified Fuji X-M1 camera on a Fornax mount. This is too heavy to balance on a simple ball and socket head and a counterweight system was used, which also features a Vixen clamp. The exposures were controlled via a handheld intervalometer connected to the camera's USB port. As it got dark, I mounted my smartphone aiming jig (described in earlier chapters) to the camera's hot shoe.*

### Image PreProcessing

Calibrating CMOS files and Fuji RAW X-Trans files in particular, is tricky. As discussed previously, some CMOS sensors have the annoying behavior of automatically doing a primitive dark subtraction using the pixel values from a masked portion of the sensor. This conflicts with the traditional CCD calibration processes that have ingrained themselves on a generation of astrophotographers and application developers. In particular, bias subtraction from dark frames and dark-frame optimization may cause the images to have black pixels and not fully remove amp glow. In the case of the Fuji X-M1, its dark frames have a higher median value than those from the bias frames, indicating that this camera, unlike the EOS 60Da, should not cause issues with traditional image calibration processes. If this had not been the case, it would have required manual intervention during calibration. I compared conventional calibrations made with DeepSkyStacker and Nebulosity 4's default calibration routines. The outcomes were similar, though the automatic registration within DeepSkyStacker (DSS) was more convenient.

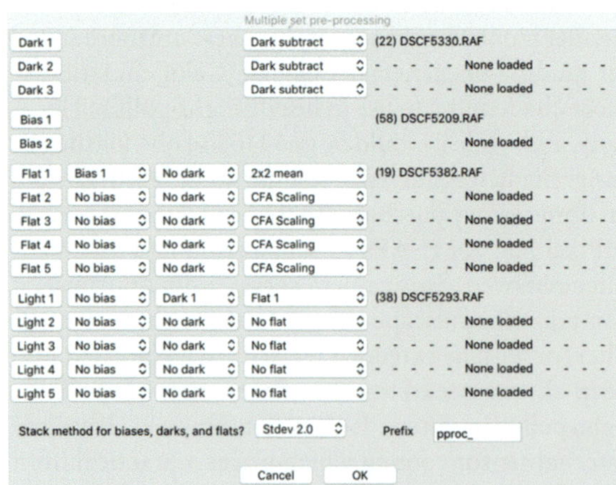


*fig.2 Nebulosity 4 set up to create master calibration files and then calibrate the image files. (Unlike the workflow in DeepSkyStacker, image registration and stacking processes are made with a second tool.)*

### Image Processing

The workflow in fig.9 was accomplished in Nebulosity 4 and Affinity Photo. Several automatic processing tools use the image data to calculate the settings. The ragged borders can interfere with the assessment and so, before proceeding, I cropped them off. While instant gratification encourages one to start with an image stretch, it is much better to correct the color balance beforehand. The reason for this is it is much easier to correct for

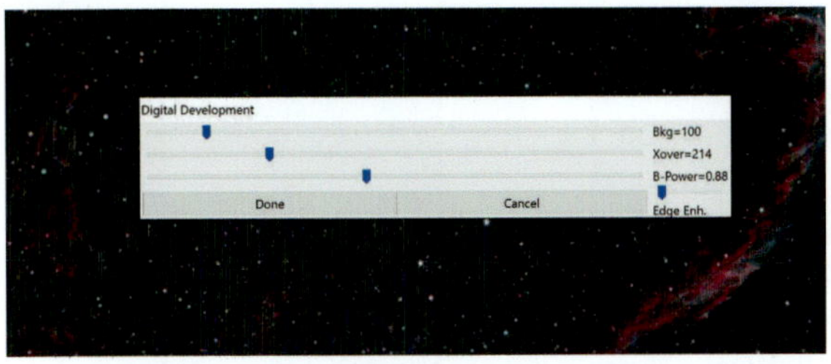


*fig.3 The Digital Development (DPP) tool combines an image stretch with a simplistic deconvolution (sharpening) process. It occurs in other applications as well. In all cases it should be used with care ,as it easy to be heavy-handed.*

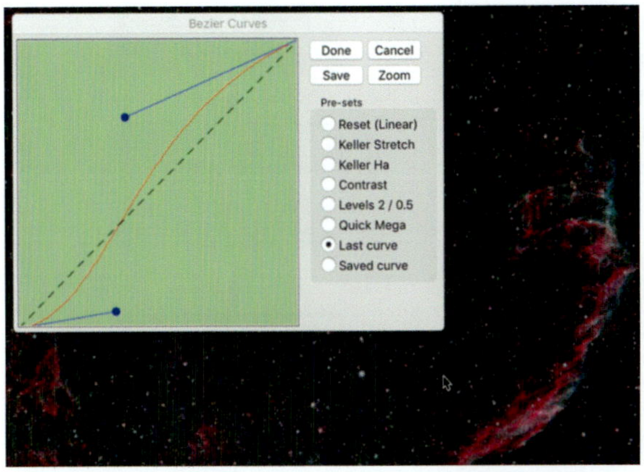


*fig.4 Nebulosity 4's Curves tool allows one to selectively increase or decrease image contrast, depending on luminosity, without blowing out bright stars.*

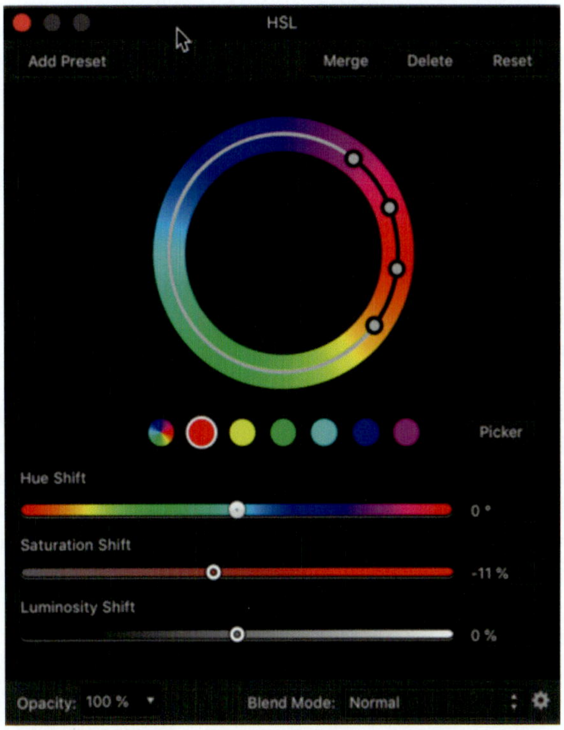


*fig.5 The Affinity Photo HSL control, here I used the picker to select some red nebulosity and then reduced the saturation slightly.*

color issues while the image data is still linear, before they are distorted further. It can be difficult to achieve a neutral background and believable color. Fortunately, an application of Nebulosity's Auto Color Balance tool does a good job. There are many ways to stretch an image and in this example, I used the Digital Development (DPP) tool with the default settings to bring the nebula to life (fig.3). DPP combines several actions that other applications do separately and while it is convenient, you need to take care with it, as it can create digital artifacts. In this image, the effect was subtle and needed a further boost, lowering shadow values and lifting mid-tones. The ideal tool for this is the Curves tool. To use it, one drags the two handles to create a spline curve, like the one in fig.4. This subdues the background but crucially boosts the appearance of the nebulosity. It also improves the appearance of shadow noise to some extent. In this case, I improved the image noise slightly with an application of Nebulosity 4's GREYCstoration noise tool, with the default parameters and scale set to 1.5. This tool has many parameters and I am certain my result could be improved upon with further experimentation. There is a useful tutorial on the Nebulosity (Stark Labs) website that suggests how to go about finding the balance between sharpening, noise reduction and keeping image details. Finally, before exporting, I reduced the star size a little with Nebulosity's Tighten Star Edges tool. The smaller stars are less dominant and allow the nebulosity to come through.

At this point, it is a good time to evaluate the image and in this case, it was apparent that the red nebulosity was more intense than the blue-green (which is a common occurrence). Transferring the image to Affinity, I tidied up the image balance using an HSL adjustment layer (HSL stands for Hue, Saturation, Luminance). Here, I slightly de-saturated the reds and enhanced the blues, balancing with the intensity and saturation adjusters (fig.5).

Finally, the image was sharpened using a High Pass Live Filter layer. With this tool, one applies the filter to a duplicate layer to reveal the structures you wish to emphasize and then change its blending mode to Soft or Hard Light. (If you are using Photoshop, the equivalent tool works the same.) The title image is a crop of a quarter of the frame, to indicate the beautiful, delicate nature of the aptly-named nebula.

I used the same Pre/Processing workflow on the images taken through the IDAS light-pollution filter. It is interesting to compare the results. My framing consistency was poor between sessions and for the comparison, I used an alternative part of the nebula, called Pickering's Triangle that was present in both image sets. The processed versions are shown in fig.6 and fig.7. These bear out the differences seen in their respective RAW images; those using the light-pollution filter have an obvious sky background and many bright stars which obscure the delicate nebula structures. Those taken with the duo-narrowband filter have a darker sky background (that measured about 8x lower for an equivalent exposure length) and the stars are less dominant in the image too.

For interest, fig.8 shows an equivalent 3-hour session, exhumed from my archive, taken with an ancient monochrome CCD camera at -20 °C using traditional single narrowband filters. All three images were similarly processed (to the extent that their image noise permitted) and show how the emphasis shifts between the stars to the nebulosity as the filter selectivity increases and how the background noise reduces as the light pollution is blocked and the sensor temperature is lowered.

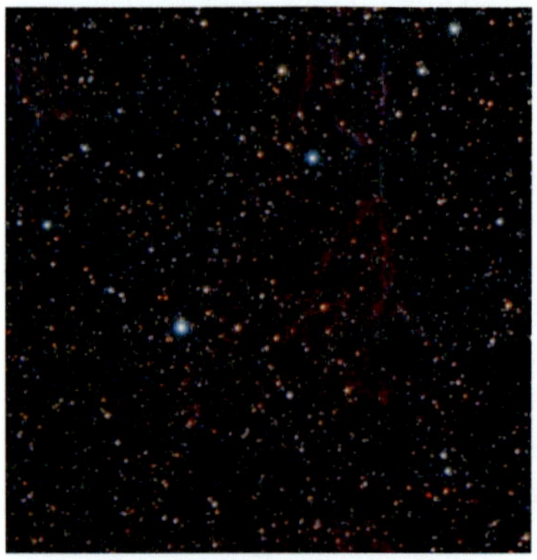

*fig.6 Pickering's Triangle, within the Veil Nebula, imaged through an IDAS light-pollution filter.*

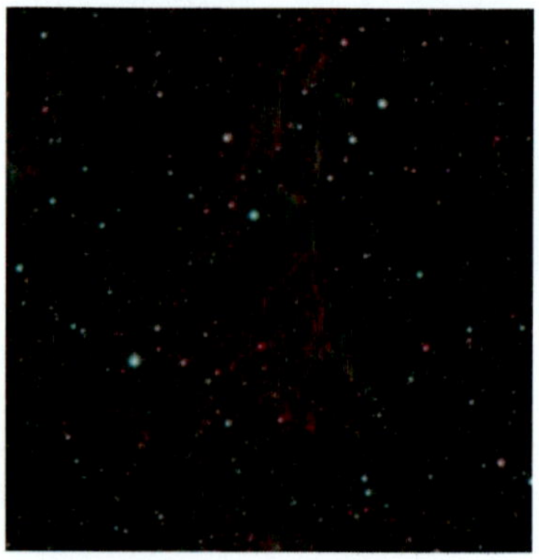

*fig.7 As fig.6, imaged through a Duo-Narrowband filter, that passes OIII and Ha emissions.*

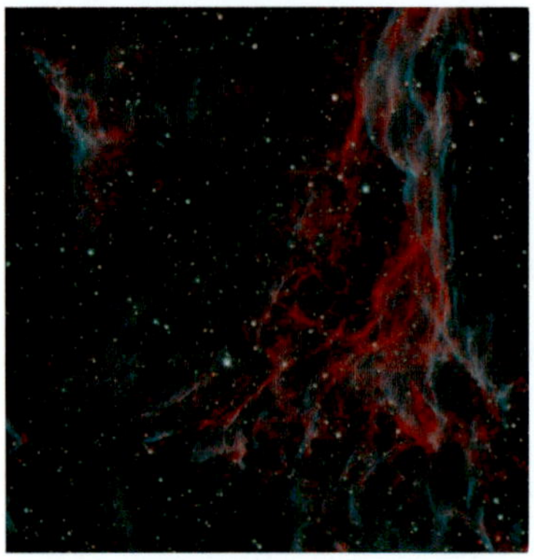

*fig.8 An equivalent comparison using an old cooled CCD monochrome camera and narrowband filters.*

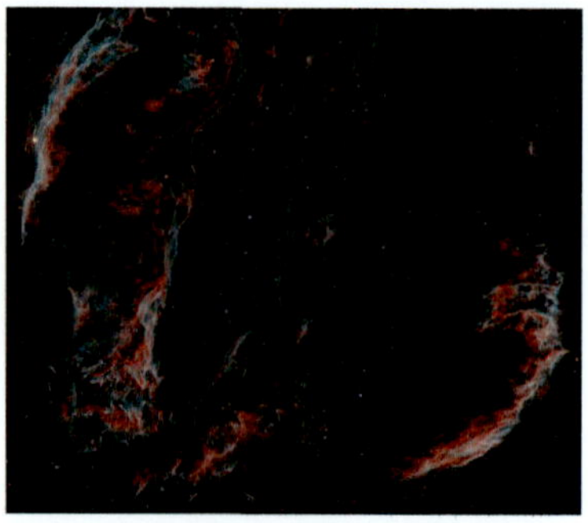

*fig.10 This shows the full extent of the nebula, about 50 light years across. It is a truly beautiful object, with delicate tendrils, belying the cataclysmic events that created it.*

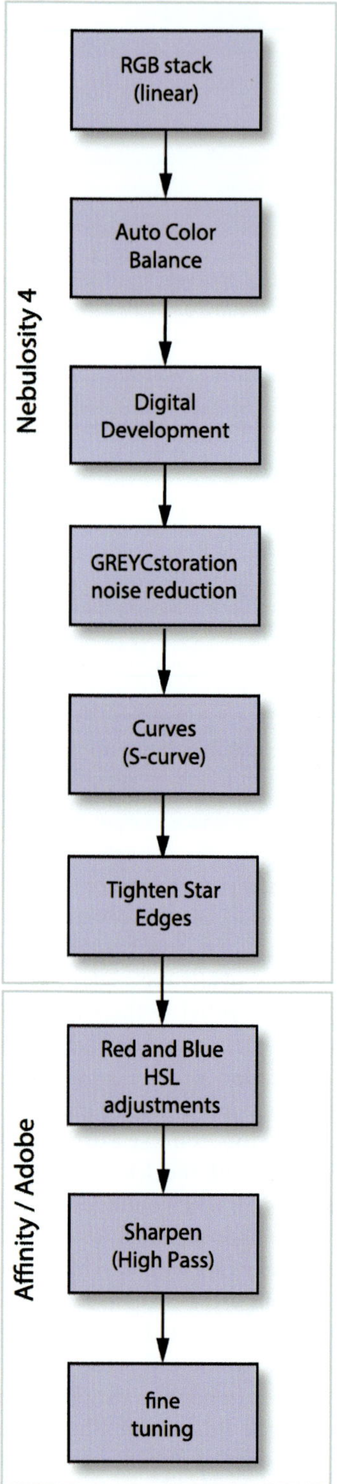


*fig.9 This is the workflow used in this example, using Nebulosity 4 for initial processing and a photo-editing program for fine tuning.*

# Image Processing (part 2)

*Image processing with empathy comes with practice.*

The introductory chapter on image processing laid out the general workflow and the broader considerations of composition, qualities and useful tools. In between, the few practical examples have provided a taster of the simpler processing steps. Not everyone has dedicated editing software and in this chapter I broke new ground to see how far we can go with a conventional photo editor on a challenging subject.

### Natural Selection

In general, the manipulations in the practical assignments have been global, in so much that they have been applied equally to all areas. This worked well, in the main, as the subjects and our expectations were less demanding. To take on more challenging subjects, not only does the image acquisition take longer but the image processing ratchets up a notch too. As you become more discerning, you notice that some global manipulations cause as many issues as they solve and require selective application, so that specific components of the image are excluded from those processing steps that do more damage than good.

It is difficult to get some manipulations correct the first time and it is easy to over-do things. In the early days, digital photographers applied adjustments directly to the "background" layer and one could only undo the steps and start over. A better approach, to retrace your steps or modify a previous change, is to apply these using adjustment and filter layers. These can be re-arranged, disabled or altered at any time to permit further tuning. Using layers also allows for endless experimentation with different layer opacities and blending modes.

Blending modes affect the way layers interact (often with unusual results). There are over 25 blending modes in Affinity Photo, of which the most commonly used are Normal, Overlay, Hard Light, Saturation, Color and Luminance. Layers also open the door to selective adjustments, through the use of layer masks. The selective adjustments usually sharpen, reduce noise, increase saturation or alter local contrast. The layer masks are usually derived from the image luminosity, color or feature scale. On occasion, it may be necessary to manually create a selection using the wand tool.

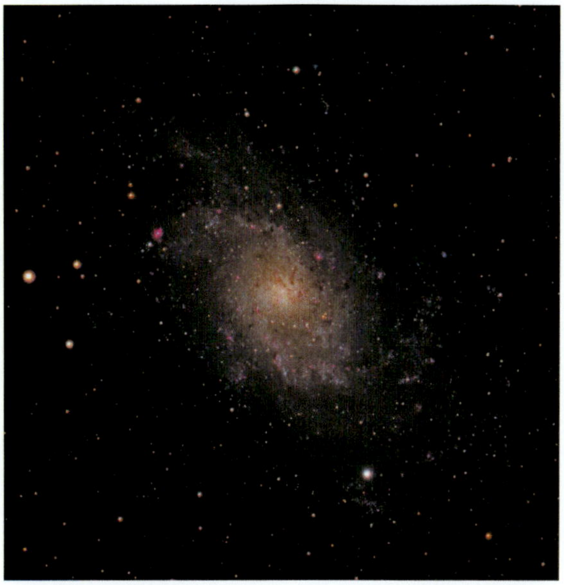

*M33, processed in Affinity Photo, using a 32-bit HDR file format*

### Editing File Format

The workflow assumes one has a calibrated, registered stack. Conventional editors will normally require a TIFF file, while dedicated astrophotography applications accept FITS files or TIFF. The stack (or stacks) should be at least be 16-bit, or better still, 32-bit floating-point. Most stacking software will optionally save to TIFF which, unlike JPEG, is a lossless format that can be 8-,16- or 32-bit. When using a color astro camera for the first time with an acquisition program, however, it will likely default to FITS. This can confuse stacking software since while a RAW filename extension shouts "I'm a color RAW file," both monochrome and color FITS files share the same file extension and the software is left guessing. In the case of DeepSkyStacker it requires the 16-bit RAW file option enabled on its FITS settings tab to tell it to de-Bayer the image and create a color image. Usefully, DSS can also convert a color FITS file into TIFF. On the Mac platform, I am not aware of a free conversion utility that will handle color file conversion.

In this case, we are using the result of 25 x 600-second exposures of M33, a large galaxy but one with low surface brightness and an extended pale periphery. This subtle galaxy will tax our manipulation skills.

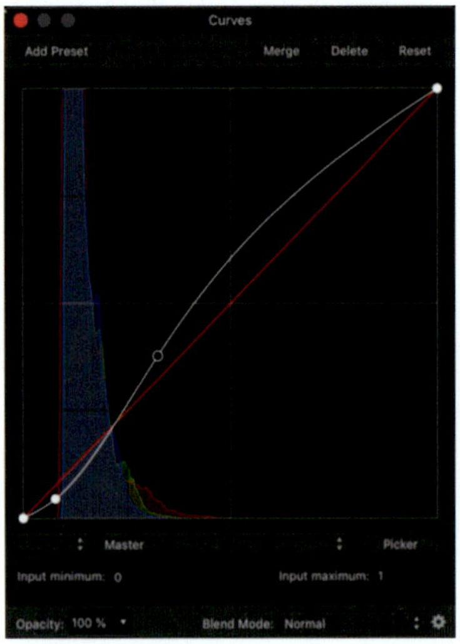


*fig.1 Curves are very useful tools for fine tuning. Where the curve is less steep than the red diagonal, contrast is reduced and similarly increased as it gets steeper. Avoid moving the right-hand inwards and only move the left hand end point if the histogram shows a clear gap (as it does here).*

### Editing Software

Going forward, I am using a 32-bit (HDR) color image of M33 with Affinity Photo, whose tools are similar to Adobe Photoshop's. This app has extensive high-bit support for most manipulations and additionally has some unique tools that are particularly useful for astrophotography. We will be using a range of common tools including layers, masks, blending modes, adjustments and filters. I'm using it on a Mac and as is typical, where I mention modifier keys, the PC version uses them too, where Ctrl, Command and the Option buttons are equivalent to Ctrl, Alt and Start keys.

As mentioned before, there is no single workflow that fits every situation. Not only can the workflow change with the image content, but it also changes with camera type (mono vs. color) and software (astro editor vs. photo editor). For instance, while it is preferred to correct color balance and background gradients before stretching, that is difficult to achieve on a linear image within a photo editing application. Astro image editors exist for a reason and that is that because they are designed for a specific purpose without compromise, working on minute pixel variations in 32-bit or 64-bit images. They uniquely having a screen-stretch function that reveals the hidden problems and image, without changing the image data itself.

## Image Preparation

After registering and stacking, the initially very dark image hides any number of issues and it requires a quick stretch to reveal the color balance, background color/uniformity and image noise (fig.3). If possible, it is better to fix these, or at least improve them, before stretching the image too far. Stretching is best accomplished with several modest actions, rather than a single extreme one. This provides an opportunity to reveal and fix some issues before the full manipulation is applied.

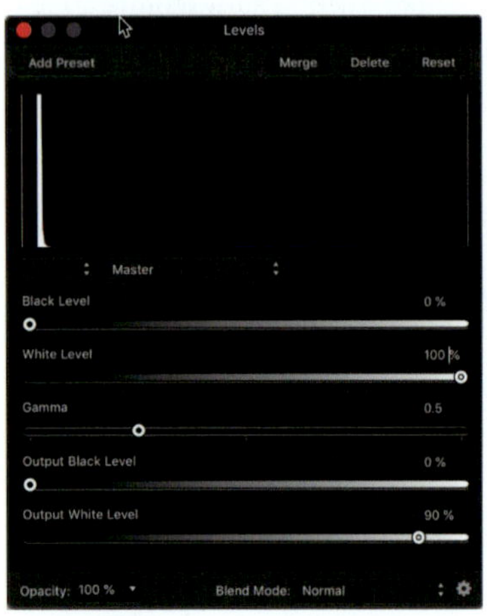


*fig.2 Another way of boosting shadows and mid-tones is to move the Gamma slider in the Levels tool. What some may think is a significant initial stretch on a 32-bit HDR image, still produces an image that is just a few white pinpricks.*

### Stretching

In a photo editor, the two classic tools that stretch an image are Levels and Curves. Of the two, the Curves tool is more flexible and duplicates the highlight/mid-tone/shadow adjustments of the Levels tool. There is one exception though, the Levels tool also has additional adjustments to give headroom for highlight and shadows tones with its output level controls. If star cores are already white, further sharpening and stretching will make them featureless white blobs. In these circumstances, it is useful to add some highlight headroom to reduce the star-core intensity and minimize clipping from later manipulations. With either tool, there are a number of things to look out for. In fig.1 or 2, moving the endpoints inwards may clip pixels to black or white. The white-point is rarely moved but after several stretches, the bulk of the histogram moves to the right and it is usually necessary to move the left-hand point inwards to bring the background level to about 90% density. Dragging a point creates a curve. It is important it is a curve, all the way to the endpoints, without horizontal portions along the top or bottom. Horizontal portions cause a range of tones to be the same and, equally unsightly, if any portion of the curve has a negative slope, it causes tone inversion. (As an aside, when imaging very faint objects, some astrophotographers deliberately invert the entire image. Human vision has better tonal resolution with bright tones and inverting an image reveals very faint objects hidden in dark tones.) The Curves tool changes image contrast according to image tone. Where the curve slope is steeper than the red diagonal, the contrast is increased and the opposite is true too. With two points an S-curve can simultaneously increase and decrease contrast, in the case of fig.1, increasing object contrast in the mid-tones and reducing the appearance of shadow noise without highlight clipping.

### Tone Mapping Persona.

There is a third tool that produces surprisingly good stretching results, especially after a mild stretch. In Affinity Photo, if you select a high-bit RGB pixel layer the Tone Mapping Persona can boost image shadows to an incredible degree, principally using its tone compression, blackpoint, exposure and brightness sliders and crucially without blowing highlights. It additionally has sliders for sharpness, saturation, vibrancy and curves. In Photoshop, a similar function is performed with the HDR toning adjustment. We shall experiment with tone mapping in our example.

After an initial stretch (figs.1,2) the background histogram is usually bunched up at the shadow end and has little effect on the visible image. It requires a more extreme, temporary manipulation to identify background defects, as in fig.3. Since the galaxy occupies a large part of the image, the common background blur/subtraction will not work. Gradients should be smooth, usually a simple linear or radial pattern and it should be possible to model closely. Dragging guides from the rulers, I identified two points in the corners that defined the color gradient. Deleting the extreme adjustment layer I Alt-clicked the color picker on the guideline intersections. These background selections appeared in the swatches palette. After adding a new pixel layer I created a linear gradient across these intersections, using the swatches as the reference colors. Changing the blending mode to Subtract I backed off slightly by adding a Brightness and Contrast adjustment layer, dragging this is as a child layer to the gradient pixel layer (fig.4). After checking with another extreme stretch there was still some coloration in the four corners. I repeated the exercise, this time with an elliptical gradient, with a local brightness adjustment and in Subtract blending mode. This reduced the gradient to manageable levels and the remaining coloration low enough to be subdued later on.

The image now consisted of a background with two blended pixel layers, ready for an initial stretch. For this, I clicked the Tone Mapping Persona button and experimented with the tone mapping sliders (fig.6). The result shows a faint galaxy (fig.5.) It is not a final image and there are few things to watch out for. The image is deliberately tuned with no shadow clipping and any highlight clipping is only on the brightest star cores. No attempt at increasing saturation or detail was attempted, since it is applied globally and would potentially interfere with the background. Once it was to my liking, I clicked Apply. Back in the Photo Persona, the next steps increased color saturation in the galaxy and stars and reduced the color and noise in the background, before enhancing the galaxy further.

### Initial Adjustments

These use the luminosity of the image itself to regulate their effect. Affinity calls them luminosity layers. In each case, a greyscale luminosity alpha channel is created, which links automatically to the next Live Filter or Adjustment layer. To create one, I Opt-Cmd clicked the image icon and

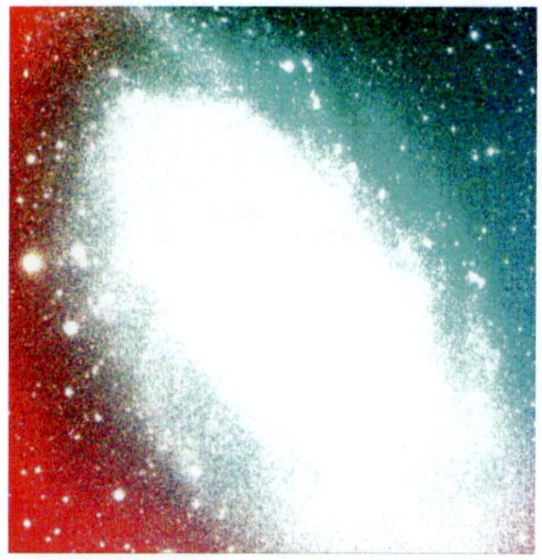

*fig.3 An extreme stretch highlights background uniformity issues and the best place to sample values.*

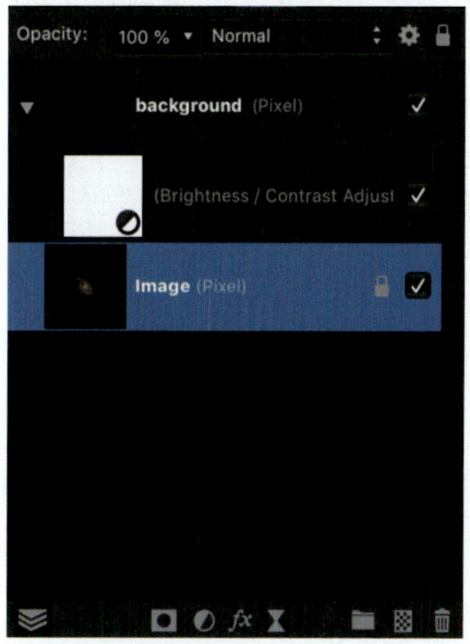


*fig.4 The layers palette, showing the background gradient layer over the image and toned down with a Brightness/Contrast adjustment.*

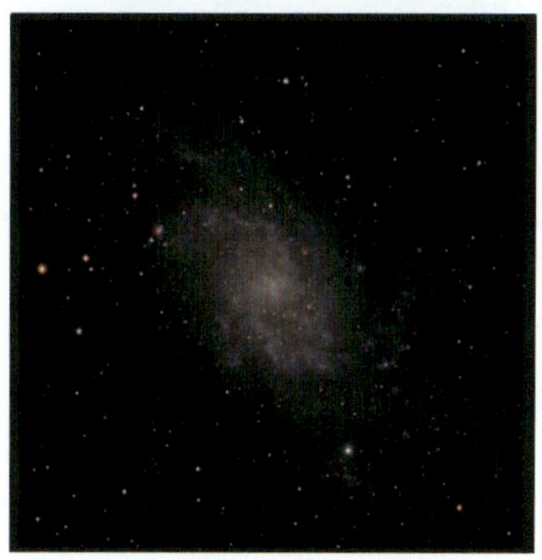

*fig.5 A faint galaxy appears after an initial stretch using the Tone Mapping Persona. It still needs some work to make it pop.*

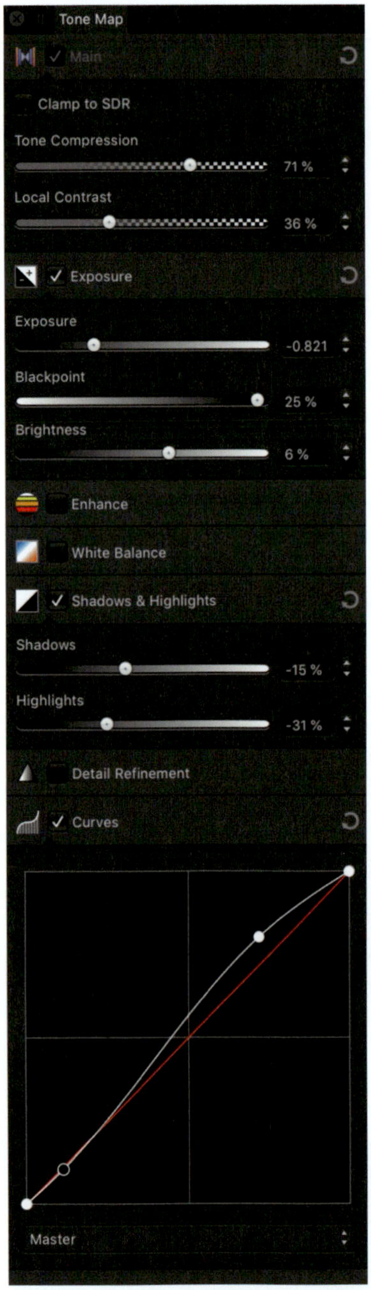


*fig.6 Tone Map is a powerful friend and it pays to take your time to experiment with the various tonal settings. In this case, I have restricted the changes to tonal values. In the Enhance section, there are additional controls for contrast, saturation and vibrance. The Detail Refinement section has tools for sharpening, which I am also ignoring for fear of accentuating noise and color in the background.*

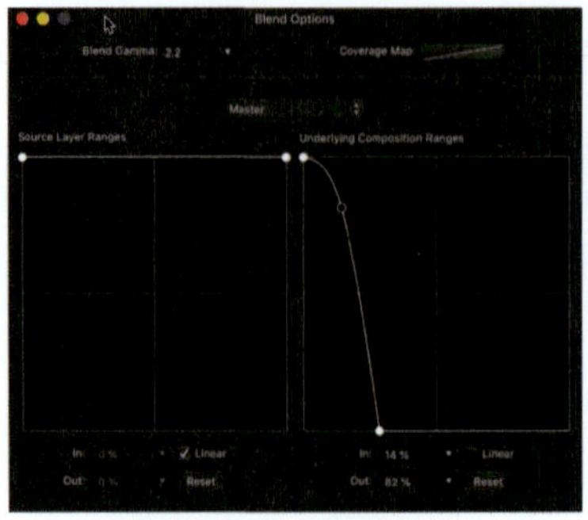

*fig.7 This blend range adjustment restricts the (de-saturation adjustment) to dark tones, effectively neutralizing the background.*

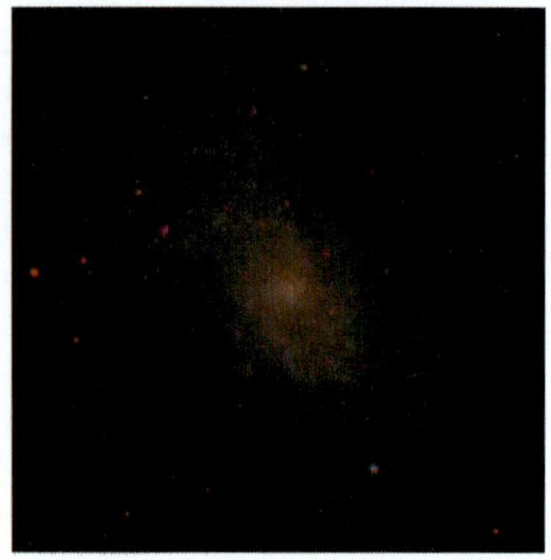

*fig.8 Slowly getting there; the galaxy is now more colorful.*

then added an adjustment layer. The layer icon changes to a miniature monochrome image and in this mode, the adjustment progressively acts on brighter tones. If I had wanted to work on the darker tones, in the channels palette, I would right-click on the alpha layer and select invert. These are linear relationships and for more control, the sensitivity to different tones can be controlled with a transfer curve in the blend options.

In our image, I added three adjustments; a minimum filter of 3 pixels on a highlight selection to shrink stars, a de-saturate background and a saturate galaxy layer. I needed more control, however, over the application of the two saturation layers. After selecting the adjustment layer, I selected the cog icon (blend options), and dragged a curve in the "Underlying Composition Ranges" graph, to dictate which tones were affected by the adjustment. In the case of suppressing color in the background, the blend curve in fig.7 shows that all the emphasis is on the bottom 25% of the tonal range, leaving the brighter tones untouched. The result is a dim galaxy, with some color and extent, on a fairly neutral background and with colorful stars (fig.8). The image is subtle, which is a good thing since it is better to work up to a more vibrant image than the other way around. To reduce background noise, I added a Denoise Live Filter layer, targeted at the 25% darkest tones to conclude the image preparation.

## Image Manipulation

M33 is a beautiful and delicate galaxy, with dust lanes and areas of intense glowing hydrogen gas. Further color and contrast adjustments are needed to enhance what is there. There is a risk that these same actions may inadvertently affect the background too so I prudently added another background adjustment. For variety, I tried a different approach; having sampled and created the simple gradient pixel layer in Subtract blending mode, I added another blank pixel layer and filled it with a neutral density of about 90% grey and changed its blending mode to Add. This sets the background level while removing more of the background gradient.

Starting the enhancements, I fine-tuned the galaxy core color with a Color Balance adjustment layer, using a blend options curve to constrain itself to the brighter tones. Next, it was the turn of the blue, newly-forming stars in the outer galaxy arms. To enhance these, I created a Hue-Saturation-Levels (HSL) adjustment layer (fig.9) and selected cyan blue and increased

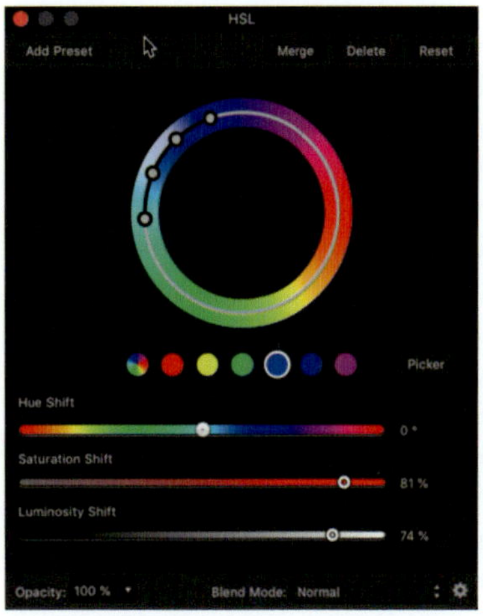


*fig.9 The HSL adjustment, ready to give a lift to blue stars.*

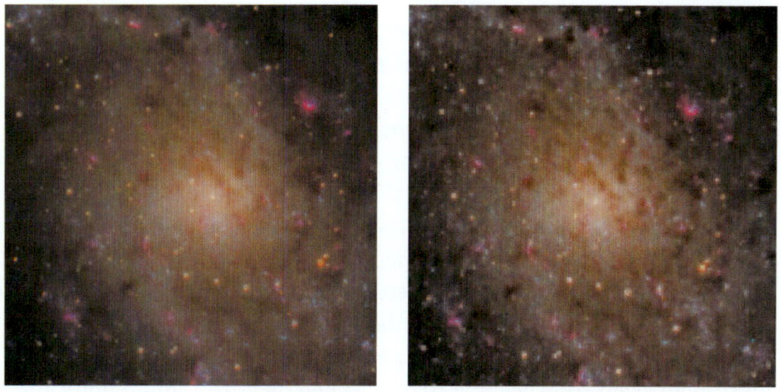

*fig.10 Before and after two applications of the High-Pass filter, set to two different pixel scales. These highly magnified crops show the absence of dark or light halos around objects. By using adjustment layers, it is possible to revisit these manipulations and tune the result, maybe for a different output scale or purpose.*

the saturation and luminosity, confining the adjustment with blend options to 20% and above of the tonal range. I picked out the red nebulosity in a similar manner, selecting the red channel and covering the orange and magenta colors. Next I needed to boost the galaxy brilliance and reduce color noise in the background before finally sharpening. Before doing so, a small digression about layers in general is useful, as you may be having difficulty making layers work in the way you want.

### Layer Mechanics and Masking

Layers are magic, there are many subtleties about them and the way in which they are created may confuse the beginner. By restricting adjustments and filters to layers (and crucially avoid using the permanent effect of the drop-down filter options in the main menu) it is easy to retrospectively tune, remove or temporarily disable any edit. In my normal photographic work, with a background in classical film photography, I rarely found the need for advanced layer techniques. Not everyone is compositing and some layer techniques may be new to a digital photographer. Fortunately, there are many video tutorials on the Internet. Our image is created from a background pixel layer with a series of adjustment layers on top (a layer stack). These adjustment layers are transparent so that we see through them. Things become more interesting when we make adjustments to part of the pixel image. In these cases, we need some form of layer mask. The effect of the mask changes its influence, depending on its position in the layer stack and how it was created. By default, creating a mask usually associates itself with the layer that was selected at the time. It is shown below that layer and slightly indented. It only influences that layer and is said to be clipped to the layer. The mask can be dragged elsewhere in the image stack, to change its behavior and influence. When Adjustment and Live Filter layers are created, they automatically include a transparent mask layer associated with them, indicated by the white rectangle in the layer palette. This also shows up as an alpha layer in its channels palette, which can be filled, cleared and inverted from the options that appear with a right-click. Masks can also be created from a selection, simply by clicking the quick mask icon in the layers palette. We have already used a luminosity mask, in which the image's luminance values are painted in the adjustment layer's alpha mask and also seen how the blend options dialog allows one to make the adjustment dependent on the image intensity. If the mask layer (or the adjustment/live filter layer) is selected, it is possible to add and subtract content by painting. To see what you have painted, Opt-clicking a mask icon makes it appear as a monochrome pixel image. Likewise, the net contribution of any layer (image or adjustment) is adjusted by its opacity slider, allowing for endless control over the final impact of the effect.

Things become more complicated as the position of a layer or mask affects its influence on the layers beneath it. The simplest stack has all its layer icons aligned on the left-hand side, acting on layers beneath it. When layers or masks are indented, that indicates they have a localized influence. This can happen by accident when dragging layers around at the point you release the mouse button. When dragging a layer up and down the stack in Affinity Photo, a blue bar appears, indicating where the layer will be inserted. When the cursor is in-between layers, it is a long horizontal bar to form its own layer, when it is over another layer's text, it is a short horizontal bar and it forms a clipping layer. When it is on the icon, it turns into a short vertical bar to mask that layer. For adjustment and filter layers, these last two options have the same effect. There are also options in the Affinity assistant that set the default location of newly created layers. These can catch you out and when you create an adjustment layer or mask, check its position in the layers palette and drag it, if required, to the desired position in the stack.

Lastly, is the subject of layer management; one or two adjustment layers soon become 20 plus. When they are created, they are given the name of the tool. It is easy to forget what each layer does and it is useful to rename the layers with something meaningful. Two other tips help manage the complexity of layers; grouping and color tags. Creating a group is as simple as selecting the layers and choosing Group from the layer menu and giving it a meaningful name. Overall clarity of purpose is restored when the groups are collapsed by clicking the little arrow button or color tags to differentiate.

### Boosting and Suppression

Editing astrophotographs is all about playing to the strengths and weaknesses of the image. It is easier to enhance contrast, color and detail in those areas with a good signal to noise ratio, that is, the brightest parts of the image. Enhancing a background is more challenging and the fortunate side-step is that we just need it to support the main image, rather than detract from it. This means reducing the noise (color and luminance) and subduing color imperfections and gradients. In our image of M33, I added some more sparkle to the image with a Color Balance adjustment for the brighter parts of the image followed by a subtle Curves Adjustment layer, lifting the mid-tones and highlights. To reduce the color in the background, I added an HSL adjustment layer, with the saturation turned down and with the blend options set to confine the adjustment to the darkest tones. To bring the galaxy to life, I added a Brightness/Contrast layer, directed at the brightest parts of the galaxy. At this point, the image was tonally correct but required some further sharpening to bring out the galaxy details.

### Sharpening

Sharpening techniques in astrophotography usually take a different path from traditional methods. The poor signal to noise ratio (compared to that of conventional images) reacts badly with standard techniques such as Unsharp mask. Over-application of this tool creates edge contrasts that do not exist and exacerbates image noise. A subtler technique is to enhance the micro contrasts that exist. This opens up another doorway in selective processing, using local contrast and tonal changes at different scales to direct adjustments to where they are needed. This new concept probably needs a little explaining.

In simple terms, a star is a tiny point in an image and is considered small scale. Boosting the contrast of small scales (typically less than 8 pixels) makes stars brighter. A large galaxy will be made of a range of image elements at different scales. In decreasing scale, the main body, the sweeping arms, dust lanes and finally, small regions of star clusters and nebulosity. If we can isolate these elements, based on scale, we can increase their contrast and boost their presence in the overall image. A nebulous cloud is made up of multiple elements too; while the overall cloud is at a large scale, its edges and details within it will be at a smaller scale. Advanced astro editors sharpen and reduce noise at multiple selected scales at the same time.

Affinity Photo has a filter that accomplishes sharpening at different scales. The High Pass filter creates a dull mid-grey image with subtle variations around feature boundaries. These variations vary with the pixel radius. At under 4 pixels, it will show a faint starfield. At larger sizes, the edges of nebula, thin branches of nebula or dust lanes and what may appear as a faint rendition of the entire image at 100 pixels. The magic occurs when you change the layer blending mode to one of the contrast group. These blending modes accentuate contrast, in increasing intensity from Soft Light through to Pin Light. These Live Filter layers can be stacked too, allowing one to sharpen feature edges and accentuate the contrast of larger features. In our M33 example, I used two; at 8 pixels and 50 pixels, using the Hard Light blending mode and restricted them to the brighter tones using our earlier blend ranges trick on each layer (figs.7, 10).

### Fine Tuning.

This is the end of the road for this image. Some of the stars show some strange coloration and a few are a little too big. Ironically, these kinds of issues are difficult to fix in the purist environment of the astro editors but present no issue to photo editors. A lasso around the offenders allows one to re-color and shrink using a Minimum Blur live filter and for those with strange color artifacts, some selective color adjustment too.

The final processing chapter looks at several specialized tools to improve quality and reduce defects in a variety of new situations.

# Orion Nebula (M42)

*Greater depth improves image appearance*

***Equipment:***
*Canon EOS 60Da*
*73-mm f/5.9 refractor (430-mm $f_L$)*
*field flattener, electronic focus*
*SkyWatcher AZ EQ5 mount*
*50-mm guidescope with CMOS guide camera*

***Software:***
*SGP acquisition,*
*PHD2 autoguider*
*DeepSkyStacker*
*Affinity Photo*

***Exposure:***
*10 x 30, 16 x 120 and 20 x 600 seconds at ISO 200*
*Dark, Flat and Bias calibration exposures*

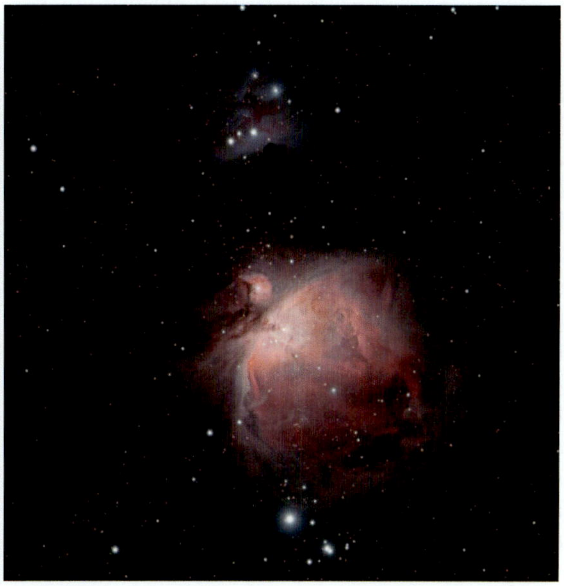

The Orion Nebula's amazing clouds are worthy of revisiting; there are many ways of approaching it, both with conventional color cameras and monochrome models, fitted with narrowband filters. At the same time, it is also a challenging subject to do well on account of its extreme dynamic range. The difference between the bright central core, around the famous trapezium star group, and the faint outer dusty perimeter is beyond the capability of a digital sensor. The only way to capture the entire dynamic range is to selectively combine images made with a variety of exposure times to capture the dim and bright nebulosity and compress the result within the range of a monitor or print. Within these covers, the first of our tracking assignments barely picked out the brightest central M42 core after a brief 1-hour sequence. At the other extreme, the narrowband version in *The Limits of Amateur Astrophotography* took over 50 hours through a much larger aperture instrument. This assignment is pitched somewhere in the middle, with an overall exposure time of 4 hours through a modest telescope, and not only introduces us to using a dual-axis mount but also to more advanced image processing techniques. These methods are the same for imaging galaxies, though usually using a longer focal length to capture the smaller objects.

## Equipment

I used a SkyWatcher EQ5 mount for this assignment. At 17 lbs, the head is reasonably portable. With a little ingenuity, it fits into a suitcase along with its folding pier and it can accompany you on vacation, without having to leave the children behind. The EQ5 range has improved over the years and this version features a belt-driven worm drive that claims, and achieves an excellent tracking error. The controls on my old EQ6 mount rotated with the RA axis. This not only caused potential wire-snagging issues but at times, the standard coaxial DC plug became intermittent, causing loss of control. This EQ5 model has a static connection panel and uses a more robust screw-in power lead. The mount receives its controls via a computerized handset, which has a second 6-pole RJ connector, through which it can receive serial commands from an external computer. I connected this to a PC, running Win 7 Pro with an FTDi chipset-based USB-serial adaptor. This mount is too small to have an internal polar scope and polar alignment was facilitated by a QHY PoleMaster camera slung under the refractor's dovetail plate and connected to a second USB port. The extended imaging time benefits from improved tracking and periodic focus updates. To that end, I modified our refractor system, by adding a 50-mm guide scope fitted with a CMOS guide camera and converting the focus mechanism for remote operation by coupling a geared focus motor. Three further USB connections connected the computer to a focuser module, EOS imaging camera and guide camera, bringing the total to 5. Since my PC only has a few USB ports, I used a high-quality 7-way USB 2.0 hub, powered by a 12-volt battery.

## Acquisition

This assignment is more demanding than the prior ones and is the first to start automating some parts of the acquisition process. I set out at the beginning of this book to use and evaluate BYE, APT, Nebulosity and SGP across these practical assignments. In the past, I started with Nebulosity and then, after a challenging period with Maxim DL, switched to SGP as its automation features and reliability worked well in portable and permanent setups . At $99, however, its price is near the top of our artificial budget. BYE and APT are less expensive and target a different user. Nebulosity is unique; at a similar price to SGP but covers basic image acquisition and processing. Using them all has been interesting and choosing between them is largely dependent upon how far you want to take the hobby. The comparison may also change with time too, as new features are added to each application and so any evaluation is time-bound. The investment is not purely monetary, it takes time and precious clear nights to fully appreciate and be proficient with any of these packages and recommendations are flavored by personal experience and circumstances.

At the top end is SGP; its unique framing and mosaic wizard, automated meridian flip and autofocusing routines accomplish long and complex imaging sequences with ease. Its controls are logical (well, to me anyway) and the developers resist feature-creep into non-acquisition territories that may interfere with the general usability of the interface. It also works at a simpler level too, with basic equipment, without being burdened by complex controls. Its strength is how the various tools and processes are integrated together.

APT works with a variety of equipment levels too and its selection of tools is in active development. It describes itself as the Swiss Army knife of astrophotography, which is an apt description. At present, its plethora of tools are less integrated with less automation. It does, however, offer several unique features including aperture and focus controls on EOS cameras as well as polar alignment and collimation tools and some specific non-ASCOM interfaces to more unusual hardware. As a consequence, the acquisition experience is more hands-on but as I said earlier, that is likely to change in time with future upgrades.

BackYard EOS and its partner, BackYard Nikon, as their names imply, are particular to those DSLR camera systems. They have been responsible for transitioning many from the world of photography to astrophotography. It accomplishes less than APT but at the same time, it uses less resources and is designed to work with a netbook's lower screen resolution. Of the three, I prefer the clean lines of the SGP interface and the way in which you can display all the information that is relevant to you. Application appearance and logic is, however, a personal thing and your tastes may differ. They all appear to be equally reliable and forum comparisons or 'bug' reports have to be read with care, as it is always tempting for users to "blame the messenger." All these applications are at the heart of a complex system and any part or parts of that system, hardware and software, may be responsible. I would be an alcoholic if I had a beer for every bug suggestion that turned out to be a user, USB cable or hub issue.

What of Nebulosity? I often feel that this program is overlooked, it started me off on a Mac and I occasionally use it on a portable PC. Its strength lies in its simplicity and for that alone, it is suitable for a novice and small screen. It works well with autoguider software and has the basic tools to aid focus and filter changes. Behind the scenes is a simple script tool to make an image sequence, if required. It also works with a wide range of cameras and in that respect, it is not made redundant by a camera upgrade. The bonus is that it also has the principal image pre-processing and processing tools that together, make a passable image.

For this assignment, I decided to run three sequences of different exposure lengths. The weather was particularly fickle and imaging spanned three nights. I chose SGP, on account of its ability to pick up where it left off and manage the meridian flip in the middle of each. The system was set to trigger the autofocus procedure for each 1 °C ambient change and continue imaging until the target altitude dropped to 35 degrees. The exposure sequence shot 10 x 30-second, 16 x 120-second and 20 x 600-second exposures to capture the wide dynamic range of the nebulosity, effectively 70x the exposure used for the wide-field view taken through the much smaller aperture. While the camera was still outside in the cold and after replacing the metal lens cap, I took 20 dark frames of the same durations and 30 bias frames, with the camera set to 1/4,000$^{th}$ second. In daylight, I shot another 20 frames of a white wall to create the flat files.

## Processing

The initial processing of camera RAW files throws some unique challenges in our path that dedicated camera files neatly avoid. For instance, just as with traditional image processing, any RAW file requires conversion into a different format for editing and, contrary to traditional image processing, the file has to remain linear. This last point confuses many newcomers; the image preview on a digital camera is deliberately boosted with a gamma correction that stretches the image mid-tones between the highlight and shadow endpoints, to match our perception of the subject.

### Linear Processing

When a RAW file is viewed in a traditional image editor, the image is similarly boosted and when it is converted to an editing format like JPEG, TIFF and PSD, it is often made permanent during the conversion and the image is said to be non-linear. In astrophotography, we also stretch files, but considerably more aggressively. More importantly, there are some initial image processing steps that require a linear image, including image calibration and stacking. The confusion arises when one views the partly processed image in one of these specialist applications, in so much that unlike the clear image on the camera's LCD, the preview looks to be entirely black, possibly with a few light dots, corresponding to the brightest stars. It is not uncommon to find forum posts from concerned users, claiming their images have disappeared after stacking. There are two potential causes; the missing gamma correction mentioned earlier and the way in which the image bit-depth is translated into a higher bit-depth file. I have noted that some applications do not scale up the stack to the full bit depth. While the image calibration works on linear files, without any attempt to interpret color, the registration and final stacking require color files. Here neighboring pixels corresponding to red green and blue photosites are combined to form a single color pixel and it is these that are aligned and combined. After this, non-linear image processing continues with non-linear stretches and manipulations. (Later on, some more advanced techniques extract the color and luminosity information from the RGB image and manipulate them separately for better effect.)

In this case, I ran Nebulosity, DeepSkyStacker and SiriL head to head and out of curiosity compared the results with my usual image processing application PixInsight. PixInsight is not for the faint-hearted and not a starter application but I consider it the benchmark, as well has having good support for many different camera RAW files. To my surprise, the results were quite different in several regards. SiriL produced a very green 16-bit file but the data only occupied a quarter of the range (i.e. 14-bit). DSS produced a serviceable 16-bit FITS/TIFF file with a blue background, PixInsight, a 32-bit file with a turquoise background and Nebulosity something different again. The most convenient for registering and onward processing, in a photography editor, was DSS's TIFF output. DeepSkyStacker (DSS) has been around for many years and has the amateur in mind. It works well with DSLR images as well as the FITS files produced from dedicated acquisition applications. Its strength lies in its pre-processing ease and high degree of automation but at the same time, offering the user some overrides to customize the calibration, stacking and rejection options. For this assignment, I ran DSS three times, using the same flat and bias frames but selecting each light frame set and its matching exposure dark frames in each case (fig.5). The resulting three TIFF files from the stacked 30-, 120- and 600-second exposures captured the entire brightness range, as shown in figs.1–3 respectively.

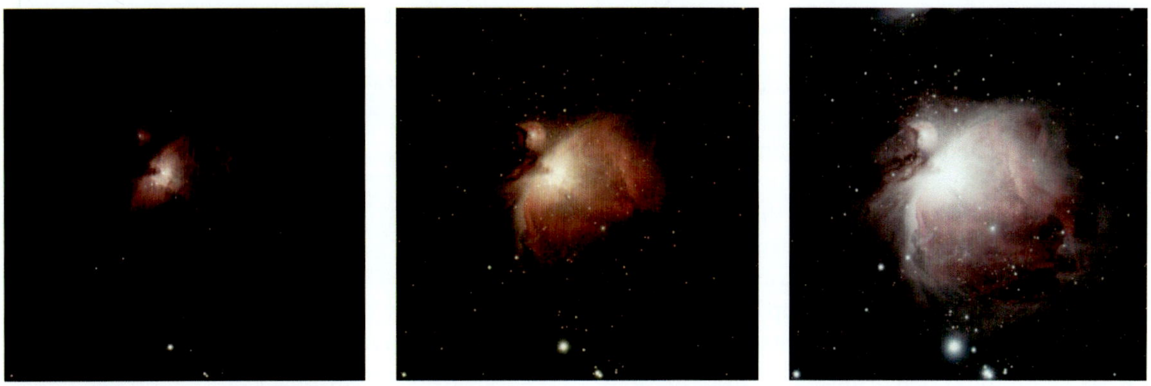

*fig.1–3 After pre-processing and a basic image stretch, the 30-, 120- and 600-second exposures reveal different parts of the nebula. These images require individual processing and then combining them as layers into a single image.*

### Non-Linear Processing

The three TIFF files were transferred over to Affinity Photo for processing and combining. (I could equally have used Photoshop's equivalent tool-set.) The general processing workflow is shown in fig.4. In the first place, the three images are individually stretched using a Levels adjustment layer which is also used to correct the background color and density. The trick is to not attempt this in one go; it is much better to make a sequence of Levels adjustments. In the first instance, simply boost the mid-tones with the master channel's gamma slider. Once the background becomes a lighter color, it is then much easier to move the Black Level slider on each of the channels to create a dark, neutral background. It is important to keep the image stretch at a modest level on the 30-second exposure to just retain the image details in the nebulosity core. The 120- and 600-second exposure images can be stretched more at the expense of clipping the image in the middle. The three images should end up with a similar color balance and intensity. (The backgrounds for the 30- and 120-second exposures are not as important since they will not be used in the final image.)

The three images are then copied as layers into a new image, in the order, 30-, 120- and 600-second. The next step requires the use of layer masks to selectively combine the images into one. There are a number ways of doing this. I chose to use the Threshold adjustment tool. In this case the Threshold adjustment is applied to a duplicate of the 600-second image to form a white blob (corresponding to the brighter core area) on a black background. This is blurred, inverted and converted to a mask, using the Rasterize to Mask option from the layer menu. Dragging this mask onto the 600-second image immediately reveals a portion of the 120-second exposure from the layer beneath. This process is repeated for the 120-second exposure, using a threshold setting that just highlights the very brightest central core.

The composite image is unlikely to look quite right at the first attempt. Almost certainly, there will be jumps in intensity and color at the mask boundaries. One way to overcome these is to then apply individual Curves and Levels adjustment layers to the three image layers and tune them so that they blend seamlessly (fig.6).

This image does not require extensive manipulation and to finish off, I simply cleaned up the background with an application of the Denoise filter and sharpened up a little using the High Pass sharpen filter, experimenting with the blending mode and transparency for the right effect. Revisiting the image a day later, I

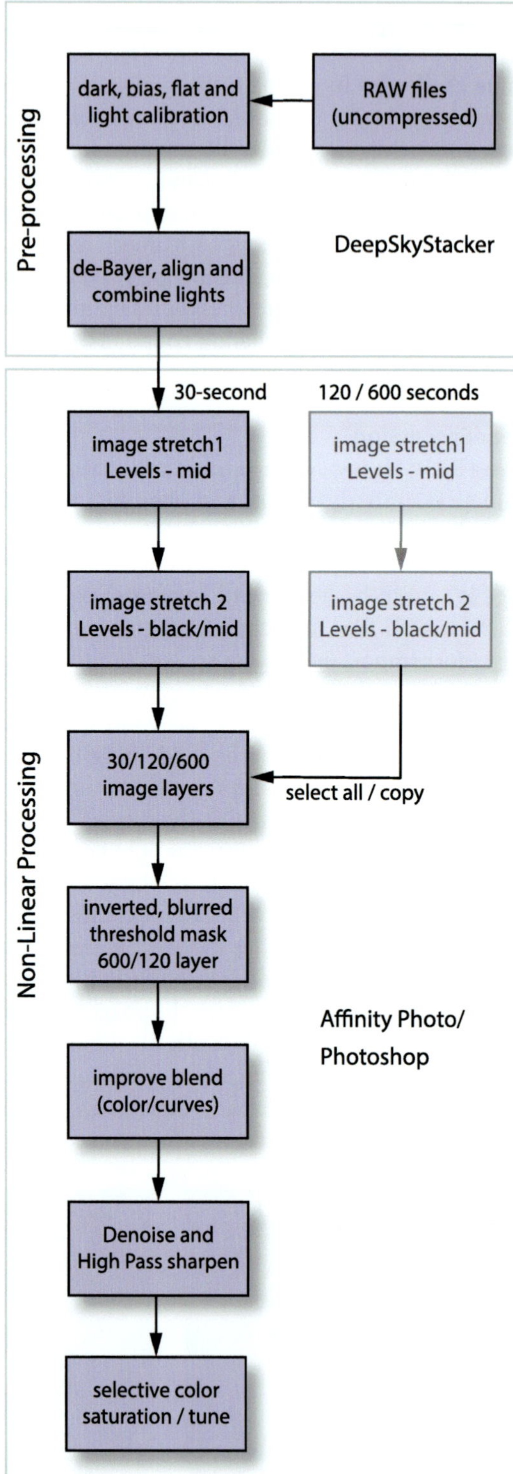


*fig.4 The image processing workflow. After calibration, registration and stacking, the processing is entirely carried out in a conventional image processing program, using image layers, masks, and adjustment layers.*

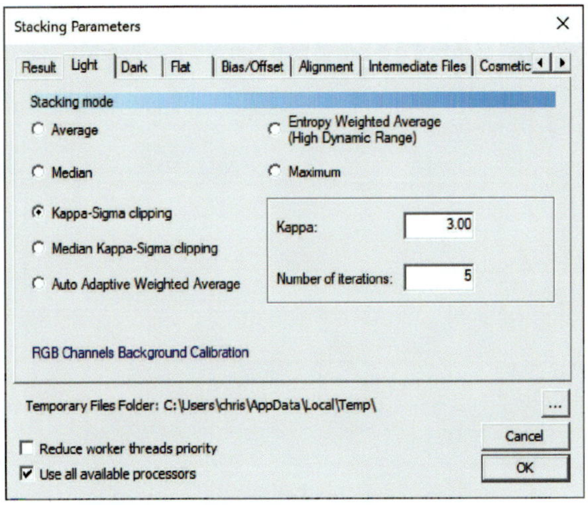


*fig.5 DeepSkyStacker automatically sets what it thinks are the best settings for calibration, alignment and stacking. It can be as simple as selecting the files and hitting one button. For those who wish to have more control, each of the image groups and processes has a control tab in the settings menu that permit altering the image processing parameters to optimize the result.*

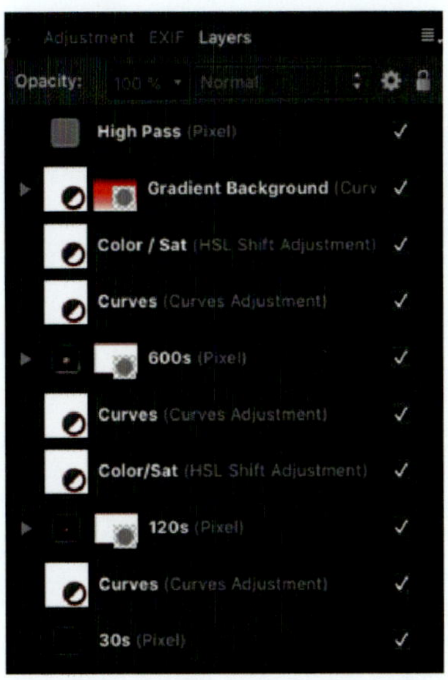


*fig.6 This snapshot of the Affinity Photo layers tab shows the three image layers, the two masks, formed by using a threshold adjustment layer on an image copy and various adjustment layers to blend the three images seamlessly. Affinity Photo and Photoshop are similar but not identical. In the case of Affinity, the threshold adjustment layer creates a black or white image, depending on the threshold value. Dragging this to a mask layer has no effect and it has to be converted to a mask format first, using the Rasterize to Mask option from the main layer menu.*

increased the blue saturation a little with the HSL tool and introduced some subtle tone changes with a Curves tool to the entire image.

In terms of further improvement, even after the close crop to the square picture you see here, the successive image stretches reveal a top to bottom sky gradient. When there is obvious nebulosity in the image, as we have here, it is difficult to remove this without specialist tools. With a simple gradient, however, it is possible to reduce its appearance by applying a Curves adjustment to a duplicate image to correct the lighter background area and then blending it with the original through a graduated mask, in the same familiar method that photographers subdue the edges of any photo. Out of interest, using PixInsight entirely for pre-processing and manipulating only delivered a modest improvement in this straightforward image (fig.7). Ultimately, the sky quality, imaging setup and modest exposure duration are the major limiting factors.

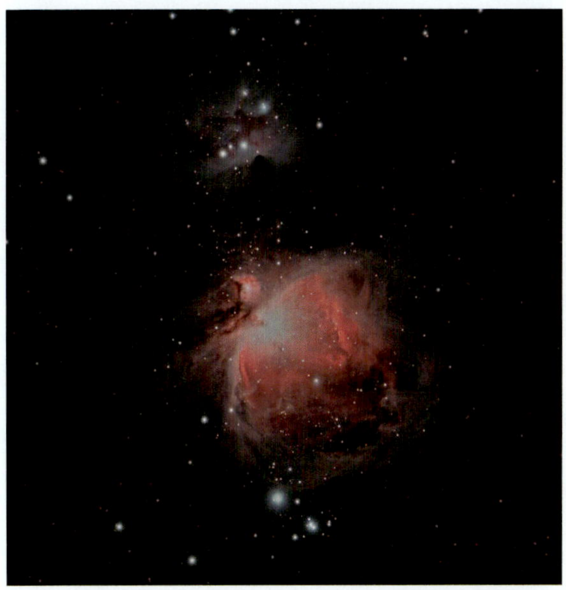

*fig.7 Out of interest, I took the original Canon CR2 RAW files and processed them entirely through PixInsight. After a few hours, I was able to produce this alternative rendering. I was able to reveal more detail in the cloud structure and extend the nebulosity further into the background too. Ultimately though, the image quality was limited by the length of the exposure and, if you look carefully, some horizontal dark lines left behind by the calibration process, arising from a small mismatch in the sensor temperature between the light and dark frames. It is fair to say that the extra effort (and expense) in this particular example, are a case of diminishing returns.*

# Extended Deep-Sky Imaging

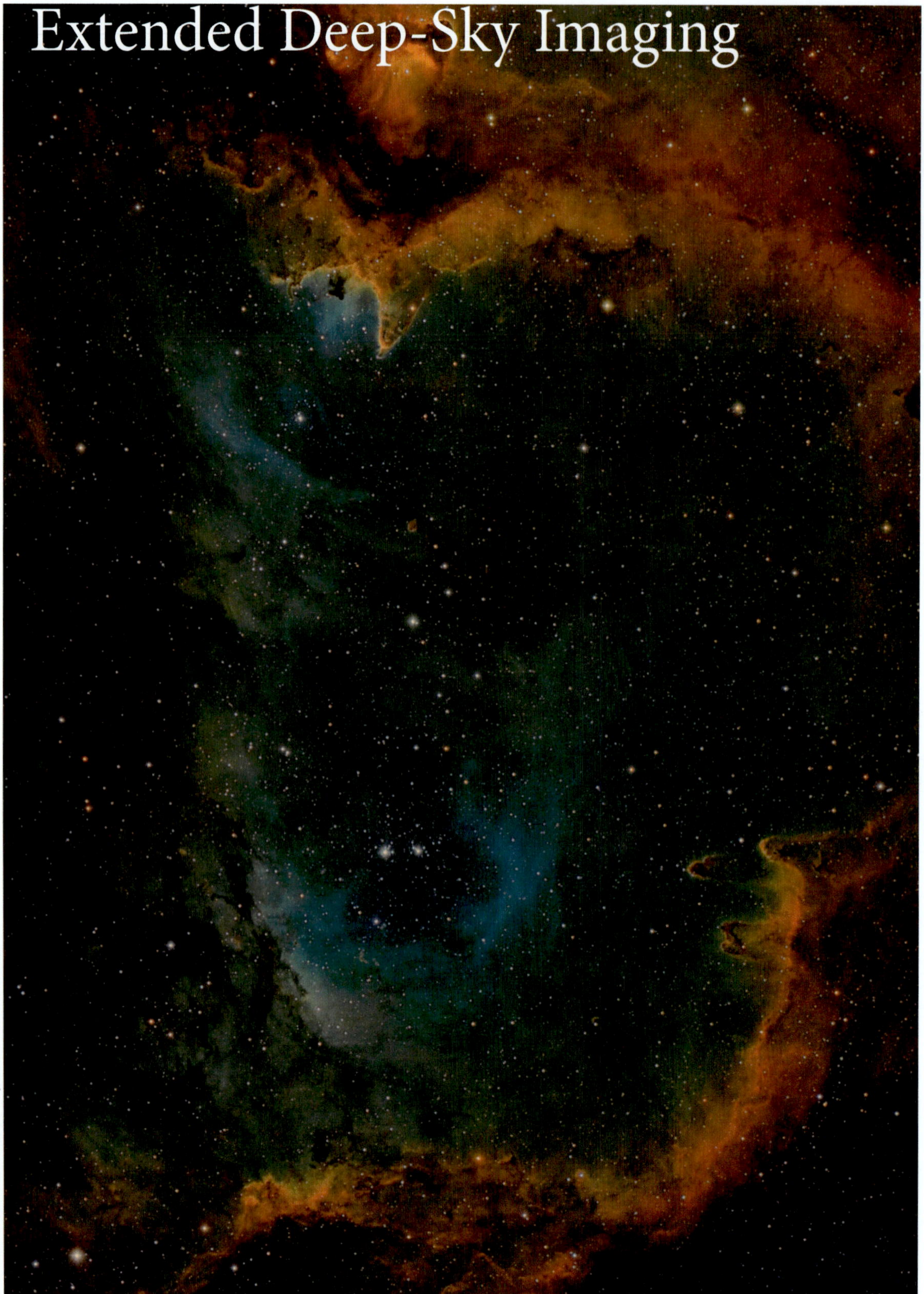

*Inner Soul Nebula (extended-exposure narrowband)*

# Pushing the Boundaries

*Fainter objects, longer focal lengths and extended imaging sessions challenge our techniques, equipment and patience.*

The targets up to now have been relatively bright and large, suitable for a camera lens or a short refractor and a digital camera. While some of the total exposures have run into several hours, that was on account of the camera operating above ambient temperature and more out of interest, to see how much background detail could be extracted.

Smaller and dimmer objects are considerably more demanding and this chapter and the ones that follow mark the watershed between casual astrophotography and something altogether more serious. These objects typically require a cumulative exposure in excess of 10 hours, together with more specialized imaging systems and processing. In effect, one has to adhere to all the good practice identified up to now, and some.

The challenges are similar to those already encountered, including optimizing focusing and tracking. These requirements are made all the more acute due to the longer exposure times and higher magnifications. The total integration time which, for some of my deep-sky images, has spanned several months (on account of the weather) is a practical and motivational challenge in itself.

This then is the realm of the dedicated astrophotographer, with a dedicated astro camera, fully integrated into an automated acquisition system that can align, focus and track for itself without constant human intervention. For all-night operation, environmental monitoring will predict or detect unfavorable conditions and even then, the output will require more extensive manipulations to tease out faint details without emphasizing noise at the same time using more selective tool application. The chapters that follow look at how we meet these more specialized needs and processes. One of the exciting features of using dedicated cameras, especially monochrome ones, is to exploit the full potential of narrowband filters. As mentioned earlier, these are dichroic filters and tuned to very precise wavelengths, corresponding to the principal emission wavelengths from excited atoms. These filters are incredibly selective, with a passband as little as 3 nm wide, excluding all other colors (and light pollution too). The most common three are Hydrogen (alpha), Oxygen and Sulphur. Less common are Hydrogen (beta), Neon and Nitrogen, usually reserved for scientific applications. These filters are expensive, increasingly so as the passband is made narrower. Even with near perfect efficiency at the chosen wavelength, many emissions are incredibly dim and individual exposures are often in the range 10–20 minutes each. Narrowband imaging is incredibly liberating and is capable of amazingly beautiful and delicate images from nothingness and gives a new lease of life to familiar objects or reveal new ones, as on the previous page.

Dark current and the noise that accompanies it is the ever-increasing enemy with extended narrowband exposure times. Fortunately, most dedicated cameras have electronic sensor cooling, achieving a 35–40 °C reduction below ambient conditions, that in turn reduces the dark current by about 100x and the random noise contribution by 10x (the square root of 100).

Astrophotography at this level is often difficult and expensive. It is hard to avoid these realities when one sets out to image the faintest or smallest targets. Deep-sky imaging tests your technique, equipment (and patience) to the limit and one should be prepared for some failures and near-misses. The flip side of that is the extreme satisfaction that accompanies a successful imaging run and subsequent processing. For many others, the invention required to overcome practical shortcomings, sustained operation, novel setups and remote operation adds another dimension to the hobby.

Extended exposures yield surprising results; French astrophotographer Nicolas Outters discovered a new nebula in 2011 that had gone completely unnoticed. You might think it was exceedingly small; to everyone's surprise, it was not. Named the Squid Nebula, or OU4, this turquoise structure spans over a degree and is an exceedingly faint OIII emission nebula. It did not require exotic equipment and was imaged with narrowband filters. Out of interest, I decided to have a try myself; after centering on its bright central star, I stretched a single 20-minute exposure taken through a 3 nm OIII filter. Although I knew it was in the middle of the image, I could not even detect its presence. That is most likely due to the less-than-perfect imaging conditions in my back yard, something to keep in mind when one is reaching for the stars!

# Dedicated Camera Systems

*Long exposures and dim subjects demand careful camera choice.*

Many astrophotographers sensibly start with their current digital camera and later on, when they wish to progress, contemplate a dedicated model. The earlier chapter *Imaging Hardware* briefly considered the generic options with which to choose a consumer or dedicated camera, including sensor size, pixel pitch and general configurations. At that point, it was too soon to go under the hood and understand their detailed workings, pros, cons and all within the context of making real images in real conditions.

Before any discussion on model choice, a few home truths are in order; even in conventional digital photography, we have reached a point that the differences between sensors are increasingly irrelevant. Equivalent 16x20-inch prints from each would be indistinguishable at a normal viewing distance. What remains are comparisons based on autofocus speed, ergonomics and increasingly, video capability. If you are coming from conventional digital photography, it is almost impossible to escape the continual rivalry between sensors and cameras and, in a book on astrophotography, there is an expectation of an in-depth evaluation of alternatives. Astrophotography is far more demanding and any evaluation cannot avoid becoming technical since it involves an understanding of the inner-workings of the silicon, to correctly interpret published specifications or amateur "reviews." The growing number of competing sensors, even from the same manufacturer is utterly bewildering, even for the seasoned astrophotographer. In writing this chapter, I hoped to reach my own purchase decision to use in a later practical chapter. It has not been an easy journey, but extremely interesting.

At the same time, any evaluation has to additionally indicate the practical (in)significance of various sensor attributes in real-world conditions. Even though astrophotography is more demanding on sensor performance, it is easy to overlook one overriding constraint, light pollution. My location is semi-rural and fig.1 shows an example of the measured signal and noise contributions from a *cooled* sensor, target and light pollution. In this case, I have squared the common electron noise figures, which converts them to noise power values, so that it is legitimate to add them in a stack graph. The interesting outcome is, in all but the darkest sites, the random noise contribution from light pollution is the overriding noise source. So, while the debate is interesting and may seem paramount, in many ways sensor performance is less important than one's environment, telescope aperture, mechanical stability, focus accuracy, tracking capability and the one item many forget about and optimistically assume is a given, hardware and software reliability.

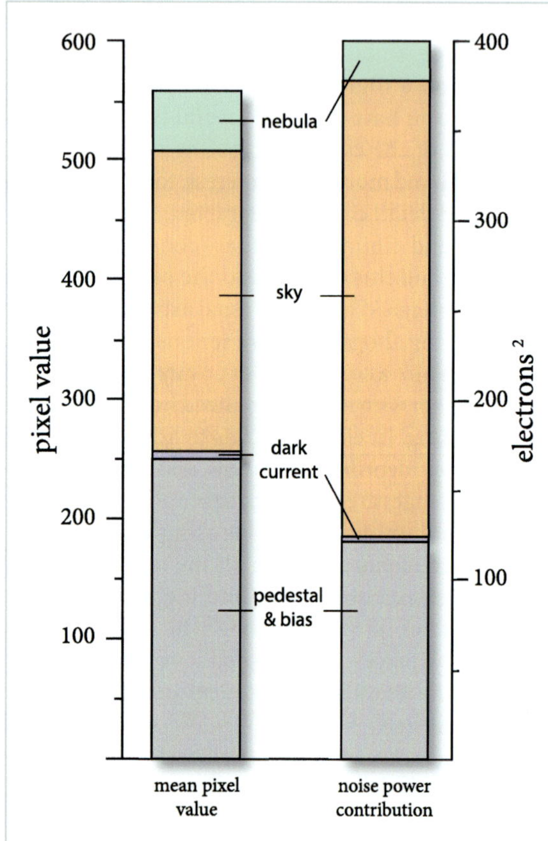


*fig.1 These real measurements show sky noise in my semi-rural area is higher than the sensor noise (KAF8300 CCD @-20 °C).*

There is nothing inherently "wrong" with old technology, most of the facer images in this book were taken with a 15-year-old CCD sensor (a youngster compared to those in the Hubble Space Telescope) and with high sensor noise (compared to modern CCD and CMOS designs). My camera system has been extremely reliable for hundreds of unattended hours and it is still a popular choice for astrophotography. It is only recently that it is being challenged by alternative astro camera designs using CCD and CMOS technology. CMOS astro cameras from Chinese companies are now mainstream and the same CMOS sensors are being adopted by the established European and American manufacturers too, to recover lost market share. The initial models underwent rapid changes (aka customer prove-out) and the earlier operational issues have largely been addressed in the latest models. CMOS designs are less expensive than the established CCD-based models and are an attractive proposition for the newcomer. With that in mind, it is interesting to compare the two technologies at this exciting time.

## CCD and CMOS

In the 2010's the overwhelming majority of astro cameras used 10-year-old CCD-based sensors that outperformed contemporary CMOS designs. At the same time, CCD sensors were to be found in professional observatories and space telescopes. The significant commercial advantage of CMOS sensors encouraged their accelerated development and, at present, I am not aware of a consumer camera that uses a CCD. Sensor design has, in general, improved over the years, with better efficiency, larger sizes, higher pixel counts, lower noise, lower power consumption and faster operation. There are now many more new CMOS sensors for the camera manufacturer to choose from than CCD models. Some new CCDs have been developed, typically in smaller sizes, for specialist applications such as low-light surveillance and some video applications.

Understandably, both editions of *The Astrophotography Manual* majored on CCD sensors and simply acknowledged that there were interesting times ahead with the gaining popularity and sophistication of CMOS sensors. It was premature at the time but now it is fair to say they have arrived, from the East, and are increasingly featured in the product lineups of the more established manufacturers. They are still not well understood and comparing the two technologies is not easy and to make it more challenging, is also a moving target. Suffice to say, CCD and CMOS-based cameras have different strengths and weaknesses.

CCD and CMOS sensors use different architectures; while both designs convert incident photons into an electrical charge, the way that each photosite charge is read and converted is quite different. CMOS sensors are more self-contained, by integrating more electronics on the sensor chip, whereas CCDs require additional peripheral components to provide the final digital output. For that reason, there may be more significant performance differences between CCD camera implementations using the same sensor, specifically in regard to noise levels and download times.

These different architectures lead to different compromises. In a CCD, the pixel charge is clocked out one pixel at a time, row by row into an amplifier that converts the charge into a voltage and then passed to an external Analog to Digital Converter (ADC). Normally, this is a high-quality 16-bit model and, as all the pixels share the same amplifier and conversion, pixel to pixel variation, linearity and uniformity are typically excellent. This serial read method has other implications too; it allows for binning, in which the charge from several neighboring photosites is combined on-chip, before being amplified and sampled. (Simply put, when high resolution is not at a premium, binning improves image noise and download times.) There are some downsides too; the process of clocking the pixel values out of the sensor is slow (for instance, my 8-MP CCD camera takes about 20 seconds to output an image in high-quality mode) and although this is not an overriding consideration for many forms of astrophotography, some techniques and subjects benefit from rapid exposure download.

In comparison, CMOS sensors have their conversion electronics integrated onto the silicon. In general, each photosite has a readout circuit on the chip, converting the charge to a voltage and each column has its own ADC, although these are more likely to be 12- or 14-bit versions. The duplication of readout circuits and ADCs allows pixel values to be read in parallel, which dramatically improves the readout speed.

> The term "pixel" is a common one with several implied meanings. A digital image is composed of pixels, which in our case, has values derived from a sensor. A photosite is one of the array of electronic circuits on a sensor that convert photons into an electrical charge. On a monochrome sensor (without a color matrix), each image pixel contains the information from a single sensor photosite. When it comes to color images, things become more confused and the term pixel is alternatively used to describe a combination of adjacent filtered photosites, a single 3-channel data element or a colored dot on a monitor that is actually made up of all three.

It is impossible to make all these duplicated readout electronics identical and as a consequence, there is more apparent pattern noise, arising from tiny differences in circuit sensitivity and linearity. On the positive side, the tight integration benefits internal sensor read noise at the expense of localized thermal noise, often referred to as amp glow, generated by the heat of these additional circuits.

Amp glow appears as a slight lightening over areas of the sensor, which after image stretching becomes noticeable and resembles a light leak in a film camera. It potentially increases with exposure time and has been linked to the image download process too. It is difficult to quantify and does not appear in specifications as such but may be mentioned in the feature list. In the latest cameras, amp glow is virtually non-existent. In any case, it is possible to calibrate out consistent pattern noise and amp glow with careful image pre-processing using matched exposure sets.

At one time, the increased real estate of sensor circuitry reduced the light capture area and efficiency of a CMOS sensor. Improvements in micro-lens design now focus more of the incident light onto the photo-active areas and, to improve things further, the latest "back-illuminated" designs increase the photosite real-estate (and well depth) by flipping the circuitry and interconnections to the dark side. Sensor efficiency is now at a level where photon conversion is close to the physical limit. CMOS sensor development is not targeted at astrophotography in particular and continues to evolve. Some of the latest designs are noisier than their predecessors, on account of the high-speed integrated electronics (and the heat they generate) required to satisfy consumer demand for high bit-rate video. As video resolutions continue to increase beyond our visual acuity, the extra speed of operation will likely be to the detriment of astrophotography applications. For instance, my small action camera becomes hot when operating at high resolutions and frame rates.

So how does one choose a camera? Specification sheets are a good starting point, but they are not the whole story. Some important aspects are not commonly reported and others require further interpretation to make meaningful comparisons between models. More recent sensors, using CMOS technology, are especially tricky to tie down.

These devices are designed to operate at different gain settings. This alters the amplification of the analog signal into the digital converter and trades the sensor noise level with its dynamic range. The published specifications commonly list the best-case values, but these do not occur at a common gain setting. It makes sense to explain and contrast the different specifications that fall into the following general categories:

- sensor architecture
- noise performance
- scale
- other specifications

## Sensor Architecture

### Quantum Efficiency (QE)

As stated in earlier chapters, QE is the conversion efficiency of incident photons into electrons. A sensor is not uniformly sensitive to all wavelengths and typical peak values are in the range 50–85%, with the peak in the 550–650 nm wavelength region. The highest efficiencies are about as good as they will ever be. Some datasheets confuse things, however, by quoting relative efficiencies for a particular wavelength. These values may appear impressive but they are quoted relative to the peak efficiency, which in some cases is not stated. Even with multi-coating, the sensor efficiency is reduced slightly with every air-glass surface (for example, UV/IR block filter) and improved with back-illumination technologies and an optimized micro-lens array. Taking the system as a whole, the color filter type also affects sensor efficiency and this is a good time to discuss the color versus monochrome sensor question as it relates to image capture efficiency.

### Color or Monochrome?

The on-line debate rages over which is best, often confused by some poor science. One way to resolve this is to consider their abilities to capture photons of different wavelengths over a set period. Color sensors have a color filter array (Bayer or similar) and a UV/IR blocking filter in front of the sensor photosites. These are optimized for conventional photography and white balance. In the case of the Bayer array versions, there are two green-filtered pixels for each red and blue. Green is not a common color in astrophotography and some consider the color filter array wastes half the photosites on the sensor, reducing efficiency and resolution. The individual RGB filters themselves are not as efficient as separate dichroic RGB filters and when one compares the peak and deep-red wavelength quantum efficiencies, there is about a 10% transmission loss. While both sensor types usually require some form of UV/IR blocking filter, a DSLR or mirrorless camera has

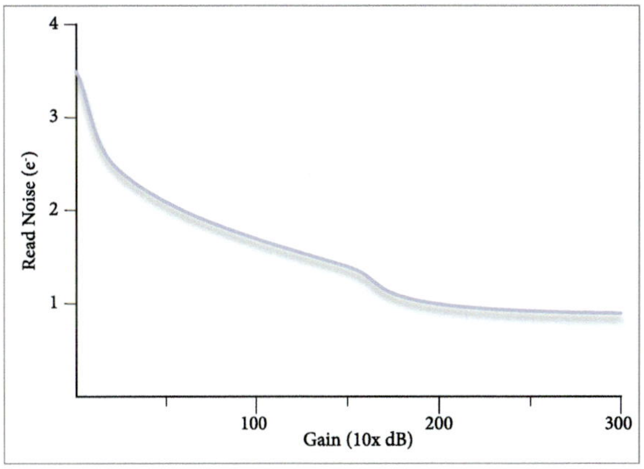


*fig.2 In common with many other CMOS sensors, the Panasonic* MN34230 *has a lower read noise at high gains, easily out-performing many common CCD designs. (A KAF8300 CCD has a read noise of about 8 e-)*

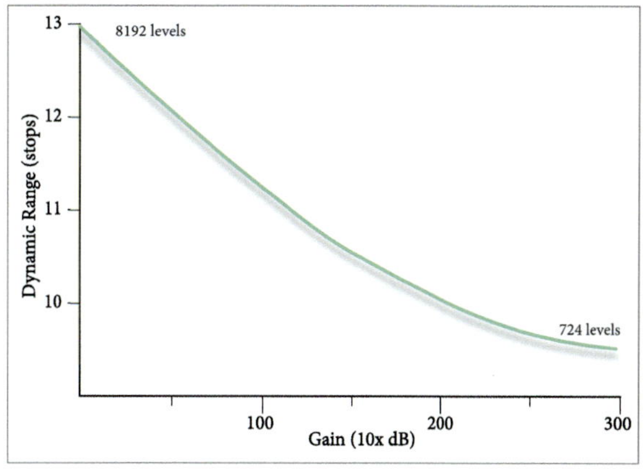


*fig.3 There is no free lunch, however, and the price for lower read noise is significantly less dynamic range. In this case it reduces by about 11x. (A KAF8300 CCD has a dynamic range of about 12 stops, or ~4,000 levels).*

an integrated UV/IR blocking filter. Consumer cameras are designed for daylight use, which unfortunately blocks those deep-red wavelengths associated with emission nebulae to some extent. In any CFA sensor, the red and green filter responses overlap and crucially, transmit the yellow sodium-vapor lamp light pollution wavelengths and their associated shot noise. (A light-pollution filter is beneficial in these situations.)

If we tackle the thorny issue of relative efficiency over a fixed period, the outcome is dependent upon the object color. If we compare an imaging session of three hours, in one case the CFA sensor exposes some of its pixels for three hours and in the mono case, with red, green and blue filters, it exposes all of its pixels for less than three hours. Taking into account the sensitivities at different wavelengths, for green light, the CFA has a 3:2 photon capturing advantage and for OIII and Hα wavelengths a 3:4 disadvantage with respect to a monochrome sensor. While some claim green pixels are a waste, they do serve a purpose, as human vision is extremely sensitive to green and the doubling-up of green pixels in a CFA reduces visible noise.

Although not part of the QE discussion per se, the imaging convenience of a color camera also comes with a further slight resolution loss, caused by the color conversion process, or de-mosaicing, that combines neighboring filtered pixels to form one RGB color pixel. Along with the selective use of pixels in the CFA is an accompanying theoretical reduction in resolution of about 50% for any particular wavelength. The key word here is "theoretical" since in many cases, astronomical seeing, tracking and focus errors are the overriding resolution-limiting consideration. For real-color wide-band astrophotography, the small real-life compromises of a color sensor are often outweighed by the benefit of their ease of use, especially for short-duration sessions. This balance changes when one considers false-color images with narrowband filters. When used exclusively, the photosite inefficiency has a more pronounced effect on nebulae, than say with stellar images, whose stars have a wide-band output from blue to red.

All is not lost, however, since the enterprising filter manufacturers now produce filters that pass multiple narrowband wavelengths. Of these, the duo-narrowband filter (passing OIII and Hα) is the most interesting and blocks more light pollution than a more conventional light-pollution filter. This, coupled with a DSLR/mirrorless camera or color CMOS astro camera make an interesting pairing, with better overall efficiency than using two separate narrowband filters, improving the efficiency of an OIII/Hα subject acquisition over separate NB filters.

## Bit Depth

The bit-depth figure in question is the resolution of the conversion of an analog voltage to a digital value. This is stated as the ADC (Analog to Digital Converter) bit depth. An ADC with x bits can discriminate $2^x$ levels. Ideally, we need any camera to discern the charge contribution of an individual electron and the higher the bit depth, the better. While most imaging cameras create a 16-bit image file, that is not necessarily the whole story. Popular imaging CCDs have a true 16-bit resolution output but their CMOS equivalents are commonly 10-, 12- or 14-bit. In these cases, a 6-, 4- or

2-bit binary shift scales up the ADC value to a 16-bit output, without any increase in tonal resolution. Where the ADC has fewer readout levels than the number of discrete analog voltage levels (the well capacity), it introduces further sampling inaccuracies called quantization noise. If the camera has a gain control, this may boost the signal, improving sensitivity and reducing quantization noise but at the expense of clipping highlights and losing dynamic range. Some of the latest CMOS sensors duplicate the readout electronics and simultaneously process the photosite voltages at two different gains and combine the results to form a true 16-bit output or have a 16-bit ADC, optimizing the sensor read noise without compromising the full well capacity. Some of these are in development and will challenge CCD dominance, especially if the sensor manufacturer releases a monochrome version.

### Gain

In a sensor, the gain is defined as the conversion rate from electrons into digital values. It is not necessarily a one-to-one relationship. Many CCDs have a low and high gain setting, often used for binned and unbinned exposures respectively. CMOS sensors have greater flexibility and have a range, just like the ISO setting on a camera. Just to confuse matters, some CMOS sensor specifications refer to gain as a general amplification setting too, in dB or an arbitrary integer value. The better models will have published graphs of read noise and dynamic range for different gain settings, like those in fig.2 and fig.3. The gain is usually set up in the driver's dialog. Along with the gain, there is usually an offset setting, either automatic and hidden, or user-selectable. It is usual to set a value that ensures the minimum pixel value is 100–500 for each gain setting.

### Full-Well Capacity (Full-Well Depth)

This defines the maximum number of electrons that a sensor photosite can hold before it saturates. When this happens, especially on older CCDs, additional electrons spill over into neighboring photo sites and cause a vertical streak, called blooming. Fortunately, most modern CCD sensors have anti-blooming circuits that mop up these rogue electrons to reduce the effect.

Astrophotography is all about extremes and many deep-sky objects have tremendous magnitude variation between the brightest and dimmest elements. Best-practice extends each exposure as much as possible, without clipping highlights, to register the dimmest elements of the sensor noise. As such, the full-well capacity is a critical parameter since the dynamic range increases with its value. Some CMOS sensor specifications are misleading if they are considered in isolation. In those models where the full-well capacity exceeds the resolution of the ADC, it requires multiple electrons to trigger a digital value change and the effective dynamic range is compromised. In addition, those sensors using a high gain setting (which reduces read noise), only use a fraction of the available full-well capacity, affecting the effective value and dynamic range. This can be seen in the comparative specifications for the same sensor in fig.4. These interactions between parameters do not make the task of choosing a CMOS camera any easier.

## Noise Performance

### False Leads

An Internet search will quickly find misleading claims about CMOS cameras (in DSLR or astro form) that confuse the novice. Take this one; "Astrophotography is all about noise. You want to use high ISO in astrophotography to capture the minimal amount of light that stars provide to you." Wrong. The ISO setting does not affect the captured light, only aperture and total exposure time do that. A high ISO (gain) setting amplifies the signal and noise by about the same amount and an increasing number of sensors are "ISO invariant" in that there is little detectable difference between a boosted ISO 200 shot with an ISO 3200 shot of the same aperture and exposure duration.

Here is another poorly-worded recommendation; "if the gain is set too high it will introduce lots of noise, so it's better to keep it around 200 to 300." It is the same argument as before, the signal to noise ratio is fundamentally controlled by the total number of photons entering the aperture and only slightly affected by the camera settings. If anything is true it is the opposite; most sensors with 12- or 14-bit ADCs have slightly *less* read noise at high ISO but as we started off this chapter, sky noise is usually the dominant noise source.

Finally, another inaccurate claim; "CMOS cameras have a lot of other benefits as well as price, such as short imaging times and the ability to capture without guiding on some imaging setups, thanks to the amount of detail that can be captured in a short exposure." This is confusing the convenience of using high ISO/gain with a short exposure to register an image on the camera's LCD screen, with something you would want to hang on the wall and show friends,

These inaccurate statements frustrate the new user, as their results fall short of various claims. So, what is all the noise about?

### Read Noise

Read noise is a loose term and is distinct from other noise types, as it is present in all exposures, with or without light and for any exposure duration. It is generated within the sensor from several physical mechanisms and can be broken down into random and consistent elements: pedestal, fixed pattern noise and random noise:

### Pedestal

To start with, the pixel values in a short exposure of nothing (called a bias frame) are not zero. The pixel values are all slightly different and the mean value is deliberately elevated with a pedestal to reduce the likelihood of clipped black pixels during pixel conversion. In a 16-bit file, they are typically around 200, give or take. In some CMOS astro cameras, it is necessary to manually set the pedestal using an "offset" value in its properties dialog, to ensure all pixels are non-zero.

### Fixed Pattern Noise (bias noise)

While a bias frame looks featureless, the average of 100 exposures reveals a faint pattern superimposed on the pedestal value. This is caused by minute but consistent differences in the electronic circuits during the voltage conversion. This and the pedestal are removed from image exposures by subtracting the average of many bias frames during the calibration process.

### Random Noise

The random element of read noise is the most important and is the one reported in the specifications. It is typically reported in electrons (e-) as a Root Mean Square (RMS) value. The term RMS may be confusing but in essence, it means you will generally have that degree of uncertainty on the electron count for any pixel. This uncertainty can only be reduced by averaging multiple image exposures. These random electrons are not the entire story. They are converted into a voltage and then sampled to create a digital value by the ADC. This digitization contributes a sampling (quantization) error to the measured read noise. When the ADC has fewer digital levels than the potential number of electrons, the quantization error is more significant. CMOS sensors uniquely have widely programmable gains, which amplify the voltage before sampling. At higher gains, it takes fewer electrons to reach the maximum output and the ADC can resolve the voltage contribution of each electron. This results in a lower read noise (fig.2) at the expense of dynamic range. In the popular Panasonic MN34230 CMOS sensor, the read noise changes from 3.6 to 1.3 $e^-$, accompanied by a 3.5 stop (11x) dynamic range reduction (fig.3). With such a difference, it is essential to know the gain setting used to specify read noise or, better still, find the read noise for the most likely gain setting (which may change with subject and application).

### Dark Current

All sensors accumulate a "dark current" during every exposure, light or dark; thermally generated electrons slowly accumulate in the photosites during every exposure. This dark current increases with temperature and the accumulated charge will eventually saturate pixels. It should be of little surprise that this current, being comprised of electrons, has a mean and random element and each pixel accumulates at a slightly different rate. In conventional photography, dark current and its accompanying noise are of little concern. During the long exposures that astrophotography requires, however, it becomes increasingly significant. The mean accumulation rate approximately doubles for every 6 °C rise in sensor temperature. In the case of a camera that can cool its sensor to 40 °C below ambient, it will lower the rate by ~100x and the random noise contribution by 10x. In the case of a sensor's dark current specification, it is reported in electrons/second/pixel for a particular temperature. Unfortunately, there is no standard temperature setting and specifications use anything from -20 °C to 20 °C. To compare results, I convert all the results to -10 °C assuming a 2x / 6 °C reduction guideline. There is some variation between sensors, which is mostly a property of the basic semiconductor materials, rather than the sensor design.

## Scale

### Sensor Dimensions

The size of a sensor for deep-sky imaging is in many ways the initial consideration. The most popular sizes are Four-Thirds and APS-C. Larger sizes provide a wonderfully wide field of view but are increasingly expensive and at the same time require larger and more expensive filters and filter wheels. With care, an APS-C sensor will work with 1.25" or 31-mm filters but the slightly larger APS-H sized sensors will require 2" or 36-mm. In many cases, larger sensors exceed an affordable telescope's imaging circle, or simply reveal the optical issues at their margins. Smaller sizes are the opposite but simply constrain the field of view and those with high pixel counts will have unnecessarily small pixel

| sensor | sensor specification | | | | | | | | | computed sensor specifications | | | | | | |
|---|---|---|---|---|---|---|---|---|---|---|---|---|---|---|---|---|
| | pixel size [μ] | gain [e⁻/ADU] | ADC [bits] | sensor size [mm] | RMS read noise [e⁻] | full well [e⁻/pixel] | dark noise [@-10°C] | QE peak [%] | QE Hα [%] | dynamic range [states] | dynamic range [stops] | full well [e⁻/μm²] | noise [e⁻/μm²] | dynamic range [/μm²] | dark noise @-10 °C [/μm²] | comments |
| **Four Thirds** | | | | | | | | | | | | | | | | |
| KAF8300 | 5.4 | 0.5 | 16 | 17.6 x 13.5 | 8 | 26,000 | 0.02 | 55 | 45 | 3,250 | 11.7 | 892 | 1.48 | 602 | 0.0037 | CCD |
| MN34230 | 3.8 | 5 | 12 | 17.3 x 13 | 3.47 | 20,000 | 0.012 | 60 | 45 | 5,764 | 12.5 | 1,385 | 0.91 | 1517 | 0.0032 | color/mono |
| MN34230 | 3.8 | 1 | 12 | 17.3 x 13 | 1.3 | 4,096 | 0.012 | 60 | 45 | 3,151 | 11.6 | 284 | 0.34 | 829 | 0.0032 | color/mono |
| IMX294 | 4.63 | 3.9 | 14 | 19.3 x 13 | 7.5 | 63,700 | 0.009 | 75 | 70 | 8,493 | 13.1 | 2,972 | 1.62 | 1834 | 0.0019 | color only |
| IMX294 | 4.63 | 1 | 14 | 19.3 x 13 | 1.7 | 14,000 | 0.009 | 75 | 70 | 8,235 | 13.0 | 653 | 0.37 | 1779 | 0.0019 | color only |
| **APS-C** | | | | | | | | | | | | | | | | |
| EOS60Da (200) | 4.29 | 1.487 | 14 | 23 x 15 | 8.3 | 24,118 | 0.068 | 38 | 30 | 2,906 | 11.5 | 1,310 | 1.93 | 677 | 0.016 | color only |
| EOS60Da (400) | 4.29 | 0.744 | 14 | 23 x 15 | 4.4 | 12,197 | 0.068 | 38 | 30 | 2,772 | 11.4 | 663 | 1.03 | 646 | 0.016 | color only |
| EOS60Da (800) | 4.29 | 0.38 | 14 | 23 x 15 | 3.2 | 7,405 | 0.068 | 38 | 30 | 2,314 | 11.2 | 402 | 0.75 | 539 | 0.016 | color only |
| IMX071 | 4.78 | 3 | 14 | 23.6 x 15.7 | 3.3 | 48,600 | 0.001 | 50 | 40 | 14,727 | 13.8 | 2,127 | 0.69 | 3081 | 0.0002 | color only |
| IMX071 | 4.78 | 1 | 14 | 23.6 x 15.7 | 2.6 | 17,000 | 0.001 | 50 | 40 | 6,538 | 12.7 | 744 | 0.54 | 1368 | 0.0002 | color only |
| IMX193 | 3.91 | 2.2 | 14 | 23.6 x 15.7 | 2.7 | 36,000 | 0.0014 | 74 | 70 | 13,333 | 13.7 | 2,355 | 0.69 | 3410 | 0.0004 | color only |
| IMX193 | 3.91 | 1 | 14 | 23.6 x 15.7 | 2.2 | 17,000 | 0.0014 | 74 | 70 | 7,727 | 12.9 | 1,112 | 0.56 | 1976 | 0.0004 | color only |
| **larger formats** | | | | | | | | | | | | | | | | |
| KAF16200 | 6 | 0.6 | 16 | 27 x 21 | 6 | 40,000 | 0.08 | 60 | 50 | 6,667 | 12.7 | 1,111 | 1.00 | 1111 | 0.0133 | CCD |
| IMX128 | 5.97 | 4.6 | 14 | 36 x 24 | 4.2 | 76,000 | 0.004 | 53 | 42 | 18,095 | 14.1 | 2,132 | 0.70 | 3031 | 0.0007 | color only |
| IMX128 | 5.97 | 1 | 14 | 36 x 24 | 2.8 | 17,000 | 0.004 | 53 | 42 | 6,071 | 12.6 | 477 | 0.47 | 1017 | 0.0007 | color only |
| KAF11002 | 9 | 0.9 | 16 | 36 x 24 | 15 | 60,000 | 0.015 | 50 | 30 | 4,000 | 12.0 | 741 | 1.67 | 444 | 0.0017 | CCD |
| KAF16803 | 9 | 1.6 | 16 | 36.8 x 36.8 | 10 | 100,000 | 0.094 | 60 | 50 | 10,000 | 13.3 | 1,235 | 1.11 | 1111 | 0.0104 | CCD |

*fig.4* *Too much data? This is a comparison of the currently popular deep-sky imaging sensors, in CMOS and CCD technologies. These are typical specifications and there will be some variation between camera manufacturers. The CMOS sensors appear twice; at minimum gain (and maximum dynamic range) and at unity gain, to give an indication of the dynamic range/noise trade-off. For reference, I have also included an EOS 60Da at two ISO settings. This camera is not cooled and so it is unlikely to ever reach -10°C and in real-world conditions its dark noise will be several electrons/second.*

*The data can be mind-boggling and it is easy to be drawn into a never-ending dilemma. The specifications and their normalized figures per square micron are just a starting point. There are other optical characteristics that make them less or more suitable, including how well they control amp glow and sensor diffraction/reflection artifacts. In general, however, this is an exciting time and there are some interesting alternatives out there.*

*At present, the choice of monochrome CMOS sensors at Four-Thirds or larger is limited. The Four-Thirds Panasonic MN34230 has just been joined by the 60MP Sony IMX455 full-frame sensor, but with an equally significant price tag. The later models usually have improved ADCs, read and dark noise.*

sizes. Having chosen a sensor size, however, the choice is reduced to a handful of models and implementations.

On a datasheet, it is the active area that is important as some sensors have peripheral photosites that are not part of the image forming area. (These are masked off and are used to implement a crude dark-frame calibration within the sensor.) Most new CMOS sensors are only available in a color version, as the demand for a mono version is too low to be commercially viable.

### Pixel Pitch and Computed Characteristics

One of the tricky aspects of comparing sensor performance is their physical pixel size. It is all very well quoting 1 $e^-$ read noise per pixel, but if two sensors of similar read noise have very different pixel sizes, the noise level per unit area is quite different. It is important to remember, at the end of the day, it is the overall picture that counts, not the individual pixels. The same is true of the full-well depth specification. Astrophotography is all about recording information and at its core, it starts with the collected and focused image photons. In conventional photography, we concern ourselves with the aperture ratio (f/stops) but in astrophotography, gathered information increases with the physical aperture diameter. A 2,000-mm $f_L$ scope with an aperture ratio of f/10 captures more information than a 350 mm $f_L$, f/4. The same is true with sensors, ultimately, their tonal resolution and noise performance is not entirely a pixel phenomenon but what it does, on the whole, regarding photon conversion and image noise. It is a tricky and fascinating concept and confuses the heck out of many. If you want to look it up, an Internet search on "Information Theory" will cure insomnia.

### Dynamic Range

In many respects, the dynamic range is a more useful measure of sensor quality than the signal to noise ratio. Dynamic range is inferred from two separate specifications, full-well capacity and read noise. While the full-well capacity specification is often considered as a dynamic range indicator, it is not the entire story. The dynamic range is degraded by the uncertainty of an pixel value caused by sensor noise. The effective dynamic range is defined by the full-well capacity divided by the minimum sensor noise (read noise and dark noise). This assumes that the full-well capacity occupies the useful range of the sensor (by which we mean the linear part, before roll-off or clipping). While this appears similar to the signal to noise ratio calculation, the difference is the SNR is highly subject dependent and, in practice, the SNR of an image may have more shot noise (from light pollution) than sensor noise. It is a point worth dwelling on. Although all noise sources are additive, there is a less compelling rationale for pursuing low-noise sensors in areas of high light pollution.

### Normalized Sensor Noise

We use sensors to get information. If we consider two hypothetical APS-C sized sensors, one with 100 pixels and another with 10 megapixels, which one has more information? Easy, the one with 10 megapixels. If we throw very different noise levels into the equation, with the 10-megapixel sensor having an awful pixel noise level, the answer is less certain. To compare sensor noise levels on an equal footing, it is useful to normalize their read noise performance per square micron or to an angle, nominally 1 arc second. The same should also be considered for dark current too, although it is often overlooked.

The noise *power* from different sources is additive and is proportional to the square of the voltage. When adding noise voltages, as quoted in datasheets, you cannot simply add them. In the case where x, y and z are individual noise contributions in $e^-$, the total noise is defined by the root of the sum of their squares:

$$\text{total noise} = \sqrt{(x^2+y^2+z^2)}$$

### Normalized Full-Well Capacity

Similar to sensor noise, using the argument of two sensors with very different pixel sizes, pixel size influences the overall ability to accumulate electrons. Again, one neat way to directly compare sensors is to normalize full-well depth over an equal equivalent area. If you consider the focused image of a star is a diffuse blob, to avoid clipping, one big pixel will need to capture more photons than a cluster of small ones.

## Other Specifications

### Cooling

Perhaps the key benefit of a dedicated camera is the potential to precisely cool the sensor and reduce the dark current and its associated shot noise. A camera's cooling specification states the maximum temperature reduction from the ambient temperature. Another key feature is the ability to set the sensor temperature and automatically regulate the cooling to maintain a constant sensor temperature. This is particularly important when one is required to image and generate calibration exposures with precisely the same conditions in order to remove amp glow and dark noise.

Cooling is achieved using an electronic heat pump (Peltier cooler) with a fan or a water-cooled jacket to remove the heat. These devices are intriguing; a single-stage cooler looks like a wafer biscuit and is thermally coupled to the back of the sensor. When current passes through the substrate, it typically achieves a 20°C difference between its surfaces. The cooler surface is coupled to the sensor and the hotter one to a heatsink. To increase the cooling amount, they are stacked in two's or three's. Most cameras achieve a 35–40°C reduction, reducing dark noise to a negligible level that is usually exceeded by read noise. (With some of the best CCD sensors, the dark current is so low that it is often more effective to dispense with dark frame calibration frames and simply use a bad-pixel map to replace a few rogue pixels with the average of their surroundings.)

In a conventional DSLR or mirrorless camera, the sensor is buried in the body making it impossible to cool it directly. Some have attempted various chilled enclosures to lower the ambient operating temperature. This helps, but unless the sensor temperature can be accurately regulated, it is difficult to precisely match image and calibration exposure conditions.

### Miscellany

Other useful specifications include the minimum exposure time (sometimes limited by mechanical shutters), anti-blooming gates, residual bulk image flash, read time and high-speed (noisy) download times. Another parameter, often overlooked, is the back-focus. This is a camera rather than a sensor specification and is the distance from the sensor surface to the camera mounting plate. Many field flatteners and reducers assume a ~56-mm distance from their mounting surface to the sensor, into which one may wish to have a filter wheel, off-axis guider and perhaps a rotator too. A large back focus on the camera restricts what can be accommodated and for this reason, some cameras combine filter wheels and off-axis guiders into a single enclosure or offer different configurations with a range of back-focus distances that permit working within the standard T-thread spacing of 55 mm (allowing for the effect of inserted filters).

The spacing of the various elements has another impact; the standard 1.25-inch filter size is just large enough to work with a Four-Thirds sensor, as the separation is usually small within a single enclosure. A separate filter wheel increases the filter-sensor distance and vignetting, requiring larger filters. Five or seven filters are a significant investment in their own right, which may easily exceed the camera cost!

### Choosing a Dedicated Camera

So, how do you go about choosing? When you take all these factors into account and lay out the performance specifications of several prominent imaging sensors, the normalized specification differences are less significant than the manufacturers would lead you to believe. In common with modern digital cameras, there are few "bad" ones. There are many more potential imaging sensors now than 10 years ago and fig.4 showcases some promising CMOS sensors with their published and computed specifications. I have picked a selection of models of different sizes. Choosing between them involves a filtering process. For instance, pixel pitch plays a significant part in the decision-making process, as does the sensor size (and hence the field of view). We know from the chapter *Imaging Hardware* that pixel pitch impacts the overall imaging resolution and there is a range of pixel sizes that are considered optimum for a particular focal length.

In simple terms, a larger aperture diameter delivers a higher resolution. As the optical resolution increases, however, atmospheric seeing plays a bigger part in the effective resolution and smaller pixel sizes have little effect. Assuming a similar pixel angular resolution, longer focal lengths achieve the same angular resolution with proportionally larger pixel sizes.

The other decision is whether to choose a color or mono camera, noting that most CMOS imaging sensors are only available in a color version. (There are a few enterprising companies that will remove the Bayer array from a sensor, but in so doing, remove the micro-lens too, reducing the sensor's efficiency.) For instance, if you live in an area with significant light pollution, a mono sensor is preferred for its efficiency and flexibility to block unwanted light using external RGB filters. On the other hand, if you have more favorable imaging conditions, natural color images are more readily obtainable and the convenience of a color sensor allows one to travel light and keep things simple, without too much resolution loss from the subsequent de-mosaic process. I prefer flexibility over convenience and frustratingly, there is only one Four-Thirds mono CMOS sensor at the moment, which has already been overtaken by color-only Sony sensors.

Moving on, if your specialty is wide-field imaging, then a larger sensor will make the most of the field of view of a telescope or alternatively, if you are imaging the smaller galaxies and clusters, it is more wasteful (and expensive). If in doubt, I recommend a Four-Thirds or APS-C sized sensor as a good starting point since it is compatible with the imaging circle of every telescope. The Four-Thirds size works with smaller (1.25-inch or 31-mm filters) without significant vignetting and the marginally larger APS-C 2:3 format with 36-mm filters, depending on the filter to sensor distance and the aperture ratio. Some CMOS sensors do not have anti-reflection coatings, which cause obvious circular reflections around bright stars, some also have more obvious diffraction spikes, from their micro-lens design.

While most cameras have USB 2.0 or USB 3.0 connections, some have built-in USB hubs too, which can be convenient for driving a filter wheel or guide camera and reduces the number of trailing USB cables. USB connections have a power delivery limit and the larger cameras and those with Peltier coolers require a separate, regulated DC-power connection. If you are using a simple guide camera without an ST4 port, a few imaging camera models include a 4-way opto-isolated guide port, though increasingly, wired guider corrections are being replaced by software commands to the mount over USB, serial or wireless connections.

A less obvious but overriding consideration is the availability and reliability of the camera's interface drivers, including PC/Mac/Linux driver availability and the implementation of its accompanying application, ASCOM, or X2 driver, as well as any applicable video standards. The best camera in the world is useless if one cannot use it reliably over extended periods. Hardware and software are best developed in close collaboration and it is sometimes the case that specialist manufacturers lack the software expertise or may out-source the driver development to third parties, with mixed results. Furthermore, in an attempt to reduce costs, they may not necessarily retain those services for timely support and maintenance.

My own experience, and that of friends, over several products and manufacturers, including several premium ones, has been patchy. Poor reliability and random failures are the most frustrating of all. Bad news travels faster than the speed of light and it is worthwhile checking the forums for warning signs that the latest model is being design-verified by its customers. Rapid development is great, so long as it does not leave behind a wake of dissatisfied users.

Lastly, one should additionally consider buying a used CCD camera. The Kodak KAF8300 CCD has been the mainstream imaging sensor in every manufacturer's portfolio for many years and is still a capable performer. It is too soon to write it off; its reliability is proven and many images in this book were made with one.

## Using a CMOS Camera/ Sensor

The fundamental differences between some CMOS and CCD cameras, that affect the way you use them, are the gain setting and the on-chip dark current suppression. As we have seen, when the gain setting goes from low to high, there is a trade-off between (worsening) dynamic range and (improving) read noise. This is most apparent in sensors whose full-well depth is greater than the number of sensor ADC levels. (Note: recently announced premium CMOS sensors have 14- and 16-bit ADCs, with sampling levels that exceed their full-well depth, as most CCDs do.) With the less expensive sensors, however, it is still the case and the stratagem is to use these characteristics to our advantage. Fortunately, in astrophotography, there is a perfect dichotomy that plays into our hands between images of stars and nebulae. On the one hand, stars vary hugely in brightness and a high dynamic range setting is ideal, using low gain to stop highlight clipping and where a top-notch signal to noise ratio has less visual impact on points of light. On the other hand, nebulae typically have a lower dynamic range but are very dim. They require long exposures for sufficient photons to register and the smooth nebulous clouds show up image noise more readily. Here, a high gain, trading off the unwanted dynamic range with lower image noise makes more sense. Stars and nebulae seldom occur independently and these two scenarios point to an imaging plan that combines two sets of exposures, optimized for specific subject matter. These are calibrated and stacked separately and are typically combined using a star mask or blending the brighter pixels in the nebula image with the corresponding pixels in the stellar image. Our last practical assignment uses this technique on a popular amateur target.

Finally, picking up on the second sensor difference concerning dark current suppression, CMOS images require a subtly different calibration process. This on-chip phenomenon may cause long exposure dark frames to have a lower mean value than a zero-length bias frame, which messes up traditional calibration methods. The chapter on calibration, explains how to avoid dark-frame scaling (optimization) during the calibration of "flats" and "lights" and additionally avoid creating zero-value pixels by subtracting a master bias from a master dark. The best results occur when the dark-frame sensor temperature and exposure durations exactly match that of the flat and image frames, which is why operating a DSLR or mirrorless camera is always a compromise, compared to a temperature-controlled astro camera with the same sensor.

# Heart Nebula (IC1805)

*Capturing faint objects works best with sensor cooling, filtering and longer exposure times over several nights.*

***Equipment:***
*73mm f/5.9 refractor (430-mm $f_L$)*
*field flattener, electronic focus*
*QHY color CMOS astro camera*
*ZWO CMOS guide camera and 50mm f/4 guide-scope*
*IDAS light pollution and duo-narrowband filters*
*SkyWatcher AZ-EQ5 GT mount*

***Software:***
*SGP (acquisition), PHD2 (guiding)*
*DeepSky Stacker, Affinity Photo*

***Exposure:***
*duo-narrowband (35 x 1200 seconds at high gain)*
*IDAS P2 (25 x 300 seconds at low gain)*
*Dark, Flat and Bias calibration exposures to match*

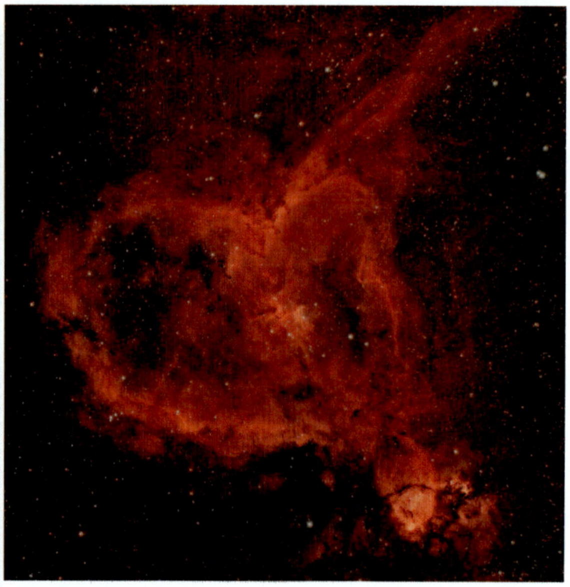

This last assignment brings together all we have learned so far and adds a dedicated color CMOS astro camera into the setup. In the scheme of things, this exceeds our self-imposed budget and propels us into more advanced territory and the limit of conventional photo editing software. The camera in question is a cooled, color CMOS model, with a 14-bit internal ADC. The choice was tricky. While this sensor is only available as a color model, its characteristics outperform the only currently available monochrome CMOS imaging sensor. I chose a large subject with a variety of elements, suitable for a short refractor; the Heart Nebula was discovered by William Herschel in 1787 and is predominantly an emission nebula, with some dust clouds and small star clusters. This nebula is more challenging and requires extended exposure and the benefit of a cooled sensor to resolve the faint clouds without introducing too much image noise. In this case, the nebulosity is mostly made up of ionized hydrogen with patches of much fainter deep-red sulfur and blue-green oxygen, that may or may not register.

### Equipment

The camera was attached to the back of the refractor's field flattener via its 48-mm thread, at the right spacing, to ensure round stars into the corners. This required a careful analysis of QHY's mechanical drawings for its camera, adaptors and spacers to match the flattener spacing and thread. Unlike a DSLR or mirrorless camera, this model does not have an integrated UV/IR blocking filter and it is necessary to use an additional blocking filter (or light-pollution/duo-narrowband type) to block the extreme wavelengths. In this system, the front end of the field flattener has a secondary 48-mm thread, perfect for mounting 2-inch filters.

While the system is light enough to run on the diminutive Fornax LighTrack mount, its 90-minute travel period is not convenient for an extended 13-hour imaging session that spans several nights and ideally requires dual-axis guiding to minimize drift over 20-minute exposures. The SkyWatcher AZ-EQ5 GT is ideal for the purpose and I had already verified my unit had excellent periodic error and backlash. For best results, an off-axis guider is the preferred option and although one is available in the QHY range, a piggy-back guider system was sufficient for this wide-field image. I used a 50-mm f/4 guide-scope, fitted with a rigid focus mechanism and an inexpensive CMOS guider camera connected by a C-mount to 1.25-inch adaptor.

## Acquisition

For this assignment, I needed three things to make the most of several nights of imaging: autofocus, autoguiding and automatic centering. For this reason, I used Sequence Generator Pro and PHD2 for controlling the acquisition process. These are purpose-built for this kind of work and for the first time user, it takes a little while to set up, starting with information about the connected devices (telescope/camera/guider/focuser) and then the information for the location, target (RA/DEC) and the exposure plan (start time/exposure length/exposure count/filter). The camera can cool its sensor to 35 °C below ambient and in this case, I chose -15 °C as the temperature set-point. This is the first occasion we are using a cooled camera and a good time to mention some best practices, to avoid condensation, as well as issues caused by repeated thermal shock.

We know that cooled surfaces attract condensation and in this camera, like most others, the sensor cavity is sealed and employs a dry-gas fill and a desiccant. Even so, if one attempts to cool a sensor too quickly, it is likely to develop condensation on its surface (or its window) that will ruin the image. This camera is no different, so, within SGP, I set the camera to cool to -15 °C over a 30-minute period. During this time I polar aligned the mount and used the camera to check focus, image alignment and experiment with different exposure times. I employed the same regime in reverse at the end of each session, gradually warming the sensor to 10 °C. In this way, there is less likelihood of repeated thermal shock causing equipment failure at some time.

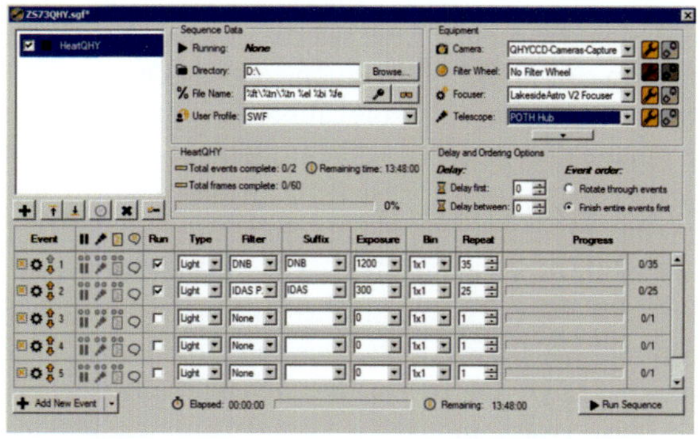

*fig.1 The sequence window from SGP shows the targets (one, in this instance) equipment and the exposure plan. The camera gain values are very different for the two filters and currently are set manually from the camera ASCOM dialog.*

(Sensors have feelings too; I know I would not like to be plunged between -15 to +10 °C.)

The exposure plan used two filters with very different glass thicknesses and which required re-focusing when they were changed over. Within SGP, it is possible to put in a pause in proceedings at the end of each filter run to allow one to change over the filter. In this case it was not required, as I only used one filter on each night.

SGP's sequence window is shown in fig.1 and its event window in fig.2, where the RA/DEC target coordinates (in J2000) are populated. These can be typed, cut and pasted from a planetarium application or, as in this case, populated from SGP's framing wizard. While SGP's slew and center options point the telescope to the rough position and then fine-tunes the aim, the catalog coordinates of any target may not give the best composition. In this case, it required a few attempts to get it just so, with small adjustments to the mount position and camera angle.

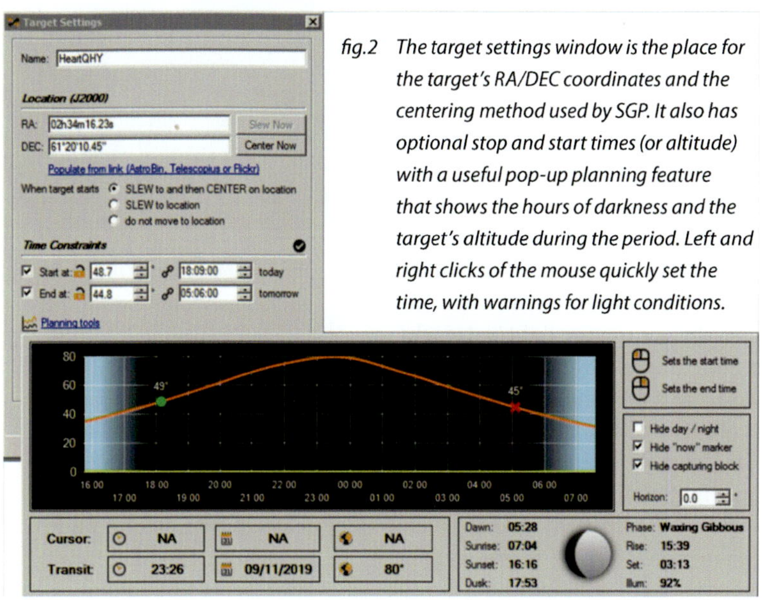


*fig.2 The target settings window is the place for the target's RA/DEC coordinates and the centering method used by SGP. It also has optional stop and start times (or altitude) with a useful pop-up planning feature that shows the hours of darkness and the target's altitude during the period. Left and right clicks of the mouse quickly set the time, with warnings for light conditions.*

Sequences are a wonderful thing but they work best with some preparation, before running. For example, SGP's focus routine works more effectively if the imaging scope is near focus. To do this, I set up the guider and imaging scope to approximate focus, with the help of a bright star and a cycle of short exposures.

### Guiding

To calibrate the autoguider in PHD2, I aimed the telescope South, at low declination and took a 1-second exposure. After selecting a guide star, I shift-clicked the guide icon to run PHD2's calibration routine. Once it had finished I checked the option to remember the calibration for next time and kept the camera permanently assembled to the guide-scope throughout the three sessions. The 5 guiding algorithms in PHD2 are designed to cope with the realities of the mount mechanism and the randomness of the star measurement. They are subtly different but all have the task of finding the right compromise between chasing seeing and keeping up with real drift or periodic error. Mount backlash is a tricky issue and the latest versions of PHD2 include a mount analysis tool that measures backlash and suggests guider settings. In this case it was minimal but if it had identified significant mount backlash on the DEC analysis, I would have selected the Resist Switch algorithm for the DEC axis, which minimizes motor/gear reversals caused by seeing noise but reacts to the underlying drift. Each of the guide algorithms have settings to vary the aggression and filtering. These can change from night to night or between high and low targets. After some experimentation I settled on the familiar Hysteresis algorithm for both axes, with aggression at 70, hysteresis at 7 and a minimum move at 0.05.

### Exposure

This assignment was "first light" for this camera and I used the information I had gleaned from our earlier experiments with the modified Fuji X-M1 body. The light pollution in my location is noticeable and I started with the duo-narrowband filter. This filter passes the specific colors of ionized hydrogen and oxygen. These are dim, so I set the camera to a high gain and took long exposures, to capture the faint signal over two nights. I was not sure if the star colors would be objectionable and on the third night, I followed with a sequence of shorter exposures, through the light pollution filter. This transmits more light and it is easy to over-expose star cores. For that reason, I set the camera to its lowest gain (which also has the highest dynamic range) and using short exposures. I made the calibration exposures for this new camera in daylight, including a series of bias and dark frames (at the same temperature, gain and exposure settings as the light frames) and some brief images with a diffuser over the end of the telescope for flats (with the same gain and filter setting as the exposures).

## Image Preparation

The image processing followed the workflow in fig.7, with calibration and stacking performed on the two different data sets, matched up to their own bias, dark and flat calibration files. Ordinarily, one would have common dark and bias frames but in this case, the image sets through each filter were taken at two different gain settings, with entirely different outcomes. In DeepSkyStacker, this was accomplished by adding the light/flat/dark/bias frames in the Main group for the duo-narrowband filter exposures and then all the image and calibration files associated with the IDAS filter/gain setting in Group 1. The stacking process requires all the calibrated images to be registered to a single file before DSS combines them into separate stacks. I chose one with small star sizes, right-clicked and selected it as the overall reference file.

The two images stacks are very different, intentionally so, for unique purposes. Saved as 32-bit TIFF files, I loaded the nebula image into Affinity

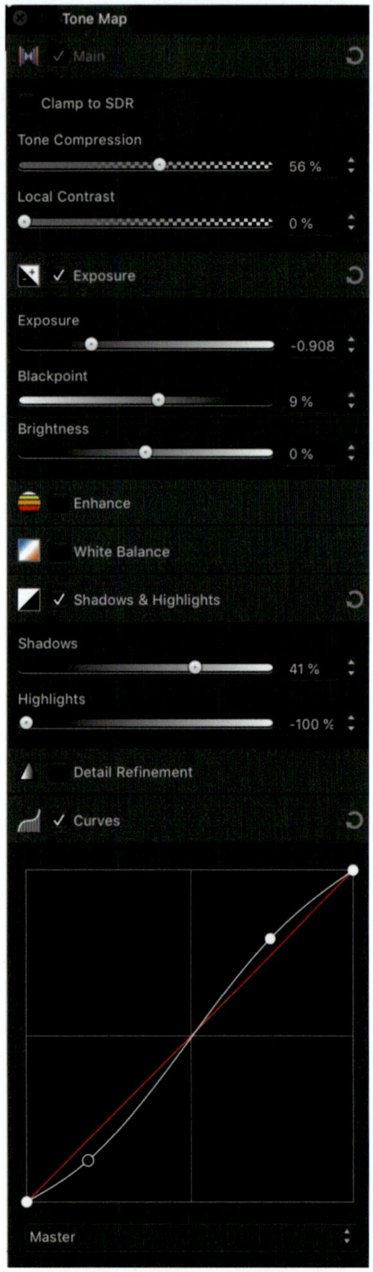


*fig.3 The Tone Map settings are worked on progressively; getting the tone compression and exposure at levels that show detail yet do not hit the extremes. The Shadows and Highlight sliders tame these extremes a little and finally, the Curves increase the mid-band contrast at the expense of a little compression in the toe and shoulder areas.*

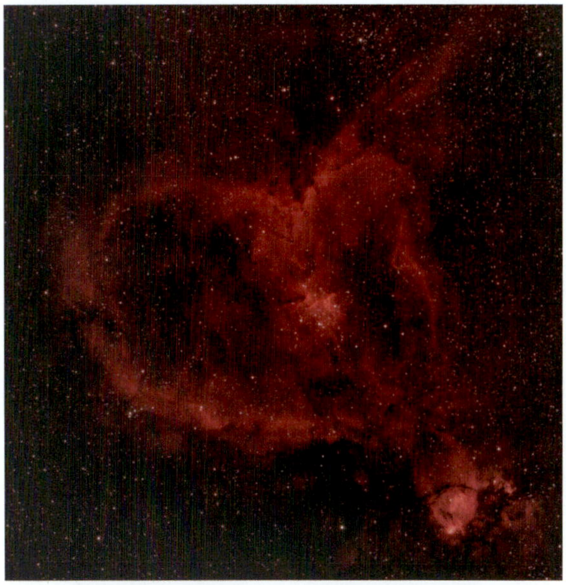

*fig.4 The result from the Tone Map persona. It is important to not over-do things. Remember, this is a 32-bit image and can always be stretched at will. Once you clip shadows or highlights, however, the bit-depth will not save the image. Less is often, more.*

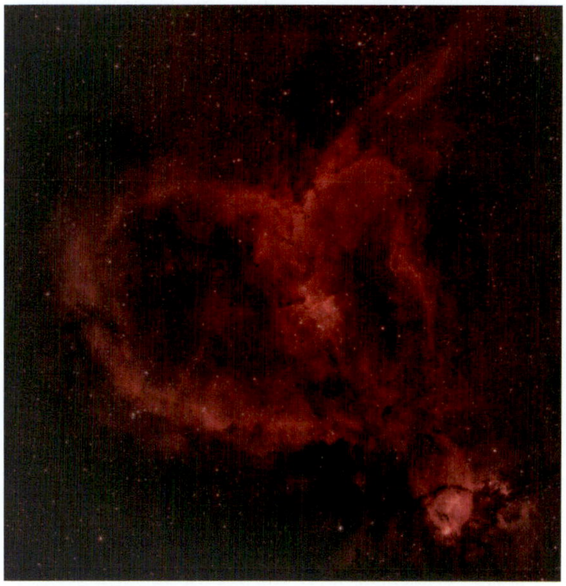

*fig.5 The result after using the Minimum live filter layer on a highlight selection (stars). A couple of things to note; this makes the nebula more dominant and even though it looks overkill, the stars will be emphasized again as the result of later contrast increases and sharpening.*

Photo and created a temporary adjustment layer with an extreme stretch. This revealed the uneven framing and background levels, dominated by an overall color cast. To remove this, I used a simple trick to neutralize the color; after disabling the stretch, I sampled one image corner, away from any nebulosity and filled a new pixel layer with that color. I reduced its opacity to 50%, inverted the layer (from the drop-down menu) and changed its blending mode to Color. I deleted the temporary adjustment layer and merged the two pixel layers to create the basis for manipulation. The temporary stretch also highlighted my less-than-perfect session to session registration. While the image clearly needed cropping, conventional photo editors have no method of pixel-perfect cropping two images and cropping was delayed until both image stacks had been combined.

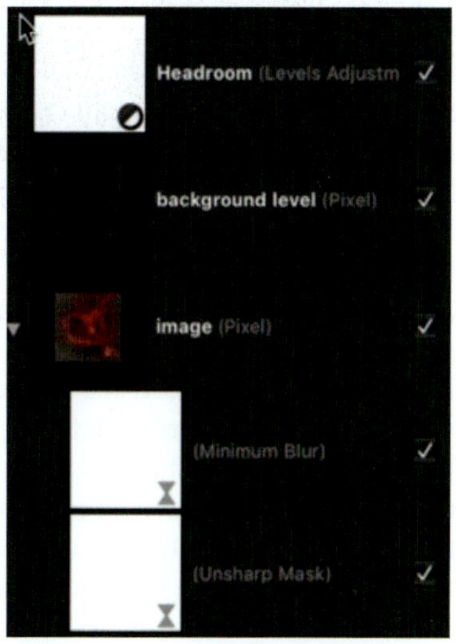


*fig.6 The layer stack with the base image and its star shrinking and deconvolution child adjustments. Above that is the sampled background elliptical gradient (in Subtract blending mode) along with a Levels adjustment that knocks back peak highlight and shadow levels to give headroom for later changes.*

## Image Processing

For the main nebula image, the initial non-linear stretch was applied by switching to the Tone Mapping Persona. I used a mixture of the Compression, Exposure and Blackpoint sliders to produce a subtle image, with no obvious shadow or highlight clipping (fig.3). Once those levels were established, I added some highlight and shadow protection and a mild S-curve to boost mid-tone contrast before committing (fig.4). Even so, after this extreme stretch, the numerous white stars dominated the image and required taming. These were selected using Select>Tonal Range>Select Highlights, immediately followed with a Minimum Blur live filter, set to 3 pixels and positioned as a child layer to the bottom image. This has a dramatic change on the image balance and one may consider it extreme, except that it is countered by later stretches and sharpening actions. One of those actions is applied to the base image as a child layer, along with *fig.7 The workflow for this faint emission nebula is complicated and like many, using multiple-filtered exposures, follows multiple paths. A similar workflow would also work on a galaxy and is close to the practical limit for a conventional photo editor Here, the long exposures, optimized for rendering faint signals, have a completely different treatment to emphasize their form with low noise and good color. The shorter exposures through the light pollution filter have more realistic star color and are designed to keep star (or galaxy) cores from clipping. At an early stage, the low-level information is discarded, to concentrate on the brighter (star) elements. The images are finally brought together by pasting the star image on top of the finished nebula image, changing its blending mode to Color and using a slightly blurred star selection as the mask.*

*The processing steps vary again in more advanced work if separate clear (Lum), red, green and blue filters are used with a monochrome camera. For instance, when the signal to noise ratio is poor, all the data from all the images may be combined into a single monochrome luminance image. This improves the signal to noise ratio slightly and gives more leeway for sharpening and stretching.*

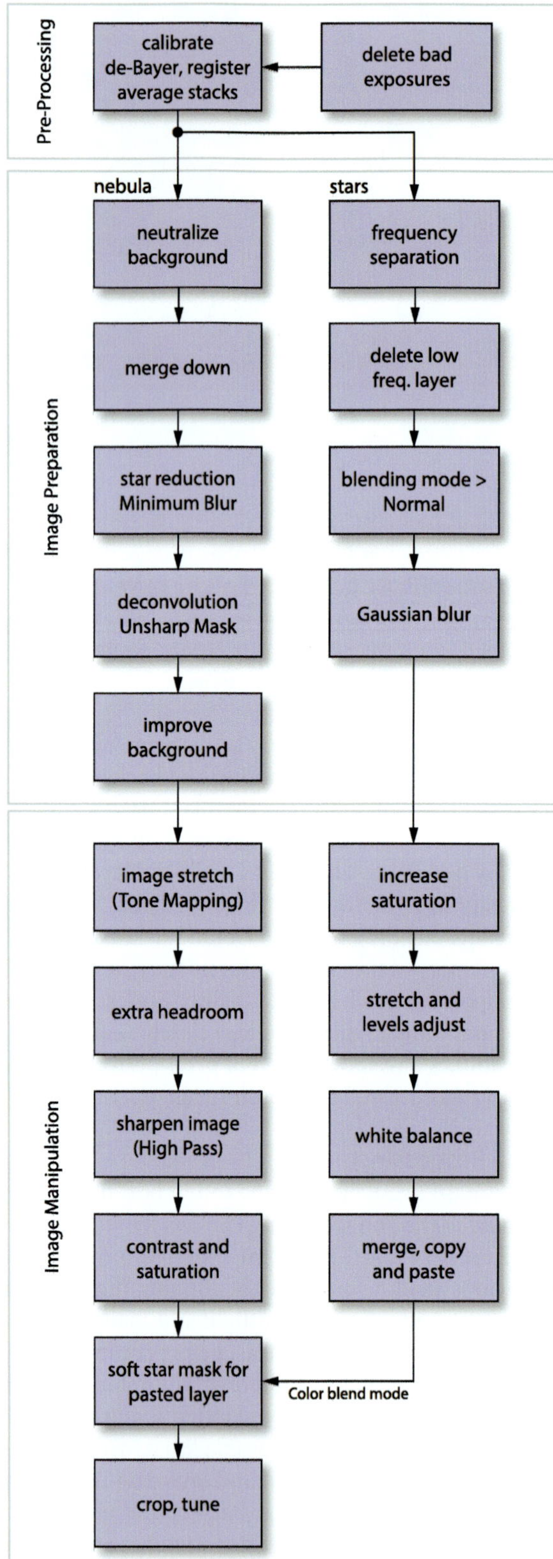


the Minimum Blur filter; image deconvolution in the form of an Unsharp Mask Live Filter, set to 0.3 pixels and Factor 4. The result is shown in fig.5.

The image at this stage is still subtle and importantly, shows the faint nebulosity surround the familiar heart outline. While there is an obvious gradient in the corners, it would be rather too easy to remove this and the faint regions with a simple subtracted blurred image. Sampling the darkest part of the image and the bottom left corner, I constructed an elliptical gradient in a new pixel layer with these values and set the blending mode to Subtract and the layer opacity to 60%. While this removed most of the color cast, the result was rather contrasty and knowing that image sharpening would increase contrast further, I rolled back the shadow and highlight values. There are a number of ways of achieving this; in this case, with a Levels adjustment layer, with the Output Black level set to 8% and the Output White Level set to 80%, providing some processing headroom (fig.6).

In the past, with a narrowband CCD image, I would be combating the background noise with various masks and filters. Up close, the noise in this image is already good and I proceeded to sharpening, tone shaping and color tuning.

*fig.8* *A High Pass filter set to Soft Light blending mode emphasizes structures of a particular scale, determined by its radius setting. The initial grey image (before changing the blending mode) indicates what will be affected. A small radius predominantly affects stars. Here, a radius of 33 pixels emphasizes some of the dark cloud structures.*

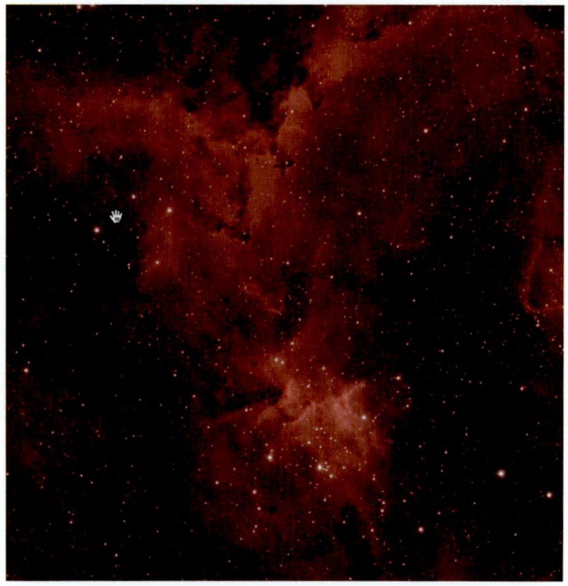

*fig.9* *After applying the High Pass filter to the image, the dark Bok globules and cloud structures became more apparent. (It is possible to alter the radius setting after the event to directly see the effect on the image. Double-clicking the High Pass filter layer will bring up the setting dialog once more and allow endless experimentation.)*

Nebulae typically do not have sharp edges but they do sometimes have well defined small, dark dust clouds called Bok globules. I sharpened these with a High Pass live filter, with the blending mode set to Soft Light and a radius of 33 pixels. The pixel size was determined by initially leaving the blending mode to Normal and examining the grey image as the radius changed (fig.8). I used the pixel value that gave the best definition of the globule structure (fig.9). I finished off the nebula image with two more adjustment layers; a slight red-hue change towards orange and an S-curve to give it some luminosity, as can be seen in the lead image.

My attention now turned to the image stack taken through the IDAS light-pollution filter. The processing for this is very different; we are only interested in the stars and most importantly, their color. To keep that color, it is mandatory to avoid over-stretching. Even if the stars are not clipped, the brighter they are, the less saturated they appear. In this 32-bit image, we can afford to compress the dynamic range without fear of later posterization. After neutralizing the background, I added a Levels adjustment layer, with a gamma boost and a healthy 70% Output White Level. Merging these layers down, I used a new tool, Frequency Separation, from the drop-down filter menu, using a radius of 25 pixels. This made no visible change but replaced the

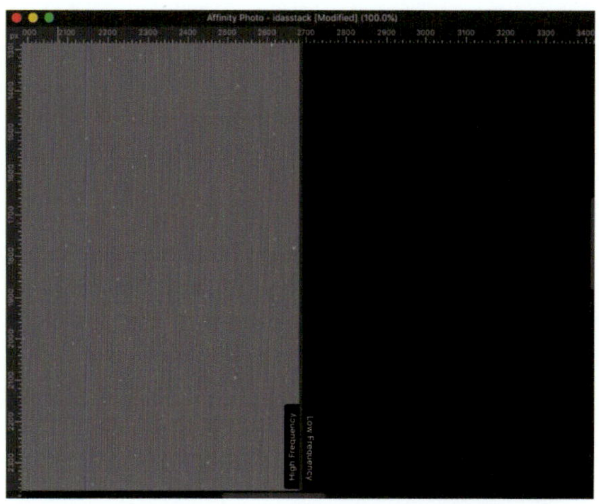


*fig.10* *The Frequency Separation filter cannot be applied as a layer. When constructed, it separates the base image into two; an underlying "soft" color image with the low-frequency information and an upper image, with a Linear Light blending mode, containing the high-frequency information. Together, they resemble the original image. When it is invoked, the screen resembles the above image, showing what is on each layer; in this case, the radius is small, to select the stars.*

image layer with two; a base fuzzy pixel layer, representing the low-frequency information and an upper layer, with the high-frequency details and a Linear Light blending mode (fig.10). The useful stellar information is in the high-frequency layer and I deleted the other. The result was an uninspiring grey image with some pale dots. A saturation and level change would made this a colorful stellar image, but, before doing so, a Gaussian blur of 1.5 pixels ensured the stars had some color information. I changed the blending mode back to Normal and used an HSL adjustment layer to increase the saturation by about 75%. I followed that with a Levels adjustment layer, carefully bringing up the Black Level slider to make a dark background and adjusting Gamma to make the stars visible. Finally, I added a small White Balance adjustment to make the stars an assortment of colors from blue through yellow, to red. After all that, the outcome was a murky background with colorful mid-toned stars (fig.11). I merged the visible layers to form a new pixel layer, selected and copied it to the clipboard to combine with the main nebula image.

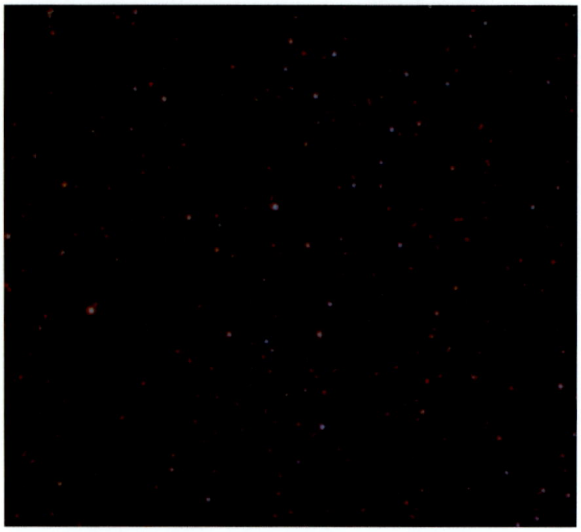

*fig.11* *A close-up of the star image, prior to combining with the nebula. In the combination process, only the star color is taken into account and here, the brightness values are kept low, to ensure good saturation.*

### Adding in the Stars

Combining the layers is fun. Going back to the nebula image, I pasted the star image as a new pixel layer on the top of the stack and made it invisible. I then selected the base pixel layer and selected the stars as before, using the Select Highlights option. Clicking the pasted star image once more, I hit the quick mask button (a dark circle in a white square) to make a mask, made the pixel layer visible and changed its blend mode to Color (fig.12). This has the magical effect of using the color information from the star image for the nebula image's stars. On close examination, the star boundaries looked a little odd and so I Cmd-clicked the mask layer and blurred it by about 0.5 pixels and experimented with the star layer opacity until I had a more natural appearance. This process of realistic star-color substitution is commonly performed with most narrowband images using a few hours of real-color acquisition.

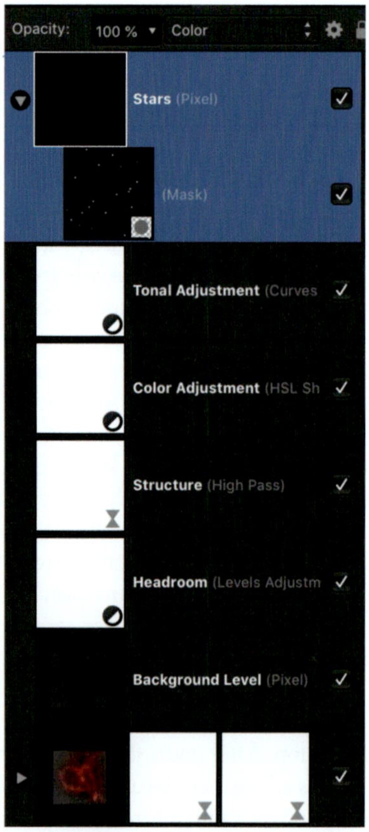


*fig.12* *The final simplified layer stack, after cropping, showing how to merge the star-color information from the IDAS-P2 filtered exposures.*

### Advanced Workflows

Red nebulosity is usually more dominant than blue-green. To balance the color we separate and boost the fainter color channels (after removing stars and blurring.) After recombining into a blurry color layer it is brought to life with a sharpened image layer, set to luminance mode. This process is common practice for those imaging with separate filters on a monochrome camera. It is, however, the subject for another book!

# Image Processing (part 3)

*Selected techniques to meet the unique challenges of deep-sky imaging.*

In this final image processing chapter, we look at some specific methods rather than an entire workflow. These consider imaging through separate filters as well as some additional and alternative tools for improving the detail and noise levels in more challenging targets. Ordinarily, I would use PixInsight and although this program is capable of exceptional results, it is more appropriate to the intermediate and advanced imager. My challenge, in writing this chapter, was to find alternatives in the more familiar world of photo editors, without spending money on plug-ins. There is a practical limit to what can be achieved, beyond which specialist plug-ins and applications give better results and are easier to use.

## Deconvolution

Deconvolution is a process of counteracting the light spread caused by diffraction and optical aberrations. The outcome is a sharper image, with smaller, brighter stars and some improvements in target details. It cannot work miracles though and I would estimate it improves enlargement potential by 10–20%. It certainly helped sharpen the initially distorted images from the Hubble Space Telescope. The star shapes themselves are a good record of the light spread and after measuring this, specialized tools use a complex iterative mathematical algorithm to roll-back the effect. It is usually the first thing you do to an image stack before it has been non-linearly stretched (think curves or gamma adjustment) and works best when the image is over-sampled. It is fiddly to do well and the process accentuates image noise too and commonly creates "panda eyes" around stars. It is not an automatic process and for best results it is used with masking to protect vulnerable areas. Two problems arise; with focal lengths under 450 mm, the pixel scale with a sensor of 3–6 micron pitch creates under-sampled images and crucially, some of the photo editors are not optimized for working with linear images. Affinity Photo has extensive linear and 32-bit file support and is an ideal test-bed for a deconvolution equivalent.

After some experimentation and research, the outcome was rather mundane; I found broadly equivalent results were obtained using the ubiquitous Unsharp Mask sharpening tool. In practice, you cannot judge this until the image is stretched, so I inserted it as one of my last actions. To implement, I clicked on the base image and added an Unsharp Mask live filter. In the filter dialog, I set Factor to maximum (4) and Threshold to zero. I found the best results were obtained with the Radius settings in the range 0.1–0.5 pixels, much lower than one would normally use and just before the onset of dark halos around small stars. Live filters in Affinity have an automatic mask, which can be modified to exclude dark background areas. A standard luminosity mask may not work here, however, as the image is almost entirely black. (The work-around is to duplicate the image, stretch it, blur slightly, rasterize to mask and drag the mask over the filter layer icon.)

## Combining Filtered Images

If you use a monochrome camera or use narrowband filters, the image acquisition process creates separate image stacks. To form a color image requires these to be associated with a color channel. There are pros and cons to combining the images before or after stretching. I find it is useful to at least view the channels as monochrome images as it allows one to assess the tonal balance more easily (fig.3). It is also common that some colors will have more of a sky gradient than others and it sometimes helps to even out backgrounds on the individual monochrome files before combining or stretching. Forming the composite RGB image is achieved by copying the three monochrome (or monochromatic color) images into a document as separate layers and changing the blend mode of the upper two to Add. Clicking on each image in the Channels palette in turn, clear red and blue for the green file, red and green for the blue file and so on. In the case of narrowband images, one can assign the images to any RGB channel for pictorial effect. The most common arrangement is called the "Hubble palette," which assigns hydrogen to green, sulfur to red and oxygen to blue. If you also have a luminance stack, taken with a clear filter on a monochrome camera, you can apply this to a color image by pasting on top and changing its blend mode to Luminosity. As we shall see later, the luminosity file carries most of the spatial information and is the one to carefully sharpen and adjust tonally. If you are combining narrowband images with RGB, then the blend mode and the Channels palette setting for that layer allows one to blend the narrowband data with a color channel. Thinking aloud, you should be able to blend narrowband or color images into more than one channel by adding them to the layer stack, selecting the color channel and rolling back their opacity.

## Green Pixels

After the complexities of layers, this is a much simpler concept. Human vision is very sensitive to green yet they rarely exist in astrophotographs. Noisy pixels, that happen to be green, are very apparent and are best replaced by a neutral tone. There is an important distinction here, this is not to say we are reducing the green content in all color pixels, only those where green is dominant. In practice, zoom in to 100% or more, and select representative green pixels with the Select Sampled Color in Affinity (or Color Range in Adobe). Adding an HSL adjustment layer uses the selection as a mask and a careful combination of the three sliders should make them melt away. (Once the mask layer is formed, the marching ants are not needed and obscure the adjustment. Cmd-d makes them disappear.) Adobe Photoshop exercises more control in its Color Range tool. It has three eyedropper tools that allow one to add and subtract from the initial pixel selection. After application, the result of targeting green pixels may cause the background to have a general magenta hue, which requires adjusting back to a more naturalistic color.

## Enhancing Color Saturation

The initial color saturation in an image is rarely sufficient and as each subsequent increase in contrast is applied, the increasing brightness or local contrast reduces what is there. A saturation (HSL) adjustment layer is a starting point but it also accentuates color noise, especially in the darker tones. In the M33 example, it was applied with a luminance mask, that favored the bright tones. This, however, potentially leaves faint nebulosity behind. There are several alternative methods that blend a more saturated and blurred duplicate layer with itself and in so doing, boost color saturation yet keeps chroma noise under control.

One of those methods is a by-product of the Soft Light blending mode. Here, one duplicates the base image twice and changes the blending mode of the top layer to Soft Light and the middle layer to Color. After merging these two new layers apply a small Gaussian blur, of about 0.5–1.0 pixels to disguise chroma noise. When you duplicate a pixel layer it also duplicates any child adjustments or live filter layers associated with it. If these are not required it is a simple matter to select them individually and delete. If you just need a duplicate of the displayed image, simply create a new pixel layer by using the Merge Visible option in the layers menu/palette. If the result is too subtle, merge down the result and repeat, or substitute the Color Burn blending mode for the Soft Light in the above process. Similarly, if the result is too intense, reduce the top layer's opacity.

A third alternative is to use a Vibrance adjustment layer. This has a subtle effect, as it increases the intensity of muted colors more than those that are more saturated. This is especially useful when there is already brightly-colored nebulosity in an image. The built-in controls that prevent it making skin tones becoming unnatural also reduce the impact on red nebulosity. To use, I duplicate the base image (and delete any adjustment hangers-on) and change its blending mode to Color. I then apply a Vibrance adjustment layer to this layer and increase the Vibrance slider. (The Saturation slider has a pronounced effect on all colors.) Again, you may need to drag the adjustment layer over the duplicate image icon to make it a child layer and, as with the aforementioned layer blending processes that use the Color blending mode, apply a small Gaussian blur to the top layer to lessen chroma noise.

A final method uses the unique capabilities of the Lab color space. While RGB color spaces have three channels, red, green and blue, Lab has a Luminance channel and two color-difference channels, a and b. If one edits the color-difference channels it alters the color differentiation. This is done with a Curves adjustment layer after selecting the AOpponent and BOpponent below where it says "Master." As normal, the dialog shows a diagonal line and if we increase the slope of the transfer function, we increase the color differentiation but, to keep the overall color balance, we need to mimic the changes in both channels (fig.1). Fortunately, in Affinity Photo, all the adjustment curves are shown in the dialog at the same time and it is easy to match them. There is room for experimentation here, one gets different results if one simply drags the endpoints inwards (like a Levels adjustment) or if one makes an S-curve.

Editing color in the Lab color space is a very powerful technique but in Affinity Photo, the color mode conversion is confused by all the adjustment layers and masks and restricted to 16-bit files. If I use this technique, I wait until the end when I can safely flatten the layer stack to a pixel layer or merge the visible layers to create a pixel layer on the top of the stack.

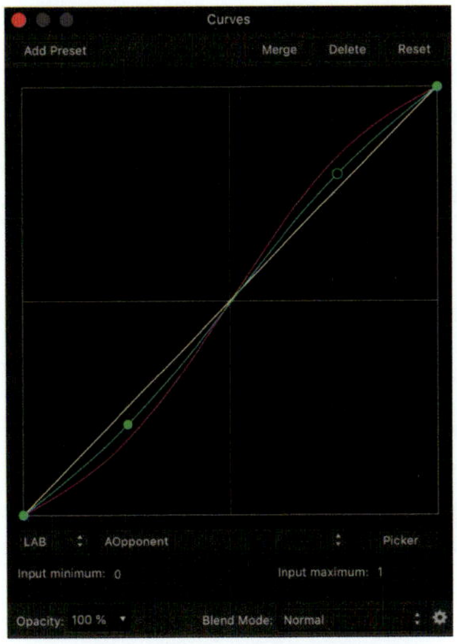


*fig.1* *In Lab mode, increasing the slope of the a and b channels increases the color saturation. Here, these correspond to the green and magenta curves and would ordinarily be perfectly overlapping, to preserve the original color fidelity.*

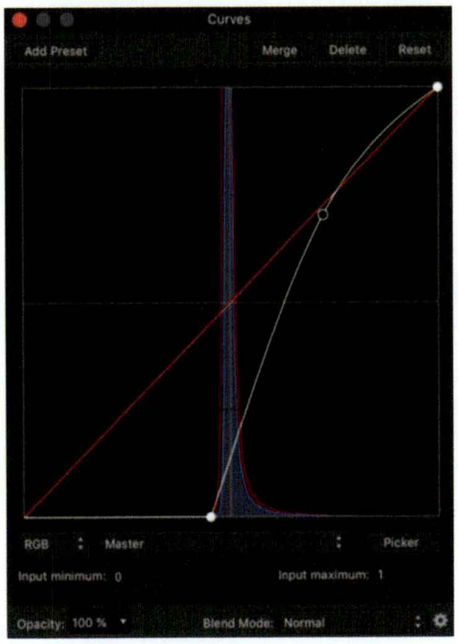


*fig.2* *If you apply a curve similar to the above on the high frequency layer (created by the Frequency Separation tool) it is easier to select most stars with the highlight selection tool. The selection is easily refined and turned into a mask, to be used in multiple places in the image stack.*

## Background Gradients (reprise)

Removing gradients can be difficult when there is a large galaxy in the middle. Yet another method of removing a gradient is to simply neutralize the color, rather than subtract the background. In past examples, we subtracted a blurred duplicate from the image and then added a nominal level. In another variation, we duplicate the image, apply a Dust and Scratches Active Filter of about 100 pixels and invert the layer. It is not looking promising but, if you change the opacity to 50% and change the blend mode to color, all the large scale color (i.e. background color gradient) is neutralized, while keeping the same luminosity. At the same time, the galaxy and stars lose some color, which is restored with an HSL Adjustment layer, with saturation boosted 50%. If the gradient occurs suddenly in the image corners, try the above but with a smaller pixel value in the Dust and Scratches filter layer. It is good practice to carefully evaluate the result; the Dust and Scratches method struggles if the object is large or is surrounded by faint nebulosity and its application partially loses this precious information. In some cases, the aforementioned inversion/color/50% blend method works well to remove the dominant background color of an image but this time by filling the new pixel layer with a plain sampled background level or linear/elliptical gradient. Failing that, the popular Photoshop plug-in, GradientXTerminator, may be your best hope.

## Stars

Stars are tricky little devils that bloat and clip at the slightest provocation, ruthlessly revealing every focus and tracking error. Keeping them under control starts with an exposure that minimizes clipping and continues through image processing. In many manipulations, we somehow have to single them out; either to receive special attention or exclude them. Even so, we may have to resort to more drastic techniques with the really bright ones. Prevention and cure consider making stars smaller, more colorful and less dominant. In all cases, we use techniques to single them out, either on account of their brightness, small size (scale) or both.

### Star Selections

In previous chapters, we have seen how the brighter stars are quickly selected with a highlight selection. That may also select bright nebula or galaxy details too. If you convert the selection to a mask (using New Mask Layer or quick mask after the selection) and Opt-clicking the mask, it is easy to exclude these items by painting over them with black. You can then drag the mask over the adjustment or filter layer that you wish to associate it with. Creating a mask also provides the freedom to modify it further after the event. Affinity Photo is also able to separate an image into two scales, using the Frequency Separation filter. As seen before, this creates two layers, a fuzzy base layer and an odd grey image on top, using the Linear Light blending mode. The real power comes by combining the two methods. Here we duplicate the image and apply the Frequency Separation filter. Discarding the low-frequency layer, we have a high-frequency layer with all the stars, no matter how bright. If we change its blending mode back to Normal and apply an aggressive Curves adjustment, like the one in fig.2, we can create a twinkly starfield with, you guessed it, a Select Highlights selection. In this way, we select most of the stars, irrespective of their original brightness. Again, we can refine the selection according to taste before turning it into a useful mask, to apply at will. Having done so, we can drag or release our star mask and delete the high-frequency layer as it is no longer required.

### Better Star Color

This technique makes use of the tiny bit of color information within each star blob. It starts by selecting the stars, pasting them onto a new pixel layer and blurring them by 1–2 pixels before increasing their color with the HSL tool saturation slider. As we are starting to appreciate, changing the blending mode to Color copies the color information across to the image, without changing the luminosity.

### Natural Looking Stars

Clipped stars are not attractive. They are characterized by hard-edged white dots of varying sizes. In a world where sharpness is over-rated, stars look more natural if they have a softer edge, with a little color. There are a couple of ways of doing this, the most obvious being to select the bright stars with Select>Tonal Range> Select Highlights and refine the selection by slightly enlarging and feathering the selection and applying a Gaussian blur of just under a pixel. (The now-familiar technique that uses the Minimum Blur filter to reduce star sizes also softens their edges.) We can also achieve a subtle effect by making the most of any transition by lessening its contrast with a kink near the top of an otherwise linear Curves adjustment layer.

## De-construction Techniques

Deep-sky images have a tremendous dynamic range across the stars, galaxies and nebulous clouds. There comes a time when even sympathetic image manipulation cannot simultaneously cope with bright and dim object matter, leading to more invasive methods. These techniques are an extension of the star selections above and deconstruct the image into constituent parts, which are then separately manipulated, before being put back together later on. Two such techniques are luminance and starless processing, borrowed from astrophotographers who routinely use monochrome cameras and combine several separately-filtered exposures.

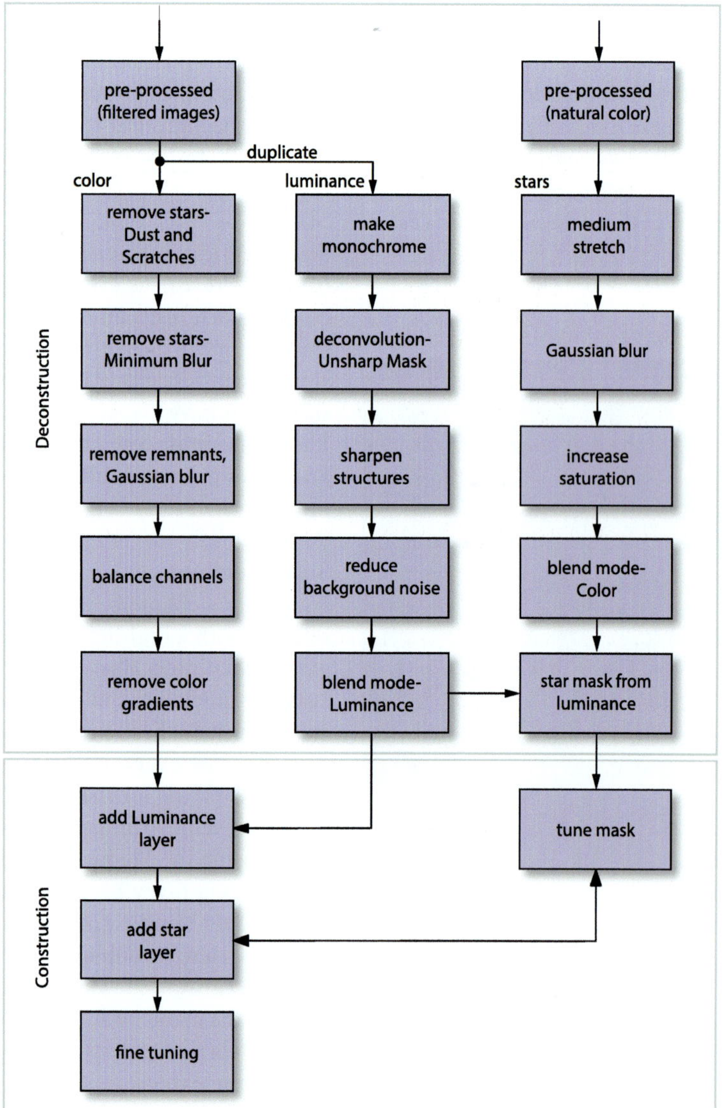


*fig.3 As your technique and equipment progress, it will increasingly be the case that better results are obtained by de-constructing an image into its constituents, separately processing according to their needs and then combining them back again. Here two image stacks, a filtered one to enhance nebula and a natural color one are used together to process color, luminance and stellar data separately and combined with different blend modes.*

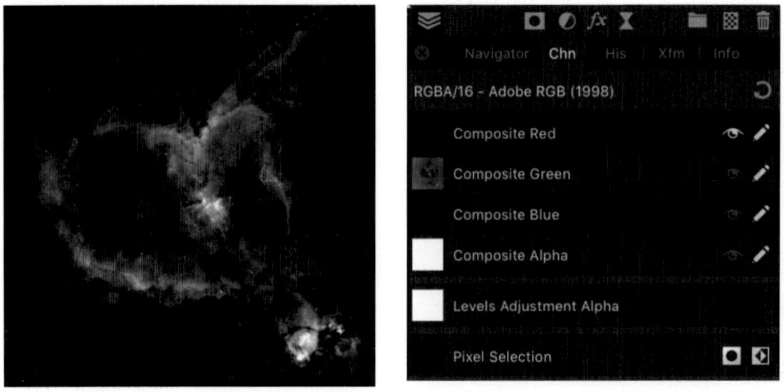


*fig.3 If you change the visibility of the color channels (right), it displays the luminance of the selected channel as a monochrome image (in this case, red). It is much easier to visually balance the relative dominance of the red, green and blue intensities from a monochrome image than trying to do it from color.*

## Starless Processing (tone mapping)

It is difficult to enhance faint nebulae without creating bloated and clipped stars. This is especially true of exposures made with narrowband filters. It is often the case that the oxygen and sulfur emissions are very much dimmer than the ionized hydrogen and to achieve any form of color balance requires some extreme stretching. Apart from bloating stars, extreme stretching puts you on a collision course with image noise. This new technique uses a visual trick that fools the brain by combining a "blurred" color image with a sharpened monochrome (luminance) image. When we combine the two, we do not perceive the indistinct color information. This is a happy coincidence since, by definition, noise reduction reduces pixel to pixel variation and unavoidably softens an image. The luminance image uses data from all the exposures, including the more intense emissions. The stronger signal requires less stretching and has proportionally less noise, making it ideal to provide the "bite" to the color image.

For those who use separate filters for each color and a clear filter for luminance, this process is a natural extension of processing the four image stacks. When using a color camera though, it is not as obvious and while creating and processing a monochrome luminance file is relatively straightforward, the bigger challenge is to make the blurred color image, on account of the wayward imbalance between the R, G & B channels and those pesky stars.

The first step is to remove the stars. We have already seen how the Minimum Blur filter can work on a highlight selection. Removing stars altogether with the Minimum Blur filter needs a little more care, as it can create strange dark holes in the image. Selecting the highlights as before, we need to expand the selection by a few pixels with the Grow / Shrink drop-down menu option. Adding a Minimum Blur active filter layer set at a few pixels reduces or eliminates many stars. What remains are some faint halos and myriad tiny dim stars (unless you are using the more sophisticated star mask mentioned earlier). A second application of the minimum filter may help but it may require other methods; these include using Dust and Scratches and content-aware painting.

Depending on the size of the stars in the image, another tack is to use the Dust and Scratches filter. In the current version of Affinity Photo, it does not work on 32-bit images, but it does on 16-bit ones. Dragging the slider up to 20 pixels or more wipes out the stars but distorts the nebula too. If it is applied with a radius set at a few pixels, however, in conjunction with our Minimum Blur application, it effectively removes many of the tiny mid-tone stars, leaving behind the halos from the larger stars. For these, select the pixel layer and paint over the ghosts with the Inpainting Brush tool, set between 20–50 pixels. This works in the same way as the healing brush in Photoshop, using content-aware fill. The outcome is a colorful low-noise image of the nebula, with no obvious distracting stellar highlights and which can withstand more stretching.

The next step is to equalize the R, G & B channels separately to produce a greater color variation. In dedicated astro editors, these files are separate (or easily separated) and various histogram adjustments equalize them in a moment. In a conventional editor, we make use of the Channels palette. Here, for instance, clicking off the eye icons for Blue, Green and Alpha we can view the red data in monochrome and so on for blue and green. We now use a Curves or Levels adjustment layer on modifying the separate colors rather than all of them at the same time (with Master selected). Using monochrome images makes it easier to match them visually (fig.4). In practice, it will usually require some strong Gamma or steep curves on the blue and green channels, to match the red channel intensity, at the same time trimming the black and white levels to suit. At the same time, it will, of course, exaggerate any background gradients. The outcome, with all the eye icons checked, is a colorful, soft image that has more emphasis on the fainter emission colors.

Adding back the luminance can be accomplished in several ways. Assuming one has already stretched the image, one could simply duplicate the base pixel layer, which places a duplicate of the layer (and its child layers) immediately above. Drag this group to the top of the stack, delete its child layer adjustment and filters, de-saturate it, shrink the stars slightly and sharpen. With its blending mode at the default Normal, the result is a detailed monochrome image with some stars. The magic occurs by changing the blending mode to Luminosity, which applies the brightness information from this image to the color image below.

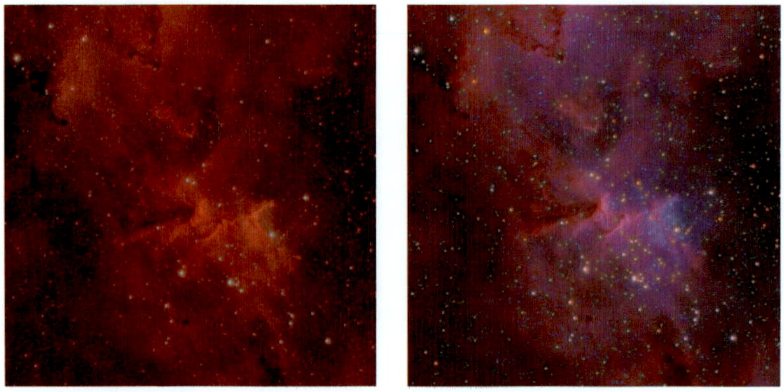

*fig.4 After using the techniques in the text with the image from the last assignment, a starless image had its color channels more closely matched and then combined back with a luminance and stellar layer. The result shows the subtle hue changes around the central feature and more varied star colors.*

This may produce the desired result with a color camera. With narrowband filtering, however, it is often the case that the stars have an odd color. Strong, unusual colors like magenta and cyan can be selected and neutralized with a sampled color selection and a hue/saturation/lightness adjustment. Another solution is to have some additional exposures to hand with natural-colored stars and process these to form a dull image with mid-tone but colorful stars. This image is copy pasted as a pixel layer above the monochrome image, immediately after a highlight selection (or by applying a star mask). The star color is then substituted into the stack by changing its blending mode to Color.

A typical result is shown in fig.4, using the data from the previous practical assignment. After star removal, the channel intensities were equalized. The intensity of the weak blue OIII emissions is now more noticeable around the central dust column feature and with a natural color star substitution, the stars have a wider range of color. It may not be entirely truthful, but these images are for visual appeal.

## Noise Reduction and Sharpening

We have already tried out some techniques for both and found simple ways to restrict their application, typically sharpening where the signal is strong and noise reduction where it is weak. There are several sharpening and noise reduction tools incorporated into Photoshop and Affinity Photo and the best way to compare them is to optimize each in a separate layer and then disable/enable these layers to quickly compare the results at 100% magnification. On top of these, there are also several third-party plug-ins, that expand the repertoire.

In this last section on processing, we can take these filters a little further, consciously deciding on whether we apply it to color and/or luminance information. We already have created the opportunity for just that with the deconstruction method described earlier. For instance, neither the background nor the color information require high resolution and both will tolerate strong noise reduction. Conversely, the luminance information conveys detail and structure and it should be singularly targeted by any sharpening process. It is interesting to compare the results of sharpening a normal color image with a similar action that is only applied to the luminance information. In the case of a normal color image, its color noise is also accentuated. Noise reduction inevitably destroys details and used indiscriminately, blurs star profiles, though some astro noise-reduction tools detect and protect stars. Each color channel usually has a different noise level too and it is useful to examine each using the arrangement in fig.3 and apply different levels as required.

Noise reduction and sharpening affect local contrast and it is sometimes easy to overlook the obvious; the Curves adjustment tool can equally differentially change tonal contrast. For instance, an S-curve lowers shadow contrast, detail and noise, accentuates details in the mid-tones and lowers the contrast in the extreme highlights. In the same way, a saturation reduction reduces color noise at the same time and is useful when applied selectively to image backgrounds. Every image is different and when it comes to image processing, the only limitation is usually our imagination, time, patience and willingness to experiment.

# Automating Operation

*The convenience of automatic acquisition and remote control.*

Automation is an advanced topic, though we have already encountered simple forms in some of the practical assignments. Once things are stable, automation (and control from somewhere warm and mosquito free) add a welcome convenience that makes long-duration sequences a practical and sociable reality. This short chapter introduces the concepts and potential opportunities. A more thorough and detailed analysis is covered in *The Astrophotography Manual,* including detailed setup instructions for the PC and network configuration.

Apart from convenience, automation also has the potential to make acquisition processes more reliable through consistency. Before automating, however, it is important to be familiar with and ensure the individual processes work reliably and without constant attention (e.g. focusing, filter changes, centering, guiding, etc.). If not, automation just makes any problems and user interventions occur more frequently! Stepping back, automation features cover a wide span and are broadly classified into several areas:

- target planning
- acquisition (including observatory control)
- remote control

### Target Planning

For deep-sky imaging, target planning is more of an assist than true automation. External planning utilities have the ability to selectively choose objects from a catalog, based on type, visibility and so on and export details into a text file format. Although an object's RA/DEC coordinates are a useful starting point, this may not be the optimum position for image composition. Some planning tools link with planetariums and project an outline of the sensor's field of view onto a sky view, that helps the user determine the optimum center, focal length and angle for the composition. An example of an alternative approach is the mosaic planning tool in SGP. This overlays the camera field of view over a Deep Sky Survey (DSS) image of the target area to plan tiled images and then populates a multiple target sequence with the center coordinates of each tile. I also use it to plan simple single-tile image compositions. Within the sequence table, the target coordinates allow SGP to automatically calculate the start and end of an exposure sequence to ensure the target exceeds your minimum altitude requirement.

### Acquisition

With the target(s) defined, image acquisition takes over. Simple sequencing is the backbone of automation and, like an intervalometer, triggers a series of exposures. Most of the acquisition applications have additional options such as filter selection, exposure length, binning, count, order and type. From here, to meet the requirements of extended sequences in real conditions, things become more sophisticated.

Other automated features include sequence scheduling, target acquisition and centering, triggered autofocusing, dithering, autoguiding, meridian flips, and monitoring. Very soon, what started off as a basic set of exposures becomes an interaction of multiple device controls and follows a prescribed logic. Even in a simple setup, SGP will do the following, automatically:

- connects to equipment and waits for the start time
- cools camera to required temperature
- slews mount to first active target at start time
- centers, autofocuses and calibrates guider
- starts imaging when the camera is at temperature
- changes filters and adjusts the focus position
- autofocuses if ambient temperature changes
- manages autoguider software (e.g. PHD2)
- manages meridian flips, with automatic centering
- ends or aborts sequence, parks mount and notifies of poor conditions

In the case of changing weather conditions, unattended operation additionally requires monitoring the environment and taking appropriate action. For example, SGP enters a recovery mode if the guider system has difficulty finding a star or settling. The recovery mode attempts to restart the guider and image acquisition several times, over a specified period. Similarly, if the autoguider reports a large tracking error during an exposure, the exposure is restarted. If, at the end of a sequence, or if the ASCOM safety monitor indicates poor weather and, depending on the selected options, SGP parks the mount, closes the observatory roof and warms the camera to ambient, before disconnecting the hardware. Some imaging applications allow access to some of their subroutines or methods which permits further customization using short programs called scripts. The market for acquisition programs has moved on greatly in the last ten years. There are several newcomers, including NINA (freeware) and Prism ($99-$450) which are gaining converts. One or another will find favor on account of being particularly good for a particular specialism.

### Remote Control

Extended imaging requires perseverance, especially when the night is cold and damp. After a year, once the novelty had worn off, I investigated operating from indoors. It is a well-trodden path and first attempts invariably consider simple USB extenders, allowing you and the computer to reside in more comfortable surroundings. Implementations vary considerably and the forums indicate many extenders are unreliable. I initially used a USB extender recommended by several mount manufacturers. This system uses two modules; a powered USB hub and a transmitter plugged into a PC USB socket. The two are linked by up to 300 foot of Cat 5 cable. This was reliable for using at home but at public events, the coiled cable was a trip hazard. I eventually replaced the hub with a mini computer located at the mount and accessed it remotely over WiFi from a PC, Mac or iPad.

This is not as extravagant as it sounds and has a lot going for it. I do not like using a laptop out in the open; they are too fragile, have limited battery life and require a table and chair to operate. A practical alternative is to use a tiny stick or brick PC, loaded with Windows and the essential applications (or one of the dedicated modules that operate over WiFi). These computers are tiny enough to site close to (or on) the mount and are powered from a 5- or 12-volt DC supply. They easily cope with the modest computing demands and store images on a memory card. In operation, they do not require a monitor or keyboard and are operated remotely. The remote operation is made possible by free applications like Microsoft Remote Desktop (MRD) or TeamViewer. MRD makes use of the built-in remote control protocol that is included with Windows Professional, Enterprise and Ultimate. This operates over a wireless or wired Ethernet connection and permits the imaging computer to be entirely controlled by a remote tablet, Mac or PC connected to the same router and running the free application. If you know your router's IP address and allow access through your firewall you can even control your system from your local WiFi-equipped bar! In the case of a dark field site, remote operation is still possible, by making the imaging PC (or a tiny portable router) set up as an ad hoc or wireless access point.

Full remote control requires further hardware, specifically power switching and reset capabilities for all the hardware. This makes things interesting; the proliferation of Internet and IP-controlled devices enable those, frustrated by light pollution or poor weather, to remotely control imaging systems sited in other countries and significantly for some, in the other hemisphere. For some amateurs, who do not want to invest in any equipment, this is also the means by which some enterprises give access to remotely-operated telescope-for-hire facilities.

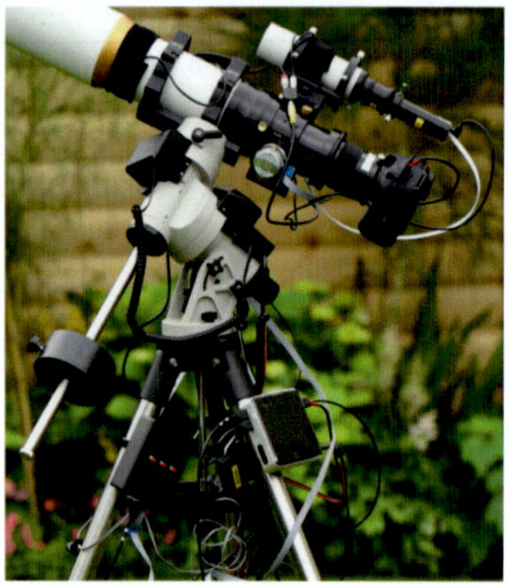

*fig.1 This iOptron mount is set up for automated acquisition. The four USB ports of an Intel NUC computer are attached to a focus module, DSLR, guide and the mount (via a USB-serial adaptor). Two small 12-volt lead-acid batteries supply power for a full night's imaging. The entire setup is controlled remotely over WiFi from an old iPad.*

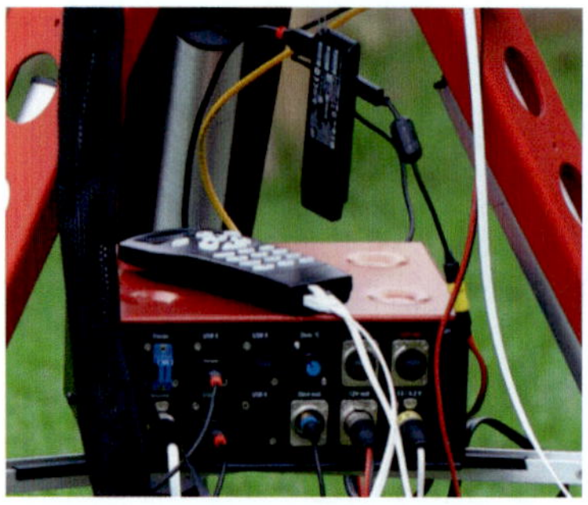

*fig.2 The tangled mess of leads and modules in fig.1 is less than ideal. Here, the USB hub, adaptors, DC converters and focus module are fitted in a box with clearly labelled latching connectors. A consistent hookup helps with automation, as there are no surprise prompts for device drivers on power-up and the ASCOM COM port settings do not need changing each night. In this case, an Intel Stick computer is being used. The box construction is a practical project in* The Astrophotography Manual.

# Appendices

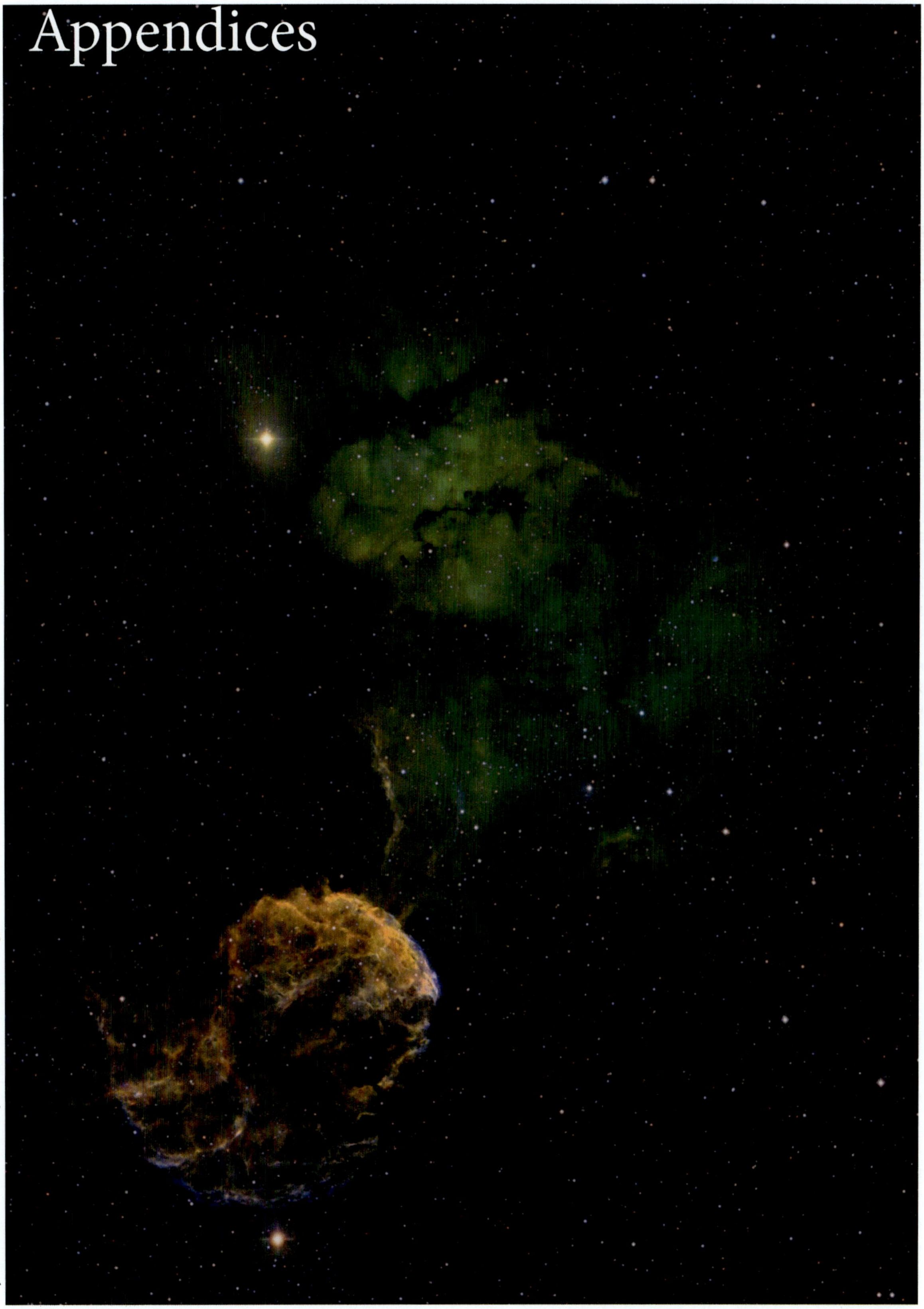

*Jellyfish Nebula (extended-exposure narrowband image)*

# Deep-Sky Lucky Imaging

*A practical experiment to evaluate the application of an imaging technique, commonly used for bright planetary imaging, on dim deep-sky objects.*

Lucky imaging is a technique that has been around for many years and is a commonly used method for high-magnification planetary and lunar imaging. These bright objects allow enterprising astrophotographers to image capture with a modified webcam or with more outlay, a dedicated high-speed color and monochrome camera, usually recording in a video format. While most of the frames are ruined from astronomical seeing, a few "lucky" ones escape, with excellent resolution. The image processing ruthlessly discards the poor frames and just aligns and combines the lucky few. With these bright objects and good conditions, image capture is completed in several minutes.

In more recent years some astrophotographers have been experimenting with this technique on much dimmer deep-sky objects. Apart from being driven by the general inventiveness of some, this novel application is increasingly plausible due to the availability of larger, faster and less noisy CMOS sensors as well as the ever-increasing ability to economically store and process large amounts of data. It may seem strange to some, however, to explore a pioneering technique in a book aimed at the beginner. This is not as wayward as one might think, as the technique has the potential to simplify the image acquisition and use less expensive equipment. The reason for this is that the typically brief exposures conveniently bypass the expense and frustration of perfect tracking and yet produce a result that has the potential for higher resolution than is possible with traditional imaging techniques.

### Feeling Lucky?

This experiment aims to eliminate the resolution-robbing effects of astronomical seeing and poor tracking and improve the image resolution to approach that of the imaging system (a little worse than the optic's diffraction limit). It is a beguilingly simple technique. In general, one takes many short exposures and then ruthlessly discards most of them, keeping only the very best. The idea is, if you take enough exposures, sooner or later you will get lucky and capture a good one. These favored few are then calibrated, registered and averaged, in a similar manner to that used in conventional deep-sky astrophotography.

### Pros

The operational settings that lucky imagers deploy additionally have other benefits, that improve image resolution, on account of using short exposures. We have not yet defined what constitutes a "short exposure" but suffice to say that it is less than a few seconds. In that time, any telescope tracking issues, from periodic gear, drift, flexure or autoguiding errors are irrelevant in all but a few extreme cases. The short exposures lessen thermally generated noise too, though the electronics do warm up appreciably on account of the constant activity. (For example, if you take one of the helmet-mounted video cameras and record for 10 minutes at a low frame rate and then at its highest, the camera runs appreciably hotter.) This then allows for some very simple setups, using a polar-aligned mount, camera and optical system. It is, however, more critical than ever to ensure the focus is precise, as poor focus will not be disguised by astronomical seeing. Of course, there is no such thing as a free lunch and lucky imaging has a few obvious drawbacks.

### Cons

The laws of noise reduction, from increasing net exposure, still apply to lucky imaging. When you consider that any astronomical target has a finite imaging window for any given night or season, the underlying concept of discarding most of the captured exposures obviously reduces the potential cumulative exposure. The brief exposures are a challenge in themselves too. Clearly, they place a demand on USB speed, storage and processing capacity. Less obvious is the effectiveness of any individual exposure. When one considers the case of a galaxy, with faint spiral arms imaged with a popular CCD sensor, the ADU value is only 100 units (in 64,000) over the background level in a 300-second exposure, equivalent to about 82 incident photons. That is equivalent to just 1 useful photon, on average, every 3.5 seconds. It is easy to predict in the case of any 1-second exposure, that it is more likely than not it will not receive a useful photon for that part of the image!

When one considers the various noise contributions during these brief exposures, dark noise and image shot noise are significantly reduced but sensor read noise remains and dominates the frame. The stretched single exposures in fig.1 and fig.2 give an indication of the difference between a 60- and 1-second frame. While this is not an issue for the much brighter solar system objects, in the case of dim deep-sky objects, the physics is undeniable and the application of lucky imaging is more gainfully deployed on brighter objects, typically star fields, loose and globular clusters and a handful of very bright galaxies and nebula features.

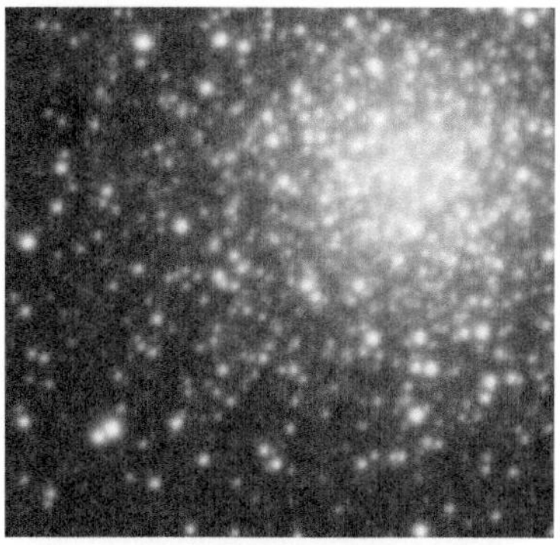

*fig.1 A single 60-second exposure taken at a low gain setting (0 dB). The image has been stretched and shows an extensive star-field with a lightly mottled background.*

## Setting Up

This chapter is an open-ended experiment to evaluate the potential of lucky imaging on a globular cluster, as it rose above my imaging horizon and with different exposure durations. At low altitude, seeing conditions are at their worst and an effective test of the technique. A globular cluster is an ideal target; the stars are relatively bright and the core definition will ruthlessly reveal poor technique. My normal CCD imager has a mechanical shutter that limits the minimum exposure time. It is also too big and slow, so I re-purposed a ZWO CMOS guide camera. It has 1936x1096 2.9 μ pixels and a rolling shutter. Like many CMOS astro cameras, it also has a variable gain setting that changes the amplification of the analog voltage before the ADC converter. If required, the USB 2.0 interface allows full resolution at 20 frames per second (FPS), though only with 10-bit resolution. This is a limit of the USB 2.0 interface and its USB 3.0 cousin is faster.

The purpose of the evaluation is to see the resolution improvement from using lucky imaging. As all the resolution-robbing effects combine, any benefit is more distinct if the prevailing seeing conditions are similar or worse than the optical resolution of the system. My 73-mm short refractor, used in the earlier assignments is not suitable, due to its small aperture and focal length. In the end, I used a 132-mm aperture refractor, with a 924-mm focal length. It has a diffraction limit of ~1 arc second and gives a 0.6 arc second/pixel angular resolution with the ZWO camera. With this, one should be able to compare short and long exposures and detect differences in their star appearance. In this case, the imaging setup was mounted on a Paramount MX, running ProTrack. This professional mount is an obvious overkill and tracks very well without autoguiding. In doing so, it effectively eliminates tracking errors as a variable in the experiment.

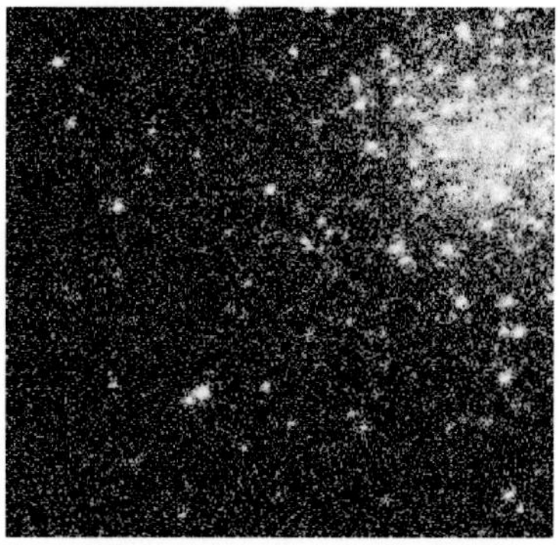

*fig.2 A single 0.5-second exposure taken at a high gain setting (40 dB) and similarly stretched as fig.1. The read noise (in $e^-$) at this high-gain setting is a third of that of that gain used for the 60-second image in fig.1. Even so, the read noise is more obvious, relative to the signal level and the fainter stars are harder to make out. Resolution is clearly better, however, both in the core and also in the case of the several double-stars in the periphery.*
*It is highly likely that, in this single exposure, there were no incident photons from the fainter stars. The probability evens out when one combines 120 of them.*

## Acquisition

The image sequence was captured in Sequence Generator Pro (SGP). Without knowing what would be effective, the sequence was set up to capture a range of exposure durations: 60, 30, 10, 1, 0.5 and 0.1-second events. With an effective exposure range of 600:1, the output was similarly scaled by changing the gain setting for each event. SGP allows the ZWO gain setting to be adjusted for each sequence event. Before I realized the gain setting was in 0.1 dB steps, I estimated the gain settings by comparing the peak intensities of the brightest stars for each exposure time, experimenting with different values, until they were roughly similar. After reading the camera's datasheet, I realized the camera has a gain range of approximately 50–0 dB, limiting the sensor gain range to about 300:1, shy of the required 600:1. As a result, the 0.1-second exposures have lower peak ADU values. With short exposures and a small sensor, there is no compelling need for dark or flat calibration files. Lucky imaging, however, involves averaging

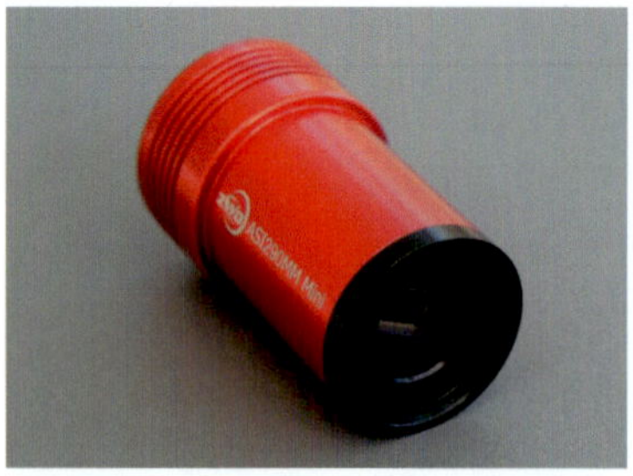


*fig.3* This 2.1 MP CMOS guide camera has a USB 2.0 interface and usefully has small 2.9 μ pixels for this experiment. It has relatively quick downloads and output video and still images.

thousands of files that will surely reveal any sensor pattern noise. For that reason, I added 50 bias frames at each of the event's gain settings to the sequence. Lastly, to make a series of comparable images, requires an increasing event counter for the shrinking exposures, so that the total exposure time is broadly equivalent. Some guesswork is required here, as it is not possible to predict how many of the short exposures will be rejected. Suffice to say that the 1,200 or so 16-bit images occupied 8 GB of drive space, and that was before I created calibrated versions of each, in 32-bit! The acquisition span was brief, spanning just a few hours. I let the system acclimatize to the ambient conditions for a few hours and then centered up the cluster as it hit 35 ° altitude and ran the autofocus routine a few times. To ensure optimum results, I paused the sequence every 15 minutes to run the autofocus routine. In practice, there is a time penalty with short exposures, as each takes a second or two to download and store before the next frame is kicked off. At the end of the imaging sequence, I replaced the lens cap and ran the bias-frame exposures for each gain setting.

## Image Processing and Analysis

The aim of this experiment was not to make a perfect image but to evaluate the effectiveness of lucky imaging and to answer some basic questions like, "what is the most effective exposure time?" and "is it worth the effort?". A such, it was important to set up the evaluation so that the pictorial and statistical performance of each exposure plan could be compared with ease. To that end, pre-processing was kept simple; constructing and subtracting a master bias frame from each of the corresponding light frames of the same gain setting. The calibrated frames were then analyzed both with SGP's free FITS image grader utility and PixInsight's Sub-Frame Selector tool to measure the star sizes in each image, noting the mean, best and spread of results for any particular exposure plan. Both utilities provide an itemized list of each file and PixInsight additionally plots the results graphically. With these, it is possible to compare the general performance of each exposure plan and its variability (fig.6 and fig.7).

From each exposure set, a selection of the best frames were registered and integrated into a single image file. These were normalized to each

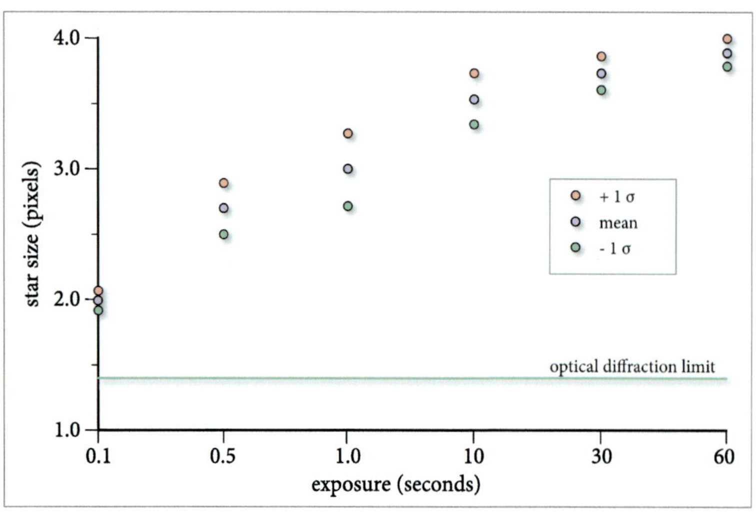


*fig.4* This illustrates the effect of increasing exposure on star size. The mean star size expands as the exposure lengthens. Interestingly, the distribution of star sizes, indicated by the ±1 σ points initially increases and then decreases. The 1-second exposure shows the most variability but the good news is that the shorter exposures have a higher proportion of keepers. Each sensor pixel subtends 0.64 arc seconds and the diffraction-limited size is shown on the graph.

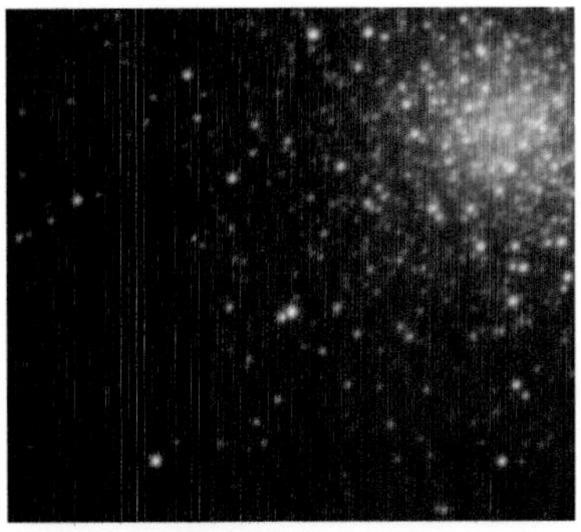

*fig.5a 1x 60-second*

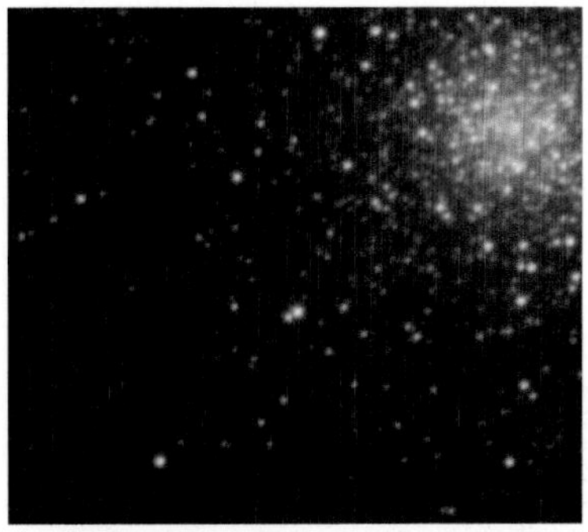

*fig.5b 2x 30-seconds*

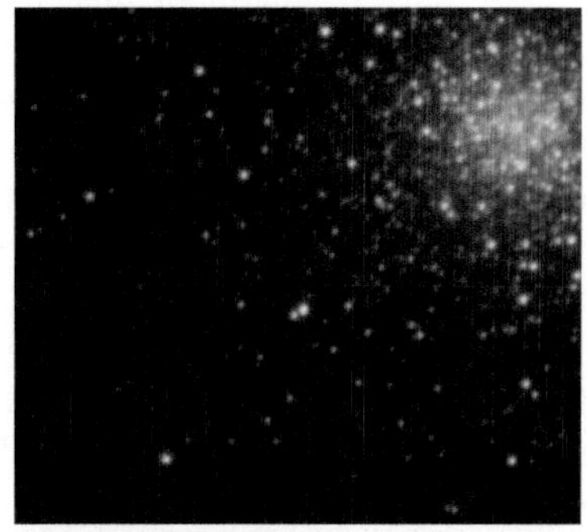

*fig.5c 6 x 10-seconds*

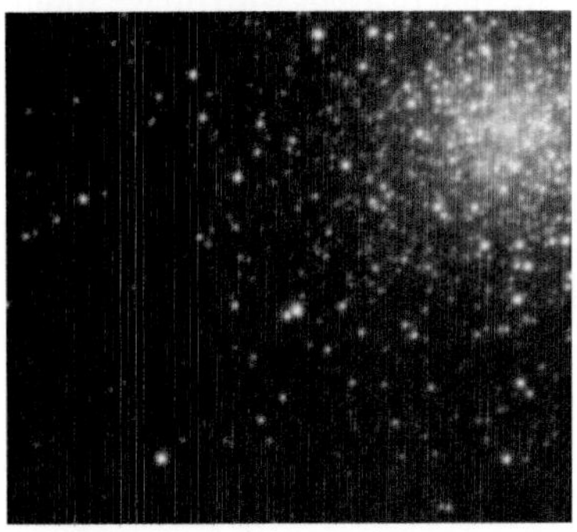

*fig.5d 60 x 1-second*

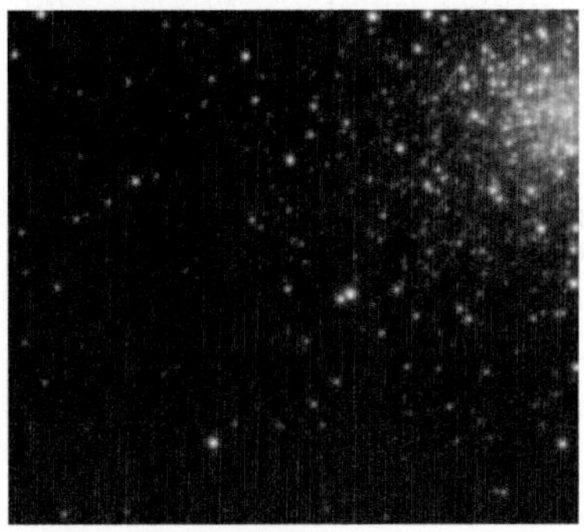

*fig.5e 120 x 0.5-seconds*

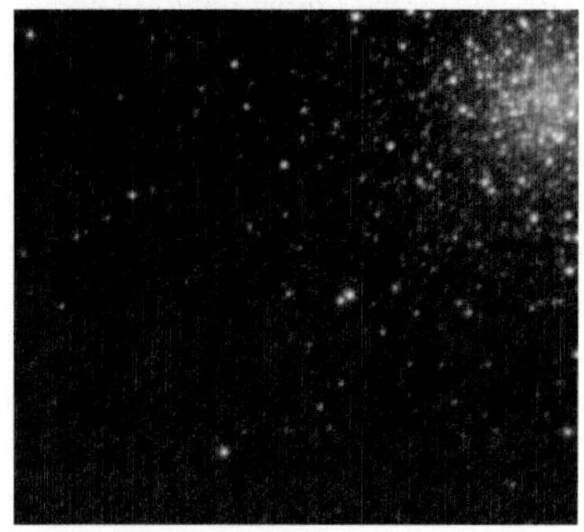

*fig.5f 600 x 0.1-seconds*

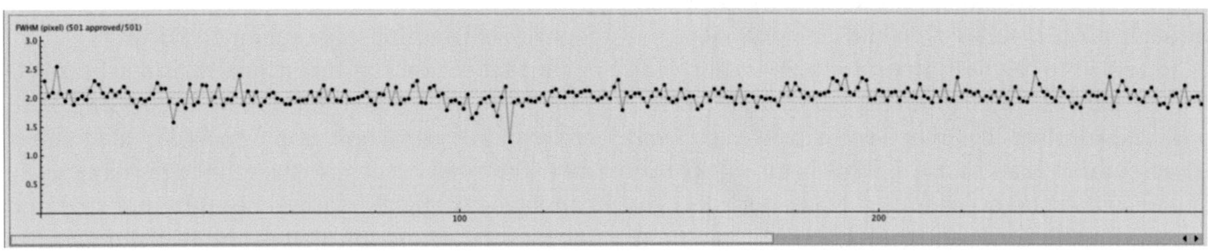


*fig.6 This is a record of the star sizes in each exposure. Contrary to expectations, the 0.1-second exposures show remarkably consistent star FWHM values to the extent that, in this case, one might safely bypass any rejection process prior to averaging.*

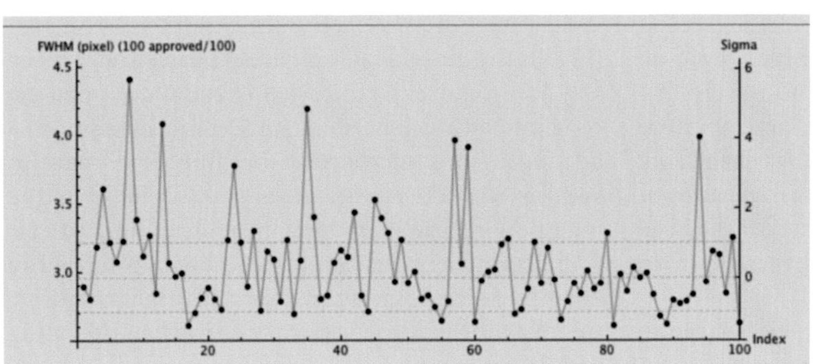


*fig.7 Interestingly, some of the 1-second exposures show clear bloated star outliers that should be rejected prior to calibration, registration and averaging. The comparison with the 0.1-second exposures suggests that astronomical seeing is not blurring per se, but a rapid image shift. The comparison suggests that the atmosphere is not altering slowly over multiple seconds but rapidly in a low single-digit second time frame.*

other (using PixInsight's LinearFit tool) to account for the slightly different effective exposures and then stretched by the same amount for closer visual examination (fig.5a–f).

## Conclusions

In truth, I did not know what to expect and the results from this simple test confirmed some general principals and provided a few surprises too. Seeing conditions are not constant and may change their characteristics with region, season or conditions. With that in mind, any tentative conclusions may not apply to another time and place. I typically use 5–15 minute exposures and assuming that the 60-second unguided result is similar to a guided longer exposure, the results in fig.4 indicate a 50% improvement in effective resolution with 0.1-second exposures. That improvement comes at a price, however, as a 1-minute duration sequence occupies 12 GB of file storage space (image + calibrated + registered image). A 10-hour sequence, as I normally would require for a galaxy or cluster, would occupy 7.6 TB, and this is with a small 2.1 MP sensor too! The mean star width of 2 pixels (1.3 arc seconds) is still a little way off the theoretical optical diffraction limit of 0.86 arc seconds but when you take the sensor resolution into account, is close to the system resolution of 1.1 arc seconds, a brilliant result for light that is passing through about 20 miles of polluted atmosphere.

In regard to the best exposure time, in this case, the improvement occurs with exposures under 10-seconds and a further 30% improvement between 0.5 and 0.1-seconds. Human nature, being what it is, one may choose a 1-second exposure out of convenience. The benefit is marginal, however, as although it has a 25% improvement over the long-duration exposures, in this experiment, it has also the widest range of results, requiring an extensive rejection process and still over 700 GB of data.

In this experiment, the CMOS guide camera was an excellent choice with low read noise (around 1 $e^-$) and high gain. Even so, the images with 0.5- and 0.1- second exposures are on the limit of its sensitivity, even on this relatively bright object. It is worth noting too that this evaluation used a long focal-length, large-aperture instrument and a small-pixel camera. Any benefit from lucky imaging diminishes with smaller aperture or focal length telescopes and to some extent if the pixel scale, in arc seconds/pixel, increases appreciably (as in the case of undersampling).

The variability of the 1-second exposure frames also has interesting implications for the mainstream astrophotographer. It is often the case we use 1-second guider exposures out of convenience. In these conditions, the results suggest that for an autoguider to be effective, it is better to use exposures exceeding 1 second, to span several random excursions. In the case of the 0.1-second exposures, even though the stars are consistently smaller, they are shifting between frames due to seeing. As we have seen before, a classical autoguider typically uses some form of low-pass filtering to discriminate the legitimate tracking error from the astronomical seeing. Some amateurs and many observatories do use very short guide exposures with fast-moving adaptive optics (a similar technology to that in a lens with image stabilization) to counteract the effect of seeing in real time.

## Future State

Apart from a few special cases, as it stands, this novel technique has some way to go before it can be used for conventional deep-sky astrophotography. It has promise though and there are a few excellent examples on the Internet. At the end of the day, however, you cannot defeat physics. While lucky imaging undoubtedly can improve image resolution, to defeat the noise mechanisms from the sensor and light itself, it still needs prolonged exposure integration times and prodigious amounts of data to perform deep-sky imaging. It has promise though, especially if one can overcome the logistics of handling large amounts of data.

In that regard, by intentionally using the ZWO camera in still-image mode and processing with conventional astrophotography tools, I created a lot of data. The pleasant surprise though was the remarkable consistency of the 0.1-second exposure star sizes and the possibility of eliminating the rejection process from the image processing workstream. This simplifies things and if one could use the camera's video mode and have an application that could acquire, calibrate, register and accumulate each frame in real time, the data storage issue would disappear.

This requirement is not too dissimilar to the live-stack feature used in electronically assisted astronomy (EAA). The individual tools are available and it probably just requires a concerted effort to optimize the code into an efficient image-processing pipeline on a fast multi-core microprocessor. This would have to keep up with the frame rate and slowly build up the image. With the introduction of faster and quieter CMOS cameras, it may only be a few years before lucky imaging becomes an automated reality that is entirely suitable for the novice astrophotographer, operating with lesser means.

# Cloudy Projects

*Even in today's product-rich society, it is still possible to find practical projects to add another dimension to this fascinating hobby.*

Astrophotography is not one hobby but many. For those with a practical nature, the summer is a great time to process images, organize things and develop various widgets that make life easier and more reliable. *The Astrophotography Manual* has several practical projects that cover a wide range of practical abilities. These are detailed on the website and are largely self-explanatory. In addition to some of my earlier gizmos for use with non-robotic single-axis mounts, here are some more that are suited for those starting out and keeping to a budget.

## Simplified Geared Head

In the early chapters, I bought a used Manfrotto 410 geared head, rather than one of the proprietary Alt/Az wedges, at a third of the cost. This model has three geared axes, ideal for minute adjustments required for polar alignment but, in position, it is a little ungainly with the off-center mass. One only needs a pan and tilt adjustment for aligning a mount and I found a post on the Internet that suggested the unit can be modified for just that purpose.

The modification invalidates any warranty, but since this was a used unit, there was nothing to lose. The whole process took 10 minutes and required a hair-dryer, a small screwdriver and a 4-mm Allen wrench. In doing so, it reduced the mass of the unit from 1.25 kg to about 900 g and improved the overall stability too. The head employs three identical gear system assemblies and the modification eliminates the middle one, joining the base plate directly to the top plate to create a pan and tilt head.

### Disassembly

This involves removing the circular logo panels that cover an Allen bolt and undoing the bolt on the tilt and level axes (fig.1, fig.2). To remove the soft plastic logo, I softened the glue by warming it with a hair-dryer and, with the aid of the screwdriver, I lifted an edge and gently peeled it off. I then loosened the exposed bolt so that the parts separated. After repeating for the other axis, I had three assemblies. A thin plastic washer, acting as a bearing surface should remain stuck to the grease and the gear should remain in place so long as you do not play with the mechanism.

### Assembly

It is now a simple case to put the middle arm to one side and assemble the other two together and tighten the bolt. Check the mechanism works fine and then warm the glue and put the circular logo back, so that the zero mark is aligned with the dot when the top plate is level. The scale on the tilt mechanism is designed for photography and for a telescope mount, where the scale is set to the user's latitude, its 0°and 90° positions are round the wrong way. For a latitude of 51 degrees, the approximate scale setting is 90–51° or 39°.

The result is neater (fig.3) and although there is a little gear play, this is eliminated by rotating the quick release ring the other way, to lock the mechanism tight. This pan and tilt conversion is reversible and it is a simple matter to re-assemble the three components to form a 3-way head once more. Polar alignment is a lot easier with this than with a conventional tripod head.

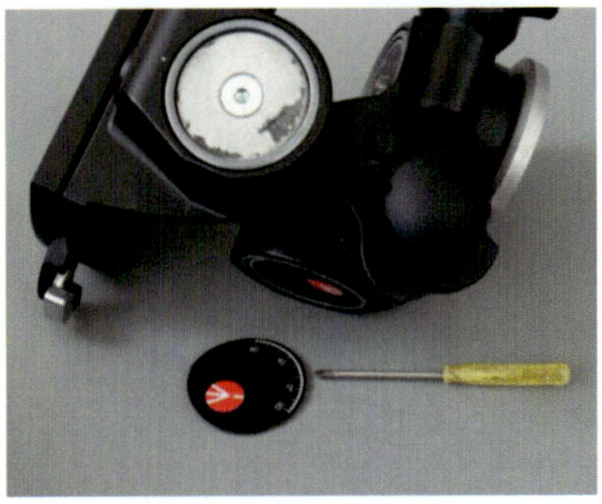

*fig.1 Gently heat the plastic logos and remove with a small screwdriver.*

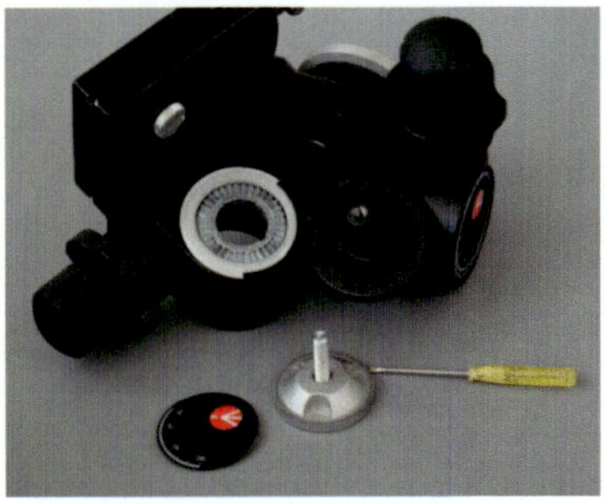

*fig.2 The three disassembled components, ready for their new role.*

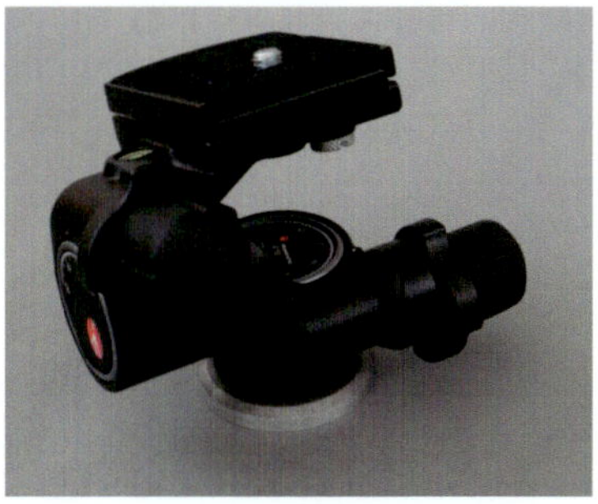

*fig.3 The re-assembled unit, with just two adjustments, is more stable.*

### Quick-release Plate Improvements

Many quick-release plates have rubber or cork mats to prevent scratching a camera baseplate. This compliance causes the mount to rock and, to improve stability further, I prised the rubber off a spare Manfrotto camera plate to reveal a metal platform with a raised edge. I filed the edge down and finished it off with some black paint. The metal to metal coupling to the telescope mount baseplate is a useful upgrade to this or any other camera plate with a cork or rubber pad (fig.4).

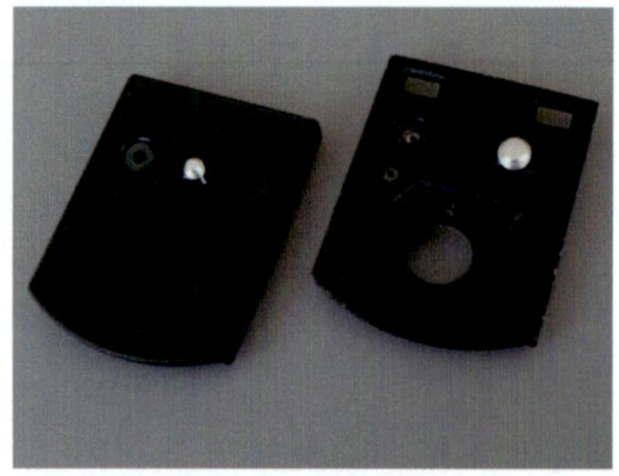

*fig.4 The original camera plate and a modified version, with a smooth metal surface.*

## ASCOM Camera Trigger

In the case of a single-axis mount an intervalometer is usually sufficient to control any camera. Once the mount becomes more intelligent and uses computer control for focusing, guiding, alignment, slewing, flipping and so on, it is useful for the camera to be remotely operated too. For those of you who use DSLRs for imaging, Canon and Nikon have enjoyed the benefit of sharing their software interface (through a published interface specification) with astronomy software developers, enabling full control over USB. Not all manufacturers are so transparent and for those cameras that have an external electronic release, it occurred to write a general ASCOM driver that allows simple camera exposure control with virtually any imaging application. The concept was inspired by the ASCOM camera simulator, bundled with the ASCOM platform, with the exception that there are no settings as such and an exposure command opens and closes a relay. The ASCOM driver installer is available for download from the book's website; what remains is a little work to connect the cable and house the unit in a small enclosure (fig.5). There are several available USB-controlled relays and although their operation is similar, their simple command protocols are not identical. It is usually not difficult to adapt the source code.

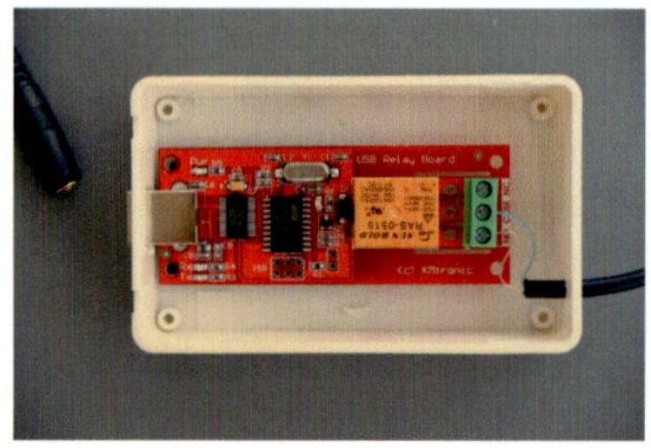

*fig.5 The KMTronic USB relay board mounted in a plastic box. The lead, terminated in a 2.5-mm female audio socket, is compression fitted through the box wall, to secure it.*

### Parts

I use several USB-controlled relays in my observatory made by KMTronic in Bulgaria. For this project, I chose their simplest one-channel model costing about $30. My Canon and Fuji cameras both have a remote release socket using the familiar 2.5-mm audio jack. The connections are shown in fig.6. To make up the connection, I bought a 1 m 2.5 mm audio extension cable (with a male and female connector). Most of my Fuji cameras and the EOS 60Da have a 2.5 mm remote control socket. For this, snip off the female connector and identify the wires connected to the plug's tip and base (1 and 3 in fig.6) and attach these to the relay terminals "COM" and "NO" (normally open). To keep connectivity options open and as my older, modified Fuji X-M1 uses its micro USB connector for remote control, I cut off the male connector instead and attached the cable with the female connector to the relay board. In this configuration, the output is a 2.5 mm socket. This is the same as my handheld intervalometer and accepts its adaptor cables, which include a 2.5 mm male-to-male lead and a micro-USB connector with a resistor dongle. The relay terminals, in this case, are connected to two USB pins via resistors. In practice, although it is possible to cannibalize a micro-USB cable, I found that not all cables, especially micro-USB charger cables, have pin 4 connected. Other connector styles are possible, including Sony and the older Canon and Nikon interfaces. The website www.doc-diy.net usefully lists several common interfaces. Some of the camera connectors

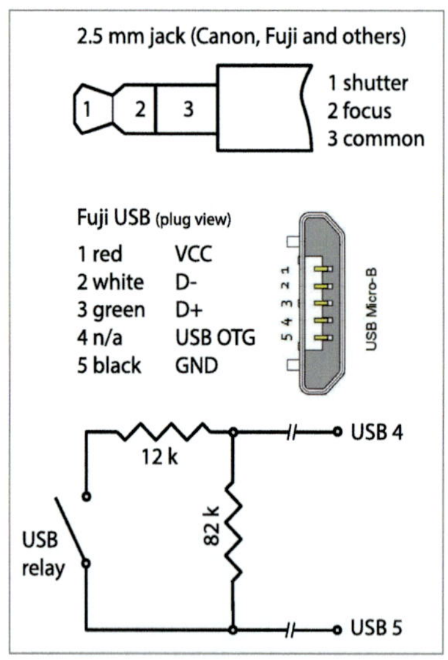


*fig.6 Two lead configurations; the jack works with most Fuji and some Canon bodies. The micro-USB connector works with other bodies. An Internet search reveals many more, for various antiquities.*

are hard to find and it may be necessary to cannibalize an inexpensive remote release.

With this relay board, the power is supplied by the USB connection and operates within the 500 mA limit. In the imaging application, for the camera device, choose the ASCOM Cable Release driver. This is not a sophisticated interface and it will not allow remote control of any camera setting; these have to be made on the camera before connecting the remote release cable to the camera. Settings include Manual exposure or Bulb mode, manual focus and manual white balance. On some cameras, the remote socket is dual-purpose and is also used for a microphone line input. In these cases, select the "remote" option in the camera's movie menu settings. Some cameras and applications have a mirror lock-up feature. Once it is set up in the camera menu and the imaging application, the relay will briefly close to lock up the mirror and then again, for the duration of the exposure.

#### Software Installation

The downloaded .exe file is an ASCOM driver installer. The installation process removes a prior version, as necessary. The relay hardware may also require a Windows driver to work. This relay board uses an FTDi USB chip-set and if this is the first time your PC has been connected to one, you may need to load the virtual com port driver from the FTDi website:

https://www.ftdichip.com/Drivers/VCP.htm

As part of the ASCOM setup, one has to select the ASCOM camera trigger as the "camera" and the relevant virtual COM port for the relay board from the drop-down list in the ASCOM driver setup dialog. You should only have to do this once, as the driver remembers the setting for next time. COM port numbers are automatically assigned by Windows and if one plugs a device into a different USB socket, Windows may be fooled into thinking it is a new device and assign the next free port number, confusing things. This is avoided by forcing Windows to use specific ports for each device. This is accomplished using the Windows device manager and setting the COM port number in the advanced properties page for the device and avoiding using COM 1 or COM 2. It may take a re-boot to straighten things out but these manual assignments are good practice with any device connected to a virtual COM port and ensure a smooth start-up. In the same manner, it also helps to keep consistent assignments through PC and operating system upgrades.

### Dew Heater Controller

A few years ago, dew heaters were expensive items and many sought out alternatives from the burgeoning sales of LED light dimmers and motor controllers from various online retailers. The opportunity was not lost on several innovative manufacturers and it is now possible to purchase a two-port USB-programmable unit for under $100. This is still significant if one is working on a strict budget and the last project tries out an inexpensive LED light dimmer. These units regulate a DC input by rapidly turning the output on and off, the technical term for which is Pulse Width Modulation (PWM). The performance of these units varies enormously, not only in reliability but also in quality too. If it is not designed correctly, the switching circuitry may cause electrical interference. I tried one unit that sent a susceptible telescope mount into a random dance. Its circuit board had a plastic enclosure and the usual suppression components were absent.

This time, I chose a dimmer with a CE rating and checked out the unit before modifying it into a more astro-friendly form. The first step is to examine the circuit board. A Google search of the chip part numbers tells you what they are. This particular one had three chips, the ubiquitous 555 timer chip, a MOSFET power transistor and an op-amp. The better units have a clamping diode (usually black, with a silver band around one end) and a capacitor across the output. With an inductive load, if one tries to suddenly chop the current, it generates a large back voltage and these components allow any back-current to flow and reduces high-frequency interference. Dew heater tapes are a resistive load but it is always a good idea to check for interference from inexpensive modules. Electrical interference is conducted through wiring and/or radiated too. A simple check is to hook up the unit to a 12-volt supply and connect the dew heater tape and place an AM/FM radio close by and see if the radio picks up any interference. In this case it did not, but for good measure, I decided to take the precaution and shield the circuit and put suppression components on the board.

#### Construction.

This $5 unit is supplied with screw terminals for input and output. While they work okay to connect a cable to a car accessory socket, they are not suitable for a fragile dew heater tape cable. Dew heater tape cables are usually terminated in an audio phono plug. The solution is to modify the casing and add a pair of phono sockets (fig.7, fig.8), salvaged from an old hi-fi project. This requires drilling two 8-mm holes for the

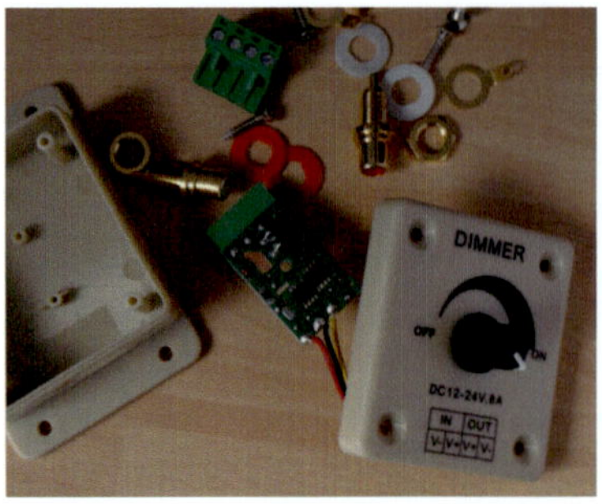


*fig.7 The LED dimmer module, opened up, along with the phono connector parts.*

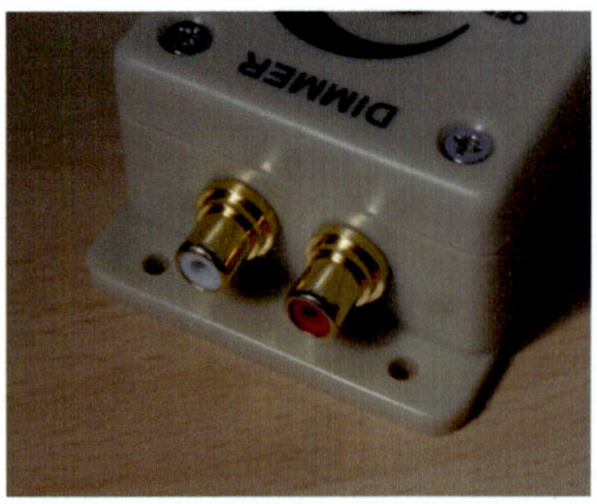


*fig.8 After drilling two holes and a little soldering, the module will now accept standard dew-heater tape connectors.*

connector body and soldering their tags to the small circuit board's output terminals (in parallel).

For shielding, I found some sticky-backed copper foil (often called guitar foil) and lined the plastic box and then lined with a layer of electrical tape, for insulation. The convention is to make the central phono contact positive. With these circuits, however, the dew heater tape is treated as a load between 12 volts and a switch to ground. This means that the outer connection will not be a true ground and is best kept away from other connector bodies. For suppression, I additionally soldered 100 nF ceramic capacitors across the input and output terminals to reduce electrical interference. The overall result is much neater and a fraction of the cost of a commercial unit. The overall schematic is shown in fig.9.

The input cable is usually a twin (figure-of-eight) cable, terminated by a cigarette lighter plug. As a general rule, I find these to be unreliable and on more than one occasion, a small movement has caused intermittent power and stopped my mount working. I have now standardized on XLR audio connectors for power. These locking connectors are more substantial and are weatherproofed to some degree. They lack LED indicators and fusing but this is taken care of within a home-made power distribution box.

### Other Module Options

Some of these pulse-width-modulation boards are particularly noisy and may cause sensitive equipment to malfunction. An alternative solution is to buy a linear power module. These use a chip that regulates the voltage output. These do not produce electrical interference as they do not chop the power. Their drawback, however, is that the regulator chip dumps the excess power as heat and often requires a small heatsink. Most use the same LM317 variable linear regulator chip and, when scouring the modules on the Internet, look out for a three-legged component on the board or in the description, with those markings. Some modules simply have the chip soldered upright without a heatsink. Heatsinks usually bolt to or clip around the metal flange of the chip. The idea is that they radiate heat away and can be made from a bent piece of tin or aluminum, so long as they do not touch other circuits. They may not be necessary for light loads (say, up to 5-inch dew-heater tapes) but anything larger will likely need one.

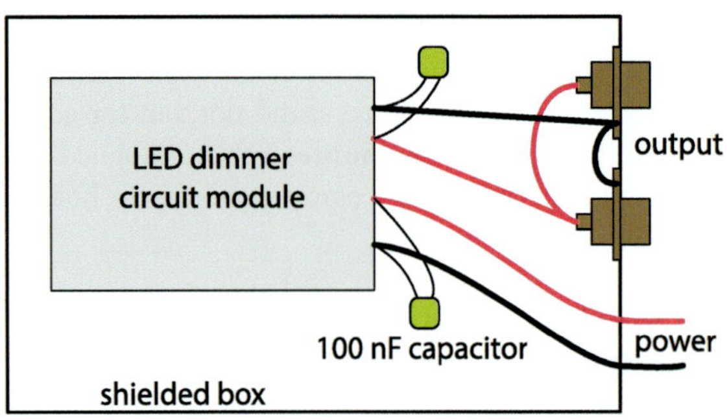


*fig.9 The completed module, with suppression components and a shielded box. There are many types of capacitors; this requires a non-electrolytic type. Ceramic types are usually a good choice for RF work with a value of about 100 nF, though this is not critical.*

The quality of dew heater cables is variable, some are thin audio designs with high resistance and drop power. In my own system, I route all cables through the mount and the dew-heater cable runs next to the more sensitive power and USB cables. As an extra precaution, I use a double-screened microphone cable to bring the modulated power up through the mount, with the outer screen connected to ground. This cable is very flexible and it terminates in a small junction box, attached to the underside of the dovetail clamp plate. The box is fitted with suppression capacitors across its two phono sockets. Each of my dew-heater tape's cables is cut down to suit and re-terminated with phono plugs that have no external metal surfaces (to avoid shorts). It may be overkill, but if there are any operational issues, I can effectively eliminate interference as a potential culprit.

# Bibliography and Resources

*Recommended reading for starting out in practical astronomy and astrophotography*

## Bibliography

Very few books on astrophotography include a bibliography. I do not know why and, for such a diverse subject, it makes no sense, as this is such a collaborative hobby.

**Astronomy and Astrophotography:**

Steve Richards, *Making Every Photo Count*, Self Published, 2011
This is a popular book that introduces digital astrophotography to the beginner. It is now in its second edition and has been updated to include modern CCD cameras. The emphasis is on using digital SLRs.

Charles Bracken, *The Astrophotography Sky Atlas*, Self Published, 2016
This atlas is targeted at astrophotographers to enable them to plan imaging sessions. Includes common and unusual objects in this well-conceived reference, organized for latitude and season.

Allen Hall, *Getting Started: Long Exposure Astrophotography*, Self Published, 2013
This is an up-to-date book which makes use of affordable equipment on a modest budget. It has an interesting section on spectroscopy and includes several practical projects for upgrading equipment and making accessories. It also features a section on imaging processing, including some of the common tools in PixInsight.

Charles Bracken, *The Deep-sky Imaging Primer*, second edition, Self Published, 2017
This up-to-date work is focused on the essentials of image capture and processing using a mixture of digital SLRs and astronomy CCD cameras. One of its highlights are the chapters that clearly explain complex technical matters.

Charles Bracken, *The Astrophotography Sky Atlas*, Self Published 2015
This is a useful reference book of astrophotography targets, some well known, some not. While most sky atlases concentrate on stars, this aims at nebulae, galaxies, double stars, clusters and nova.

Ruben Kier, *The 100 Best Astrophotography Targets*, Springer, 2009
This straightforward book lists well- and lesser-known targets as they become accessible during the year. A useful resource when you wish to venture beyond the Messier catalog.

Thierry Legault, *Astrophotography*, Rockynook, 2016
This book provides a general overview of astrophotography, touching upon most of the available equipment options with an emphasis on solar-system photography, for which Thierry is highly regarded.

Robin Scagell, *Stargazing with a Telescope*, Philip's, 2009
A good general book on choosing and using a telescope, aimed at the first-time purchaser.

Chris Woodhouse, *The Astrophotography Manual*, second edition, Routledge, 2017
This book is the natural follow-on from this one. While there is some overlap in the introductory chapters, it quickly moves on to intermediate and advanced topics, especially narrowband imaging with monochrome cameras. It is written in a scientific and practical style.

## Internet Resources

### Less Common Software (or use Internet search)

| | |
|---|---|
| PHD2 (guiding software) | *www.openphdguiding.org* |
| Nebulosity (acquisition / processing) | *www.stark-labs.com/nebulosity.html* |
| Sequence Generator Pro (acquisition) | *www.mainsequencesoftware.com* |
| NINA (acquisition) | *www.nighttime-imaging.eu* |
| EQMOD (EQ6 ASCOM) | *www.eq-mod.sourceforge.net* |
| APT (acquisition) | *www.ideiki.com/astro/default.aspx* |
| Cartes du Ciel (planetarium) | *www.ap-i.net/skychart/en/start* |
| C2A (planetarium) | *www.astrosurf.com/c2a/english* |
| Polar Drift calculator | *www.celestialwonders.com* |
| PlateSolve 2 | *www.planewave.com/downloads/software* |
| SiriL | *www.siril.org* |
| Deep Sky Stacker | *www.deepskystacker.free.fr/* |
| ASTAP (astrometric stacking) | *www.hnsky.org/astap.htm* |
| Local astrometry.net plate-solver | *www.adgsoftware.com/ansvr/* |
| ASCOM downloads | *www.ascom-standards.org* |
| Affinity Software | *https://affinity.serif.com* |
| StarStaX | *https://www.markus-enzweiler.de/software/software.html* |
| Startrails | *http://www.startrails.de/html/software.html* |

### Popular Forums

| | | |
|---|---|---|
| Stargazer's Lounge | *www.stargazerslounge.com* | (UK) |
| Cloudy Nights | *www.cloudynights.com* | (US) |
| Ice in Space | *www.iceinspace.com* | (AU) |
| Astro buy and sell (regional) | *www.astrobuysell.com/uk* | |
| Sequence Generator Pro | *www.forum.mainsequencesoftware.com* | |
| EQMOD (EQ mount software) | *www.groups.yahoo.com/group/eqmod* | |

### Weather

| | |
|---|---|
| Metcheck (UK) | *www.metcheck.com* |
| The Weather Channel | *www.uk.weather.com* |
| Clear Sky Chart (N. America) | *www.cleardarksky.com* |
| Scope Nights (App for portable devices) | *www.eggmoonstudio.com* |
| FLO weather (also iOS app version) | *www.clearoutside.com* |
| Dark Sky (also app versions) | *www.darksky.net* |
| Windy (also as iOS App) | *www.windy.com* |

### Interesting Websites

| | |
|---|---|
| DeepSkyColors | *www.deepskycolors.com* |
| Harry's Astro Shed | *www.harrysastroshed.com* |
| J-P Metsavainio | *https://astroanarchy.blogspot.com* |
| Astrometry | *astrometry.net* |
| Astrosurf | *www.astrosurf.com* |
| Bahtinov mask generator | *http://astrojargon.net/MaskGen.aspx* |
| Wilmslow Astro | *http://www.wilmslowastro.com/software/index.htm* |
| SiriL tutorial | *https://pixls.us/articles/processing-a-nightscape-in-siril* |
| Deep Sky Browser | *https://telescopius.com* |
| Star Circles | *https://starcircleacademy.com* |

# Glossary and Index

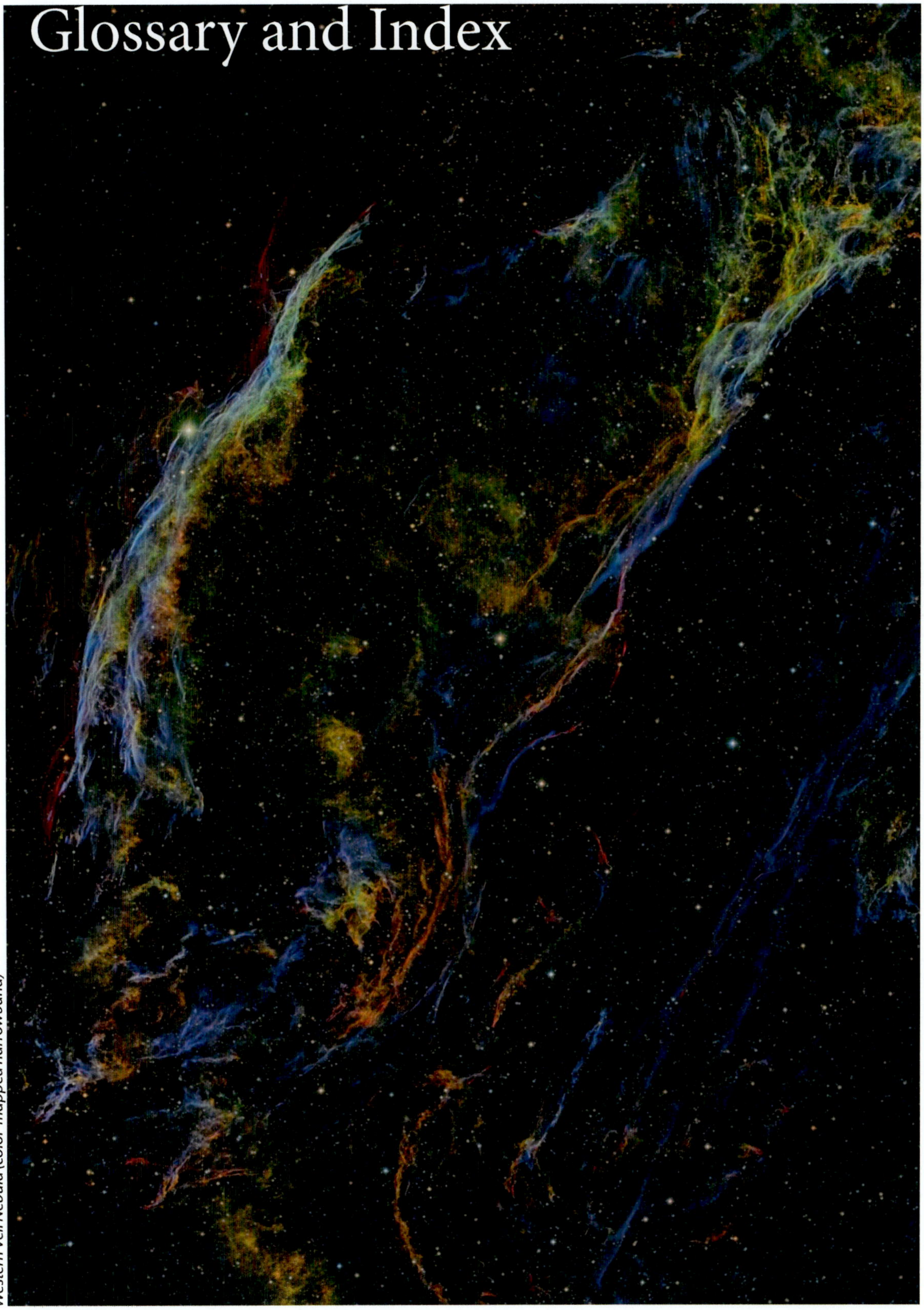

*Western Veil Nebula (color-mapped narrowband)*

# Glossary

*A selection of common technical terms and acronyms*

This is not an exhaustive list but a selection of unique items that occur in the main text.

Achromat: A refractor made of objective lenses of different materials to bring two colors of light to almost the same focal point.

Adobe RGB (1998): A popular color space (profile) for photographers and image editing. It has a moderate color Gamut.

ADU: Refers to Analog to Digital Units, the digital pixel values from a digital sensor.

Apochromat: A refractor made of several objective lenses of different dispersion characteristics that minimize spherical and chromatic aberration.

ASCOM: A non-profit initiative to create an open-source standard for interfacing astronomy software, and hardware, on the Windows platform.

Astigmatism: This is an optical defect that renders stars as ovals. More common with eyes than optics!

Astrometry: This is the measurement of a star's position and motion.

Bahtinov Mask: A focus aid that looks like a drain cover, which, when placed over the front of a telescope, creates 3 diffraction spikes that intersect when the system is in focus.

Bias Current: Random electrons that occur in every exposure, irrespective of temperature or duration.

Blooming: The unsightly effect of an overexposed photosite leaking electrons into its neighbors.

CCD: (Charged Couple Device) A semiconductor architecture used in sensors that stores electrons and transports them serially to an external circuit for conversion into digital values.

CMOS: (Complimentary Metal-Oxide Semiconductor) a semiconductor technology that is also applicable for digital sensors in consumer and astro cameras.

C-Mount: A thread standard often used on cine lenses but also used on small CCD cameras: 1-inch diameter, 32 threads per inch and with a flange distance of 17.5 mm.

Centroid: The position of a star's center. Calculated during autoguiding and astrometry.

Chromatic Aberration: In glass optics, the optical elements refract light to different degrees, depending on wavelength. The aberration arises as the different color components of white light do not focus to the same point.

Clipping Mask: In photo editors, a clipping mask associates an adjustment layer to the layer below.

Cosmic Rays: These are random high energy particles from space. They trigger electrons in a sensor and leave small white streaks. They are normally processed out during image calibration.

Dark Current: This is the ongoing thermally induced accumulation of non-image electrons, the number of which increases with exposure duration and temperature.

Diagonal: A mirror or prism that deflects the light path to enable more convenient viewing. Often fitted into the back of a focuser on a refractor or SCT.

Dovetail: A metal rail with angled edges that clamps onto a telescope mount. Popular standards are Vixen (~43 mm flange) and Losmandy (~75 mm flange).

Half Flux Density (HFD): Often used by autofocus algorithms. A measure of image diameter of a star within which half the energy or flux occurs. Similar to Full Width Half Max (FWHM) measurement but more robust in poor seeing conditions.

Field Rotation: The effect if a mount is not accurately polar aligned, in which, during a long exposure, stars will appear to rotate around the guide star.

Gamma: Is a non-linear transform applied in imaging systems using a simple power-law expression. Some color spaces, such as sRGB and Adobe RGB(1998) are based on gamma 2.2. A linear image has a gamma of 1.0.

German Equatorial Mount (GEM): Most commonly used for imaging, especially with Newtonian and refractor designs. The design requires flipping on both axes to track past the meridian.

Narrowband: Refers to a style of imaging that uses filters to select one principal emission wavelength, with a passband typically only 3–9 μm wide.

Over-sampled: When the sensor resolution exceeds that required to resolve the projected image.

Off-Axis Guider (OAG): A small mirror, normally before the filter wheel, that deflects peripheral light to a guide camera, set at 90 degrees to the optical path.

One-Shot Color (OSC): A term used for conventional digital cameras or sensors fitted with a Bayer color array (or similar) and produce a color image with a single exposure.

OTA: Optical Tube Assembly. Some telescopes are sold as systems with mounts and tripods. An OTA is just the optical telescope component.

Parfocal: Refers to different optical elements that have the same effect on focus position. Applies to filters and eyepieces. It is often an approximation.

Periodic Error Correction (PEC): Software-based system that measures and corrects for worm-gear tolerance issues, in real time as the worm rotates, using a look-up table in the mount or in a computer application or driver.

Peltier: A semiconductor that exhibits a thermoelectric effect. A heat difference between surfaces generates a voltage and likewise, an applied voltage generates a heat difference. When sandwiched between a sensor and a heatsink, the practical upshot is that it transfers thermal energy from the sensor to the heat-sink.

Pixel: An ambiguous term that refers to the sensor's light-sensitive cells (photosites) as well as to composite RGB picture elements in an image or the elements of an image.

Plate-solve: The process of calculating an image's position by matching the image star pattern with a star catalog database.

Pulseguide: An autoguiding system that uses software rather than hardware to control the mount movements. Often combined intelligently with PEC.

Quantum Efficiency (QE): An expression of the efficiency of incident photon conversion into electrons in a sensor.

RJ45: 8-way connector system used for LAN / Ethernet communications. Simple robust locking connector system also used in 6-way (RJ12) and 4-way (RJ10) for serial communications, autoguider ST4 and focuser systems.

Seeing, or Astronomical/Atmospheric Seeing: The effect of the constantly varying optical properties of the atmosphere, caused by moving air pockets of different densities.

sRGB: A color space (profile) that is used extensively as the default for consumer imaging devices and Internet use. It has a restricted color gamut.

ST4: The name derives from an early SBIG autoguiding system and now adopted to mean the "standard" electrical interface for autoguiding inputs into a mount, based on opto-isolated switch closures.

T-Mount: Sometimes also called T2 thread, this an M42x0.75 metric thread for optical systems, designed for a 55-mm flange spacing (thread to sensor or film). T-thread adapters for various cameras are deliberately sized to maintain this 55-mm flange spacing to the camera's sensor. (Confusingly, some refer to the latest M48 as T-mounts too.)

Transparency: Not to be confused with atmospheric turbulence or seeing, this is the clarity of the air and the absence of mist, dust and pollution.

Under-sampled: When a sensor's pixel pitch is too large to resolve the projected image,

# Index

*Where possible, index entries are grouped logically and indicate primary references.*

For Product Safety Concerns and Information please contact our EU representative GPSR@taylorandfrancis.com Taylor & Francis Verlag GmbH, Kaufingerstraße 24, 80331 München, Germany

Batch number: 10722693

Printed by Printforce, the Netherlands